Tools of Chemistry
Education Research

ACS SYMPOSIUM SERIES **1166**

Tools of Chemistry Education Research

Diane M. Bunce, Editor

The Catholic University of America
Washington, DC

Renée S. Cole, Editor

University of Iowa
Iowa City, Iowa

Sponsored by the
ACS Division of Chemical Education

American Chemical Society, Washington, DC

Distributed in print by Oxford University Press

Library of Congress Cataloging-in-Publication Data

Tools of chemistry education research / Diane M. Bunce, editor, The Catholic University of America, Washington, DC, Renée S. Cole, editor, University of Iowa, Iowa City, Iowa ; sponsored by the ACS Division of Chemical Education.
    pages cm. -- (ACS symposium series ; 1166)
  Includes bibliographical references and index.
  ISBN 978-0-8412-2940-2 (alk. paper)
  1. Chemistry--Study and teaching. 2. Academic achievement. 3. Curriculum enrichment.
I. Bunce, Diane M. II. Cole, Renée S. III. American Chemical Society. Division of Chemical Education.
  QD40.T64 2014
  540.71--dc23
                                        2014027945

The paper used in this publication meets the minimum requirements of American National Standard for Information Sciences—Permanence of Paper for Printed Library Materials, ANSI Z39.48n1984.

Foreword

The ACS Symposium Series was first published in 1974 to provide a mechanism for publishing symposia quickly in book form. The purpose of the series is to publish timely, comprehensive books developed from the ACS sponsored symposia based on current scientific research. Occasionally, books are developed from symposia sponsored by other organizations when the topic is of keen interest to the chemistry audience.

Before agreeing to publish a book, the proposed table of contents is reviewed for appropriate and comprehensive coverage and for interest to the audience. Some papers may be excluded to better focus the book; others may be added to provide comprehensiveness. When appropriate, overview or introductory chapters are added. Drafts of chapters are peer-reviewed prior to final acceptance or rejection, and manuscripts are prepared in camera-ready format.

As a rule, only original research papers and original review papers are included in the volumes. Verbatim reproductions of previous published papers are not accepted.

ACS Books Department

Contents

Strategies for Qualitative Research

Analyzing Quantitative Research Data

Cognitive-Based Tools for Chemistry Education Research

Practical Issues for Planning, Conducting, and Publishing Chemistry Education Research

Application

Indexes

Preface

The origins of the *Tools of Chemistry Education Research* really began with the creation of the *Nuts and Bolts of Chemical Education Research* (DOI: 10.1021/bk-2008-0976). Both editors (Diane and Renée) had heard from readers of the *Nuts and Bolts* book that there was a need for information on more in-depth resources for those interested in doing chemistry education research. Thus, we turned our attention to developing the *Tools* book as a continuation of the dialogue started in the *Nuts and Bolts* book. With this *Tools* book as a companion volume to the *Nuts and Bolts* book, we believe that both new and experienced researchers will now have a place to start as they consider new research projects in chemistry education

The creation of both books was a great way to bring together a group of talented researchers to share their insights and expertise with the broader community. We intentionally included both early career and more established chemistry education researchers as authors. This was done to promote the growth and expansion of chemistry education by drawing on the expertise and insights of both junior faculty and more experienced researchers, each of whom has unique insights to offer other practitioners in chemistry education research.

The reader should also note that we changed all references in this book from "chem*ical* education" to chemi*stry* education". These terms were used interchangeably in the *Nuts and Bolts* book but as our field has matured, it has become obvious that the research we are engaged in deals with chemistry education just as physics education research deals with physics and biology education research deals with biology education. Although "chemical" has been used historically, it makes sense that if we are dealing with the teaching and learning of chemistry, rather than with chemicals per se, the switch is a logical one.

The journey to create this book has had a few twists and turns, but along the way we have had the opportunity to work with very talented and dedicated people, including the authors and ACS staff. It will be interesting to see where the future takes the field of chemistry education research and our efforts in it. We welcome your thoughts and opinions on the book as well as your suggestions on what should come next.

Thanks

Renée and Diane sincerely thank the authors of these chapters for their diligence in reporting and generosity in sharing their talents and expertise with the readers of this book. We thank Bob Hauserman and Tim Marney of ACS Publications for their encouragement and support of our efforts to both develop and bring this book to fruition. We thank Jasmine Suarez of ACS Publications for her relentless devotion to the completion of this project and her seemingly unending patience with us as we worked on the book. And last but definitely not least, our thanks to Mackenzie Cole for her talent and generosity in spending countless hours copy editing each and every chapter of this book.

Chapter 1

An Introduction to the Tools
of Chemistry Education Research

Renée S. Cole[*,1] **and Diane M. Bunce**[2]

[1]Department of Chemistry, University of Iowa,
Iowa City, Iowa 52242, United States
[2]Department of Chemistry, The Catholic University of America,
Washington, DC 20064, United States
*E-mail: renee-cole@uiowa.edu

This chapter provides an overview of the issues and tools associated with chemistry education research projects and introduces the different chapters. The intent is to highlight the information available that may prove useful to individuals engaged in chemistry education research projects. To facilitate use by readers, the material has been organized into four sections: strategies for qualitative research; analyzing quantitative research data; cognitive-based tools for chemistry education research; and practical issues for planning, conducting, and publishing chemistry education research.

Introduction

As the field of chemistry education research matures, more researchers are using more tools to answer a greater range of questions. In the *Nuts and Bolts of Chemical Education Research* (*1*), we provided an overview of the field and discussed how chemistry education research questions could be addressed. The intended audience for that book was quite diverse, including those who wanted to learn about aspects of chemistry education research (CER) from many perspectives: novice researchers, scholarly teachers who wanted to improve assessment of practice, grant writers, and chemists who want to be more informed about chemistry education research. In this volume, the audience has been more narrowly defined as those who wish to learn more about specific techniques used in chemistry education research, although many aspects may still be useful for a broader audience.

Many active areas of research in chemistry education were described in the 2013 National Research Council report on Discipline-Based Education Research (2), hereafter referred to as the DBER report. These areas included student conceptual understanding, the use of technology to support student learning, analysis of student discourse and argumentation patterns, the use of heuristics in student reasoning, and the development of assessment tools to measure student thinking about chemistry. More detail into many of these studies is provided in a review of the peer-reviewed literature conducted by Towns and Kraft (3). The review includes research with many different areas of focus and that use a variety of research designs (qualitative, quantitative, and mixed methods). The review also summarizes several instruments that have been used by the chemistry education research community.

The DBER report (2) also includes a series of recommendations to advance DBER, which includes CER, as a field of inquiry. These recommendations include a research agenda that emphasizes the following: exploring the similarities and differences among different groups of students, research in a wide variety of course settings, research that measures a wider range of outcomes and that explores the relationships among those outcomes, research that includes more nuanced aspects of instructional strategies and their implementation, and longitudinal studies.

The areas of research and specific tools selected for this book represent a range of approaches (including qualitative and quantitative). The selection of topics to be included in this volume was based on interactions with members of the audience for this book. We received several requests from readers of the *Nuts and Bolts* book for more information on some topics, while other topics were chosen based on new directions and opportunities for growth in chemistry education. For example, we selected *R* for particular attention due to its growing use in many disciplines. It is an open-source program that is more easily available to many researchers and has capabilities to address data analysis for some areas of research (such as eye-tracking) in ways that are not yet easily available in other programs. The book is not intended to present an exhaustive list of tools and strategies that can be used in investigating chemistry education research questions, but should present a starting point and encourage broader perspectives of what can be done.

Strategies for Qualitative Research

Qualitative research methods are necessary to address "how" and "why" questions. Bretz (4) broadly described qualitative research methods in the *Nuts and Bolts of Chemical Education Research*, including issues related to data collection and quality, theoretical perspectives, data analysis, and other practical considerations. Towns (5) also addressed qualitative research methods in her chapter on mixed methods research designs. In this volume, the emphasis is on highlighting particular qualitative research methods that can be used to answer chemistry education research questions.

2

The qualitative analysis section starts with a chapter on using classroom observations as a tool for investigating chemistry teaching and learning. Yezierski (Chapter 2) describes reasons to use classroom observation protocols as a component of research projects as well as guidelines for collecting and analyzing data. The table of research-based observation protocols that can be found in the literature provides a starting point for any researcher that would like to take advantage of existing instruments. This is extended into a more complete discussion of observation protocols that have particular promise for use in chemistry education research.

This is followed by a discussion of using student interviews by Herrington and Daubenmire (Chapter 3). They focus on using open-ended and think-aloud interviews, including several examples from the chemistry education research literature. They provide guidance on developing interview protocols, constructing questions/tasks, selecting participants, conducting the interview, and analyzing the data.

The chapter by Cole, Becker, and Stanford (Chapter 4) introduces the area of discourse analysis as a tool for research in chemistry education. They begin by defining discourse and discourse analysis and then describe the types of questions that can be addressed through discourse analysis. They continue by providing an overview of methodological considerations, including data collection and analysis. The chapter concludes with a summary of some CER studies that have used discourse analysis.

The section concludes with a chapter by Talanquer (Chapter 5) describing how computer assisted qualitative data analysis (CAQDAS) programs can be used to facilitate organization and analysis of data in qualitative research. He describes a variety of programs that are available, but focuses on how they can be used to support research activities. This ranges from handling and organizing data to assisting in the coding annotation of data to visualizing data.

Analyzing Quantitative Research Data

Quantitative research starts with the collection of data, but statistical methods are required as part of the analysis. There are a number of articles that describe problems with how statistical analysis are conducted and reported in educational research (6–9). Sanger (10) provided an overview of inferential statistics in the *Nuts and Bolts* book, including the steps in conducting a quantitative research study and common misconceptions. In this volume, we add to the previous discussion of statistics and extend it by including chapters on additional statistical techniques and on the software package R.

Pentacost (Chapter 6) builds on the work of Sanger by describing how analysis of variance (ANOVA) techniques can be used to support claims in chemistry education research. He begins with a discussion of what it means to determine a difference in data sets and the assumptions about the data that must be met to use these techniques. He then helps readers decide which type of ANOVA is most appropriate for their study and provides examples of how each approach would work using examples from the CER literature.

The majority of the discussion of statistics in the *Nuts and Bolts of Chemistry Education Research* focused on parametric statistics, which make certain assumptions about the normality of the data. Many of the studies in chemistry education research do not meet these assumptions and require the use of non-parametric statistics. In the second chapter of this section, S. Lewis (Chapter 7) presents an overview of nonparametric statistics. He begins by discussing the different data scales and comparing nonparametric and parametric statistics. He then summarizes a number of nonparametric statistical tests that are useful in chemistry education studies.

The section closes with a chapter by Tang and Ji (Chapter 8) on the statistical program R. They begin with reasons that researchers may want to learn how to use R, including a description of some of the advantages while acknowledging the disadvantages of this particular environment. This is followed by a discussion of the program itself and it's capabilities. They also describe some areas of research where R has some functionality that makes it better suited to completing the data analysis as compared to other programs such as SPSS.

Cognitive-Based Tools for Chemistry Education Research

As chemistry education research has developed as a field, more sophisticated tools and methods have also been identified and developed by researchers to address research questions. There are a number of tools that are emerging as being particularly useful for chemistry education research, many of which have foundations in cognitive psychology. The examples included here highlight some methods that are showing increased use in chemistry education research.

The first chapter in this section is a discussion of concept inventories by Bretz (Chapter 9). She begins with a discussion of a variety of design and development methods. This is followed by an extensive discussion of the validity and reliability of the data generated by these instruments as well as their limitations. The chapter ends with a discussion of how concept inventories can be used to measure what and how much chemistry is learned and provides recommendations for chemistry education researchers.

In the next chapter, Neiles (Chapter 10) explores the use of tools that can be employed to measure students' structural knowledge of chemistry. After defining what is meant by structural knowledge, she describes two approaches, concept mapping and proximity data techniques, that have been used to create representations of the connections. She then presents a detailed discussion of how to analyze the resulting structural knowledge networks, particularly with the aid of computer programs such as Pathfinder and GEPHI.

The third chapter in this section focuses on eye-tracking technology and its use in chemistry education research. Havanki and VandenPlas (Chapter 11) begin with a discussion of how vision works and how eye movements relate to cognitive processes. They describe different types of eye tracking instruments before describing the types of research that are amenable to eye tracking studies. The heart of the chapter is a discussion of considerations that guide experimental design, data collection, and data analysis.

In the last chapter in this section, Cooper, Underwood, Bryfczynski, and Klymkowsky (Chapter 12) present a short history of how they have used technology to model and analyze student data. Of particular emphasis is how to use tools to both support student learning and capture data that can be analyzed for research purposes. They describe the features of IMMEX, Organic Pad, and beSocratic from the perspectives of how they were designed to support student learning and as research tools to provide insights into how students develop knowledge and science practice skills.

Practical Issues for Planning, Conducting, and Publishing Chemistry Education Research

Important areas that are rarely addressed in other forums are the practical issues of how to plan, conduct, and publish chemistry education research. While the importance of careful design and planning are emphasized for data collection and analysis, this is generally done in the context of ensuring the quality of the data. Several chapters are included in this volume that provide valuable advice on how to ensure that a project goes as smoothly as possible and culminates with results that are publishable.

In the first chapter in this section, Bunce (Chapter 13) describes a two-pronged approach to dealing with nonsignificant results. She begins by describing the specific issues in chemistry education research that make this topic important. She then explores two ways of ensuring that studies with statistically nonsignificant results are still valid and contribute to the body of knowledge about teaching and learning in chemistry—planning and post-hoc analysis. She uses examples from her own work to demonstrate how these two approaches can result in quality, publishable studies even if the results indicate there are nonsignificant effects.

In the chapter on doing chemistry education research in the "real world," J. Lewis (Chapter 14) describes the challenges of conducting research projects that involve collecting student data in real classrooms at multiple institutions. This chapter takes a more conversational tone and provides valuable guidance to researchers whose research or evaluation activities involve this type of project design. She focuses on two phases of projects that involve data collection in multiple classrooms, involving multiple instructors, and often times multiple institutions. The first phase addressed is that of planning. After exploring many aspects of planning that make it more likely that the project will result in usable data, she moves on to discussing the monitoring and controlling phase of a project. She ends with discussing how a well-designed and executed project can result in publishable results, even if events do not unfold as planned.

The topics of the ethical treatment of participants and Institutional Review Boards (IRBs) should be of concern to all chemistry education researchers (as well as to any one sharing data collected from students or other people). Bauer (Chapter 15) presents an overview of the fundamental principles, purposes, and process of obtaining appropriate IRB oversight of studies involving human subjects. He describes how to get started and some of the expectations for completing applications for IRB approval. He provides several examples to help

Taber, Towns, and Treagust (Chapter 16) present an overview of
publishing chemistry education research, including what makes a manuscript
suitable for publication as well as the practical issues of how the process works.
Many aspects of what make research publishable have been addressed in previous
chapters, but the authors reinforce the importance of thinking about what makes
research publishable during the design phase by describing a variety of factors,
including following ethical guidelines for research with human participants. The
majority of the chapter focuses on the practical aspects of publication including
preparing the manuscript for submission, the submission and review process, and
what happens after approval or acceptance.

Application

The final chapter of the book (Bunce and Cole, Chapter 17) illustrates how
the resources provided in the book can be used to assist a researcher in planning,
conducting, and publishing their research. Two different research questions are
explored to provide comparisons of how the nature of the question drives further
decisions.

Final Thoughts

The book has been designed so that readers can either focus in to learn about
a particular topic or read through the entire book for a broader view. While the
information in each chapter should provide enough information for a reader to get
started in an area, additional reading is likely to be needed to gain further expertise
in a specific area. Our hopes are that the material here will facilitate conversations
with colleagues or other researchers to continue to develop and expand the field of
chemistry education research.

References

1. Bunce, D. M., Cole, R. S., Eds.; *Nuts and Bolts of Chemical Education Research*; ACS Symposium Series 976; American Chemical Society: Washington, DC, 2008.
2. *Discipline-Based Education Research: Understanding and Improving Learning in Undergraduate Science and Engineering*; National Research Council: Washington, DC, 2012.
3. Towns, M.; Kraft, A. *Review and Synthesis of Research in Chemical Education from 2000-2010.* Paper presented at the Second Committee Meeting on the Status, Contributions, and Future Directions of Discipline-Based Education Research, 2011.

4. Bretz, S. L. Qualitative Research Designs in Chemistry Education Research. In *Nuts and Bolts of Chemical Education Research*; Bunce, D. M., Cole, R. S., Eds., ACS Symposium Series 976, American Chemical Society: Washington, DC, 2008; Chapter 7.

5. Towns, M. H. Mixed Methods Designs in Chemical Education Research. In *Nuts and Bolts of Chemical Education Research*; Bunce, D. M., Cole, R. S., Eds.; ACS Symposium Series 976; American Chemical Society: Washington, DC, 2008; Chapter 9.

6. Keselman, H. J.; Huberty, C. J.; Lix, L. M.; Olejnik, S.; Cribbie, R. A.; Donahue, B.; Kowalchuk, R. K.; Lowman, L. L.; Petoskey, M. D.; Keselman, J. C.; Levin, J. R. Statistical practices of educational researchers: An analysis of their ANOVA, MANOVA, and ANCOVA analyses. *Rev. Educ. Res.* **1998**, *68* (3), 350–386.

7. Lewis, S. E.; Lewis, J. E. The same or not the same: Equivalence as an issue in educational research. *J. Chem. Educ.* **2005**, *82* (9), 1408.

8. Henson, R. K.; Hull, D. M.; Williams, C. S. Methodology in our education research culture: Toward a stronger collective quantitative proficiency. *Educ. Res.* **2010**, *39* (3), 229–240.

9. Kirk, R. E. Promoting good statistical practices: Some suggestions. *Educ. Psychol. Meas.* **2001**, *61* (2), 213–218.

10. Sanger, M. J., Using Inferential Statistics to Answer Quantitative Chemical Education Research Questions. In *Nuts and Bolts of Chemical Education Research*; Bunce, D. M., Cole, R. S., Eds.; ACS Symposium Series 976; American Chemical Society: Washington, DC, 2008; Chapter 8.

Strategies for Qualitative Research

Chapter 2

Observation as a Tool for Investigating Chemistry Teaching and Learning

Ellen J. Yezierski*

[1]Miami University, Department of Chemistry and Biochemistry,
Oxford, Ohio 45056, United States
*E-mail: yeziere@MiamiOH.edu

This chapter provides an overview of observation in teaching and learning settings with the purpose of guiding novice chemistry education researchers in planning for, executing, and reporting on studies which rely on observational data. Critical design considerations, logistical concerns, data collection guidelines, data analysis suggestions, and reporting recommendations are presented along with exemplars from the chemistry education research literature.

Introduction

Observation is a primarily qualitative research method with roots in cultural anthropology. Over the course of the late nineteenth and early twentieth centuries, the method of observation developed as ethnographic studies employed participant observation to study and document complex sociocultural phenomena. Observation was popularized by anthropologists such as Margaret Mead (1) and has been used in sociology, education, and other fields interested in characterizing and understanding human interaction.

Like most research methods, the decision if and how to use observation should be driven by the research questions. There are numerous chemistry education research (CER) studies for which observation would have been inappropriate. Such studies often have characteristics of clinical techniques used in psychology where the participants can be studied outside of learning environments. To illustrate, consider multiple studies that have explored students' and teachers' understanding of chemistry concepts. Data contributing to such studies range from scores on concept inventories given to an entire class to transcripts of participant interviews recorded as they answered questions or interacted with chemical phenomena. Likewise, learning outcomes of interventions have been evaluated by examining the efficacy of students in solving problems and explaining chemical concepts before and after the intervention. Some research questions, however, warrant the investigation of phenomena that only can be studied while participants are immersed in the learning environment. In such cases, observation may be the most crucial of research methods. The dynamics, interactions, actions, and reactions of teachers and learners being studied, along with the taught curriculum require observational data if these types of phenomena are the units of analysis.

Observation has been used for decades to evaluate teaching performance. It is important to note that this chapter frames observation solely in a research context in which observation is used to gather data to inform basic and applied research questions regarding the teaching and learning of chemistry. Although some of the content may be aligned with the practice of evaluating chemistry teachers in secondary and university teaching environments, and some observation tools presented have been developed for such purposes, the practices and strategies that follow do not substantially address the large body of scholarly work on teacher evaluation nor have explicit implications for the scholarly field of teacher evaluation.

Observational data have been used for decades in CER, but there is not a wealth of observational studies relative to other research methods. The extent to which observational data have contributed to the overall findings of studies has certainly varied; however, with the inexpensive and easy access to tools used to collect video data, for example, we are likely to see a rise in the number of studies that have some observation component.

This chapter presents an overview of observation in teaching and learning settings with the purpose of guiding novice chemistry education researchers in planning for, executing, and reporting on studies which rely on observational data. Although observation is presented as a tool of CER, the chapter focuses on collecting and using different types of observational data in the form of field notes, video, and documentation via observation protocols. Critical design considerations, logistical concerns, data collection guidelines, data analysis suggestions, and reporting recommendations are presented along with exemplars from the chemistry education research literature.

Planning for Collecting Observational Data

Participant–Observer Continuum

Considering the role of the observer provides a useful way to begin thinking about collecting observational data in a study. The role of the observer can be characterized by a continuum born in the social sciences called the *participant-observation continuum* (*2*). The nature of the research questions, the theoretical framework of the study, and the researcher role lend key considerations to where data collection should fall on this continuum. Data collection in the tradition of psychology is on the *observer* end of the participant-observation continuum. Here, the researcher has little or no contact with the participants being observed. An example of this might be an observer sitting in the back corner of a general chemistry classroom taking field notes on teacher-student interactions. Glesne (2) refers to the next point on the continuum as *observer as participant* in which the researcher has a little bit of contact with the participants. An example might be a researcher visiting an analytical chemistry laboratory to videotape students conducting experiments. Although the researcher does not teach the students, she might interact with them to learn more about what they are doing and thinking or chat briefly with students to request access to their notebooks or apparatus to add to the video record of the laboratory experience. If the interactions are significant, it might qualify the researcher to be on the next point on the continuum – *participant as observer*. In this case the observer has a more significant role in the experience of the people being observed. The *full participant* functions as the investigator and as a full-fledged member of the community that is being studied. Conducting research in one's own classroom while in the teacher role is a good example of this most extreme point on the participant-observer continuum – *full participant*. It is important to consider what is best for the study in terms of the researcher role. Is it more appropriate for the research to have the "eye of the uninvolved outsider" at the observer end of the continuum rather than opportunities for seeing the environment under study through the eyes of its participants in the full participant role? Where the researcher may be situated in this continuum depends entirely on the research questions, the context of the study, and the theoretical perspective of the study.

For example, two studies may aim to evaluate the effect of a reformed teaching strategy on teacher-student interactions in physical chemistry. One study conducted by a teacher-researcher may focus on understanding the types of verbal and nonverbal cues given by students which lead to productive teacher questioning behavior. The perspective of the teacher (full participant) is critical; therefore, this teacher would videotape her interactions with students using a headset camera while interacting with students in large and small groups. The teacher would capture what she sees and hears from students during question and answer sessions to address the research question. On the observer end of the continuum, an outside researcher may videotape in the same classroom with the goal of identifying which types of interactions lead to correct student elicitations of content. Here the researcher would capture student and teacher discussions and evaluate the students' conceptual understanding of physical chemistry while attempting to link the more scientifically accurate elicitations to specific teacher-student interactions observed.

Observational versus Self-Report Data

The theoretical framework of the study and the theoretical perspectives of the researchers must be considered throughout any qualitative study. The following ideas guide where the observer is situated on the participant-observer continuum as well as the type of data that should be collected. In considering the participant-observer continuum, the observer role at one extreme aligns with the positivist view that the researcher can observe and analyze events from a purely "objective" perspective. Consider the opposite end of the continuum in which the participant experiences the phenomena and reports on it using field notes. The validity of the data coming from these two extreme perspectives depends solely on the research questions and what perspectives are warranted to effectively investigate and answer these questions. It is also clear that the promise of pure objectivity is likely impossible. Although video as a data source, a common method to document observation, feels just like being there, it is necessary to recognize that the videographer makes choices about what to capture that confine the viewed experience to his or her perspective. This is akin to the observer taking field notes, in which the experience is filtered entirely through the observer. Nevertheless, a full participant reporting on an experience is much like other self-report data collection methods found in education research. Self-report data are generated by a questionnaire, survey, or other instrument in which participants select and/or generate responses without a researcher intervening. Prior work has determined that self-report data may be accurate for some features of reporting on teaching and learning, but not for others. For example, Mayer (3) found that while self-reported survey data from teachers provides a fairly accurate accounting of the amount of time spent on particular practices in classrooms, it does not, however, accurately capture the quality of interactions between teachers and students. If the research questions aim to explore teacher-student interactions, observational data collected by an outsider may yield the most valid data.

Design Considerations

The participant-observer continuum helps researchers to think about the role and perspective of the observer. A related continuum-style set of frameworks that assist researchers in determining the parameters of observational data collection is proposed by Patton (4) and offers four dichotomies along with two other criteria. Three dichotomies align with the participant-observer continuum – participant versus onlooker (or both)? Insider versus outsider? Overt versus covert? – while the fourth brings in a new parameter – solo versus team observation? The other two criteria are duration of observations and focus of observations. Patton (4) and similar qualitative methods resources do not provide guidelines about the "ideal" number of observations or any of the other considerations. What is recommended is that each of the parameters is carefully considered in light of the research questions, setting, resources, and personnel during the design of the study. In addition to Patton (4), Phelps (5) offers a succinct chemistry-centered guide for key considerations in qualitative research design which aligns well with qualitative studies relying on observational data.

Human Subjects and Related Logistical Considerations

When planning to collect observational data in a classroom, museum, or other location, the most important provision is for consent from the stakeholders involved. In the case of a college or university researcher, the institutional review board (IRB) affiliated with the researcher's institution will review the application requesting to collect observational data. An excellent overview of IRB related processes, regulations, and advice is described in detail in this volume by Bauer (6). Considerations specific to planning for observational data collection are described here. It is important to note that each institution's IRB is somewhat idiosyncratic. The guidelines provided here are meant to be helpful but may not be applicable to all institutions' IRBs. It is good practice for investigators to work closely with their own boards. As an example, the age of majority (associated with adulthood) when participants can sign consent forms to participate in research is typically 18; however, this varies by state.

The method of observation and the ages of the participants primarily guide the human subjects' considerations. Any person whose likeness may be captured on video must consent by signing a form. In the cases of college classrooms in which all participants are 18 years of age or older, observations of public behavior that are not recorded (e.g., classroom or laboratory observation) are considered exempt (7). Exempt status means that the research activity is not monitored by the IRB and is not subject to continuing IRB oversight. It is still the responsibility of the research team to obtain informed consent, minimize risks, secure data, and protect the confidentiality of research participants. If the public behavior is recorded, by video for instance, the proposed research is not exempt but will likely be eligible for expedited review. If the research participants are under 18 years of age, the application to conduct research will not meet the criteria for exemption; however, it is typical for applications involving the collection of data in K-12 schools, for example, to be eligible for expedited review. In the case of minors, no observational data of any kind may be collected without parental consent and usually child assent. Bauer (6) focuses on IRB consideration for adult participants. As such, IRB considerations for observational data collected across all levels will be discussed here with special consideration for K-12 teaching and learning settings.

In the case of larger K-12 school districts, the districts often have their own review board affiliated with the assessment and evaluation arm of the upper administration. It is important to consider that urban school districts are continually approached by investigators to partner on grant-funded research projects. Developing relationships to gain access to certain environments is critical to exploring particular educational environments. Researchers should contact the district of interest and work with district personnel to prepare applications for the district's internal review. It is also useful to contact the research team's IRB and determine which approval is contingent upon the other, so that the timing of application completion and submission aligns with the various partners in oversight of the research.

In addition to the students in chemistry classrooms, informed consent must be obtained from the teacher along with permission from the teacher's supervisor. In the case of K-12 schools, successful IRB applications describing research including observational data collection includes a signed letter from the school principal stating that s/he has agreed for the research to take place in his/her building. For college classrooms, it is wise to speak with the teacher of the class to determine if supervisors such as department chairs or deans need to be alerted for research to occur in his/her class. In the case of museums, planetariums, and other informal science venues, the director or head curator of the institution should provide a permission letter for the research team to submit with the IRB application. If observational data are to be collected in community outreach chemistry programs (e.g., National Chemistry Week activities), program coordinators and the venue's director would be appropriate authorities who could provide permission letters to submit with the IRB application for the study.

Much like research participants have the right to withdraw from a study at any time, participants involved in a study in which video data, for example, are being collected may ask for the camera to be turned off at any time. A description of this right must be part of the informed consent document (and assent, if applicable).

After informed consent (and assent, if applicable) forms are collected, the researchers can begin visiting the research sites and collecting observational data. It is important to note that most classrooms contain consenting and nonconsenting students. Capturing students on video who have not consented is problematic for some IRBs and not for others. As such, several strategies are presented here in an effort to describe the range of solutions used in CER for coping with classrooms containing consenting and nonconsenting students. The following highlights the importance of ensuring that the study's protocols align with the institutional IRB.

In a classroom setting, the instructor usually does not know which students are participating and which are not. However, if the IRB does not allow non-consenting students' likenesses to be captured on video, it is necessary for the teacher to alert the videographer to students who cannot be videotaped. This can be done without the students knowing who is and is not participating. Alternatively, some boards may ask the investigator to provide seats in the classroom for students who do not wish to participate. The video camera is then placed such that these students are never captured on video. In this case, the teacher and students will know who is participating in the study. On the other hand, the IRB may insist that others not know who is participating. In this case, the entire class is videotaped and contributions from the students who did not give consent will not be transcribed or coded. All of the aforementioned approaches can make data collection challenging. There are two ways to optimize the situation, and ideally, both are employed. First, increase the number of consenting participants. Students (and their parents, if students are minors) are more likely to consent to a study that they take interest in and understand. Working closely with the instructor to have adequate recruitment time helps to increase the number of participants. When researchers give a brief recruitment presentation followed by a question and answer period, they can share their excitement for the work as well as key information about data security and the balance of benefits and risks. Second, select the class section for which there is the highest percentage of students with signed consent forms. It is optimal to recruit instructors who teach more than one section. The human subjects considerations should not be a barrier to the research. They should, however, motivate research teams to plan ahead, reach out to key stakeholders, and work diligently to ensure that the study is well designed, such that the benefits outweigh the risks for the participants.

Capturing Observational Data

Field Notes

Field notes are descriptive accounts of observations and experiences made by the researcher (8). These accounts are not free of the perceptions and interpretations of the researcher and are not simply facts about what occurred. The researcher's inscriptions filter out parts of the experience while focusing on others. This is why the focus of the study must be carefully planned along with the specific aspects of the target phenomena which should be attended to during observation. For example, the research may be focusing on how students interact with one another or with the teacher; how students use particular resources such as models or laboratory apparatus how students respond to a particular instructional activity; or how teachers frame questions to students. If the researcher has multiple points of interest, it will be impossible to capture all of the complex interactions, actions, behaviors, gestures, and verbalizations in a typical chemistry classroom. What is essential is that, before the observation, the researcher plans what types of interactions, behaviors, words, and actions need to be captured so the data will contribute to answering the research questions. Furthermore, the researcher must capture a rich description of the research setting. Lastly, the researcher should plan the most efficient way of capturing such classroom phenomena. It is helpful to plan shorthand-like representations to improve efficiency over narrative field notes. However, planning ahead should also include cues for when capturing a detailed, word-for-word record of the classroom phenomena is necessary.

Regardless of the points of interest for the researcher, the field notes will serve as important data sources for the observation. It may be useful to have multiple researchers conduct observations, even simultaneously, to lend reliability to the findings. It may also be helpful to conduct multiple observations of the same classroom to ensure that the phenomena observed are somewhat representative of the overall class environment in the setting of the study. This illustrates the importance of decisions regarding the frequency of observations during the planning stage of the study.

Audio and Video

Because classrooms are homes to complex interactions among people, things, cultures, and school environments, it may be challenging to capture the necessary data using field notes. The rapid improvement of electronics has made devices which capture high quality audio and video widely available at a reasonable cost. This accessibility of good equipment combined with the capacity for storage of large data files has made observation captured by audio and video devices commonplace in education research.

Audio recordings of a whole classroom are unwieldy; however, capturing classroom interactions or teacher words using audio recordings has been successfully done in CER. To understand analogical models when teaching and learning about equilibrium, Harrison and de Jong audiotaped lessons while observing and taking field notes (9). Video has also been used in CER to capture classroom environments and archive them for later analysis (*10–14*). Although audio remains a useful means to capture data in formal and informal educational settings, the rest of the chapter will focus on video as a data source and provide key considerations regarding its use.

The temporal and sequential structure of videos' multimodal record – talk, gestures, interactions, facial expressions, and actions – serve two key purposes not best served by field notes. First, the detailed record reminds and informs the observer. Second, the video record extends the observation experience to other researchers. Jewett (*15*) characterizes video data as sharable, durable, and flexible. Video data can be analyzed by multiple researchers repeatedly, even if one researcher served as the videographer. Moreover, the record of the events can last as long as it is needed. Not only does this provide the ability for more researchers to be involved, but also for replicate studies or deeper analyses of the events captured in the observation. Multiple passes through the video provides the research team the opportunity to notice things that might have been missed during the initial observation in the field. Although the archive of the phenomena does not change, the aspects of the video that are analyzed will vary according to the research questions and theoretical framework underlying the research. For example, a general chemistry course in which students interact with a simulation could be analyzed for multiple research questions. What might researchers notice with the sound turned off versus listening to the audio only? Furthermore, segments can be sped up or slowed down to provide another perspective on particular classroom actions. In this sense, the video can be "changed" to help researchers attend to different elements of the action.

The advantages of video are simple—the observer has an archive of the classroom phenomena that looks and sounds much like it did during the time it was captured. It is important to recognize that the setting, as it was experienced by the observer, is not the same as the video. If the research team has the ability to have multiple cameras in the room capturing actions, as did Bond-Robinson and Rodriques with their remote audio video observation system (*16*), then the video archive can more closely resemble the observer's experience.

Frequently, one camera is used to capture chemistry classroom phenomena. While the video feels just like the "real thing," it is necessary to recognize that the observer/videographer made choices about what to capture that confine the viewed experience to his or her perspective. For example, the videographer decides which students are captured, for how long, how wide/tight the shots are, which lab phenomena are captured, when the focus is on the teacher, etc. Much like the observer taking field notes, the experience is filtered through the observer (in this case, the videographer). As such, the choices made by the videographer should be made explicit to the rest of the research team who may be analyzing the video. The strongest design yielding the most robust observational data is generated when the research team articulates to the videographer (who should be a member of the team) what needs to be captured as prescribed by the research questions. For example, in Yezierski and Herrington's study (*14*), student-teacher and student-student interactions were important to the analysis of the secondary chemistry classroom phenomena. As such, videographers on the research team were certain to capture as many of these interactions as possible. Alternatively, if teacher actions were the focus of the study, the videographer would then follow the teacher around the classroom and ensure that all of the video documented teacher actions only.

Video Tools

Ensuring that technically high-quality observational data are captured on video can be challenging. Most research teams do not have the support to hire professional videographers. Moreover, setting up professional video equipment complete with a crew in a classroom would make the observation disruptive and lead to participants behaving unnaturally. Fortunately, it is relatively easy for research teams to collect high-quality video data in classrooms using tools such as handheld video cameras and small ear-mounted cameras. These tools have electronic image stabilization (EIS) made possible by mini-gyros in the lens which detect motion and programs which correct the images altered by a shaky camera. For example, this camera feature makes it easier to follow participants around in an active laboratory classroom setting. Of course, there are limitations to this technology, and videographers should pay attention to the basic principles for quality video capture. Derry's *Guidelines for Video Research in Education* (*17*) is a comprehensive reference.

Fortunately, gone are the days when high quality video required tripods, giant obtrusive cameras, and professional videographers. The small and easy to use devices that capture audio and video make observation more inexpensive for researchers who have been using it for decades while also inviting more researchers to use observation as a means to understand teaching and learning. Handheld digital video cameras with large internal hard drives are affordable and allow easy capture of video data as well as downloading of data. Another technology which offers a different vantage point for video is the ear-mounted camera. These small devices, no larger than a Bluetooth® earpiece, attach to the ear of a participant and capture audio and video. In a chemistry laboratory class, an instructor can wear the camera and comfortably interact with students and apparatus around the lab. The video captures the laboratory experience from the teacher's perspective and can be quickly downloaded to a local or remote site. Multiple students in a classroom working with ball-and-stick models could be wearing these small cameras to offer the observer an inside look into student-student interactions with chemistry models. The tools a researcher selects should align with the perspective necessary to best address the research questions.

Role of Artifacts

The video record of an experience in a chemistry lab, classroom, museum, or other learning venue is sometimes not enough to capture the key aspects of the phenomena under study. To carefully recapitulate a chemistry learning experience to analyze it in response to research questions, artifacts of the experience are critical. Artifacts may include student handouts, teachers' lesson plans, and samples of student work such as lab reports, textbooks, or laboratory procedures. In CER, most artifacts are print materials that help to document the teaching and learning activities that took place during the observation. Collection and analysis of the artifacts, like capturing the observation, should be planned for as with any other task in executing the study. Although artifacts are discussed here as enhancements of the video record, they can more generally be thought of as another data source that can contribute to the triangulation of findings.

Special Considerations in Chemistry Classrooms

Chemistry observations frequently involve some type of laboratory component ranging from a teacher demonstration to students carrying out procedures in small groups. As such, researchers visiting settings where chemistry experiments are conducted should subscribe to the strictest laboratory hygiene protocols including closed-toe shoes, long pants, and safety goggles. It is helpful to understand any other protocols in the setting that participants are asked to follow to ensure that the researcher follows them to the letter as well. For example, some high school chemistry settings have combined spaces for laboratory and classroom activities. The researcher carrying a water bottle into the classroom space might be acceptable to one teacher while violating a serious rule in another teacher's space. The key is to understand that every environment may be slightly different and violating the rules associated with a particular environment could jeopardize the researcher's access to that environment in the future.

Analyzing Observational Data

Applying General Qualitative Procedures

After video data are collected, much like interviews and other qualitative data, it can be transcribed and a host of analytic techniques applied. Researchers can organize the data around themes, break the data into smaller units such as codes, or count occurrences of behaviors or statements. Wu (18) investigated how teacher knowledge, real life experiences, and interactions with students helped students to construct meanings of chemical representations. Wu employed a variety of analytic techniques to answer the research questions. Techniqes include coding video transcripts for (i) chemical concepts addressed, (ii) links between chemical concepts and real situations, and (iii) the length of teaching events. Further analytic steps included the construction of maps to further connect related events and themes from the data. The technology that has improved the accessibility, quality, and storage of video has also resulted in tools to assist with the analysis of rich and complex qualitative data. An excellent example of software that aids in the analytic tasks necessary to process and explore qualitative data is described in this volume by Talanquer (19).

Although it is impossible to create a list of analytic procedures that researchers might use to unpack and understand video data, it is possible to capture the essence of many techniques in two frequently used methodological categories--discourse analysis and observation protocols.

Discourse Analysis

Because observing and videotaping teaching and learning settings provide a rich account of interactions among participants, discourse analysis is a frequently used technique to understand the phenomena captured in such observations. With the aim of determining how discourse supports or constrains opportunities to engage in experimentation and making sense of new experiences in secondary chemistry, Bleicher, Tobin, and McRobbie (20) used discourse analysis. Authors videotaped instruction and broke down transcriptions of classroom interactions into smaller analytic units (shorter than sentences) called message units. They describe how communication occurs in small chunks including words, gestures, changes in inflection/intonation, speed of delivery, etc. and the characteristics of the message units encode meaning. Discourse analysis as a procedure has been widely employed in sociology, and methodology scholars have devoted a great deal of work to describe this complex analytic method. In chemistry, an outstanding example of how discourse analysis can be used to understand the nature of students' understanding is contained in this volume. Cole, Becker, and Stanford (21) provide a detailed account of how to plan for and engage in discourse analysis and have carried this out to investigate the process of argumentation in the physical chemistry classroom (22). An alternative approach to analyzing observational data will be presented here: using research-based observation protocols.

Observation Protocols for Analyzing Teaching and Learning

Observation protocols are research tools, which may be paper or electronic, that enable an observer to rate, score, or count particular observed events. Unlike field notes which can be used to document any observed event, observation protocols preselect for specific observables that are aligned with the theoretical framework underlying the protocol. In this chapter, using observation protocols is classified as a data analysis technique because of the emphasis on capturing observations on video. It is important, however, to note that using protocols can be synchronous with the observation (i.e., researchers fill out the protocol during a classroom observation), which can be considered more aligned with data collection.

Using an observation protocol or other instrument to analyze observational data captured in teaching and learning settings acts like any other measurement tool. It selects for and probes a particular characteristic of the data and excludes other characteristics. For example, a mass spectrometer can determine the masses of molecules found in a sample, but NMR is required to determine the structure of the molecule fragments in the same sample. Likewise, one observation protocol designed to characterize the interactions among students in a chemistry classroom may not serve the researcher in understanding the content knowledge of the teacher. As a result, the protocol must strongly align with the characteristics of interest in the teaching and learning setting. Most observation protocols used in high school and college classrooms not only aim to characterize the learning setting but also evaluate its quality in comparison to best practices. This framework significantly narrows the scope of the analysis. Although it makes the process seem more straightforward, it is essential to be aware of the instrument's selection criteria and what aspects of the teaching and learning experience are excluded from the analysis by virtue of using an analytic protocol.

A search using the Google Scholar database yielded a large number of research-based observation protocols for science classrooms that included documentation of their development and use as well as reports of the validity and reliability of data generated by such instruments. These protocols are shown in Table 1. Although the search was not exhaustive, the protocols represent a wide range of instrument structures (e.g., qualitative versus quantitative; inclusive versus exclusive of artifacts; scaled versus unscaled items). The following discussion of the protocols highlights their important characteristics and suggests key considerations for using any protocol for observational data analysis.

Table 1. Existing Research-Based Observation Protocols for Science Classrooms

• *Classroom Assessment Scoring System(23)*	• *Oregon Collaborative for Excellence in the Preparation of Teachers* (OCEPT) *Classroom Observation Protocol (30)*
• *Electronic Quality of Inquiry Protocol (EQUIP) (24)*	• *Reformed Teaching Observation Protocol* (RTOP) *(31)*
• *Expert Science Teaching Education Evaluation Model (25)*	• *Science and Engineering Classroom Learning Observation Protocol(32)*
• *Framework for Teaching (26)*	• *Science Management Observation Protocol* (SMOP) *(33)*
• *Horizon Analytic and Observation Protocol (27)*	• *Secondary Science Teaching Analysis Matrix (34)*
• *Individualizing Student Instruction Classroom Observation Protocol (28)*	• *Science Teacher Inquiry Rubric (35)*
• *Inquiring into Science Instruction Observation Protocol (ISIOP) (29)*	• *Teaching Dimensions Observation Protocol* (TDOP) *(36)*
	• *UTeach Observation Protocol (37)*

The instruments and related published papers on their development and implementation were analyzed to determine their common and unique characteristics. The results of this analysis are summarized in Table 2 and demonstrate that many of the instruments have characteristics in common. For example, all 14 observation protocols focus on student engagement. As a result, student-centered classrooms would earn higher scores on particular items than would teacher-centered classrooms. The protocols also emphasize the content knowledge of the teacher. Although they are not specific to any science discipline, they are aimed at quantifying or qualifying teachers' understanding of the subject matter for the lesson observed. Additionally, observation protocols in Table 1 generate data that are independent of student achievement. These instruments are intended to characterize and measure the quality of the instruction that the teacher implements without necessarily taking into account student learning gains. What is most significant about the group of instruments in total is that they use national science education reforms documents (*38*, *39*) to frame what is considered high quality science instruction. Since these documents provide a synthesis of research on how people learn and instructional practices aligned with such theories of learning, higher scores on these instruments are associated with lessons, environments, and instruction aligned with the best practices for teaching and learning.

Table 2. Common and Varying Characteristics of Classroom Observation Protocols Analyzed

Common Characteristics	*Varying Characteristics*
o Student Engagement	o Conceptual Framework
o Questioning Techniques	o Aims (Evaluation vs. Description)
o Content Knowledge	o Pedagogical Emphases
o Independent of Student Achievement	o Classroom Management
	o Data Type (Likert-type Scale vs. Descriptive Indicators)
	o Nature of Data (Qualitative, Quantitative, or Both)
	o Observer Training

Many characteristics vary among the observation protocols listed in Table 1, particularly the conceptual frameworks and aims of the instruments. The *Inquiring into Science Instruction Observation Protocol* (ISIOP) (*29*) and *Electronic Quality of Inquiry Protocol* (EQUIP) (*24*), for example, are designed for evaluating inquiry-based lessons, but the *Reformed Teaching Observation Protocol* (RTOP) (*31*) is framed more generally around instruction that aligns with a constructivist theory of learning. Alternatively, the *Teaching Dimensions Observation Protocol* (TDOP) (*36*) does not aim to evaluate instruction but rather to describe it.

Another varying characteristic of the protocols is whether or not they seek to evaluate classroom management practices. The *Science Management Observation Protocol* (SMOP) (*33*) allows observers to focus on specific teacher behaviors and classroom characteristics that influence how well an inquiry-based classroom is managed; however, the other protocols have few, if any, indicators related to classroom management. This may be due to the absence of such managerial issues in the national documents providing the foundation for the protocols.

The protocols also vary in the method of measurement (quantitative, qualitative, or both). EQUIP and the *Horizon Analytic and Observation Protocol* use both quantitative and qualitative data. They contain rubrics with descriptive indicators for each possible numerical rating, but they also have several places for the researcher to provide open-ended responses about the lesson observed. The RTOP and SMOP are primarily quantitative observation protocols with Likert-type scales.

One important varying characteristic among the observation protocols pertains to their ability to generate quality data when used with raters with differing levels of experience--specifically, whether they require observer training before use. TDOP and UTOP have training manuals, CLASS training seminars can be purchased, and EQUIP and RTOP have training manuals and online training modules. The modules are free and include videos for trainees to rate and then compare their scores to those of expert raters. Negotiation among raters in a research team further improves the quality of implementation of the protocols. The benefit of using instruments with training modules is clear--calibration allows for the generation of valid and reliable data.

Observation Protocols with Particular Promise for Chemistry

Researchers have yet to publish an observation protocol specifically for use in chemistry classrooms; however, the two protocols with extensive training and research bases – RTOP and EQUIP – are strong contenders for CER studies relying on an observation protocol that measures instructional features in relation to practices aligned with national science instruction reforms. These protocols have commonly been used while viewing video data but can also be used while observing in classrooms.

The RTOP is designed for evaluating instruction in mathematics and science classrooms from middle school through college. The score on the RTOP describes the extent to which the class observed aligns with reforms (*33, 34*). High scores are associated with student-centered classrooms where student discourse drives the lesson, students build conceptual understanding from carrying out investigations, and frequent communication occurs among students. The 25 RTOP items are divided into three scales, namely, lesson design and implementation, content (propositional and procedural knowledge), and classroom culture. Each item is rated on a 5-point Likert-type scale from 0 to 4 (never occurred to very descriptive). An overall RTOP score ranges from 0-100, based on the total of all 25 items. The RTOP is the most widely reported observation protocol in the science education literature. In CER, Yezierski and Herrington (*14*) used the RTOP in a five-year study of teacher participants in Target Inquiry, a professional development program in high school chemistry. Yezierski and Herrington modified the RTOP using expert rater descriptions to improve inter-rater agreement. Use of the RTOP in this study also included a minimum of three trained raters per video and provisions to minimize instrument sensitivity and calibration decreasing over the course of a longitudinal study. For CER, the RTOP contains items which map onto dimensions of teaching and learning chemistry that distinguish chemistry from other science disciplines and are valued in CER. For example, RTOP item 9 [*Elements of abstraction (i.e., symbolic representations, theory building) were encouraged when it was important to do so*] and RTOP item 11 [*Students used a variety of means (models, drawings, graphs, concrete materials, manipulatives, etc.) to represent*

24

phenomena] address important dimensions of chemistry teaching and learning; however, with only two out of 25 items, the chemistry-specific aspects of the observation are minimized.

An instrument with great promise that has yet to be used in CER is EQUIP. This protocol was designed to measure the quantity and quality of inquiry-based instruction in K-12 math and science classrooms; however, it has promise for implementation in student-centered college and university chemistry classrooms as well. It is a qualitative and quantitative rubric-style instrument that consists of descriptive indicators and a numerical rating scale. With 19 indicators that are divided into four constructs—instruction, discourse, assessment, and curriculum – each indicator is rated on a 4-point scale from 1 to 4 (pre-inquiry to exemplary inquiry). Each construct (4 to 5 indicators) is then given a summative rating based on the essence of the indicators within that particular category. After all four constructs are rated, the lesson is assigned a summative score based on the essence of the overall lesson, which is not necessarily an average of the individual construct scores. Recent unpublished work in the Yezierski research group demonstrated that EQUIP produced reliable data when implemented with lessons ranging in their extent of alignment with quality inquiry teaching. A chemistry-specific supplement for EQUIP has been developed and is currently under study in the Yezierski group.

The majority of the published observation protocols were designed specifically for use in formal science education environments. Chemistry learning in informal environments can be likewise studied through observation and analyzed with a protocol. Christian and Yezierski (*40*) created an observation rubric to evaluate the quantity and quality of viewers' interactions with a chemistry museum exhibit focused on chemical and physical change. The protocol for informal chemistry education research was informed by the contextual model of learning (*41*) – a theory that describes the parameters of learning outside of formal environments.

Validity and Reliability Considerations

As with all CER, demonstrating the validity and reliability of findings is essential. Although the commentary on the validity and reliability in the research report of a study is often associated with the results, researchers generating high quality studies using observational data have considered validity and reliability throughout the study, beginning with the design. How many observations of a particular modeling technique under evaluation are necessary to accurately characterize the method? How many different raters using an observation protocol with video data are feasible given the time frame of the study? These example questions have serious implications for the quality of the findings. There are no hard and fast answers, since the decisions are contingent upon the research questions, theory, setting, and other contextual considerations. What can be generally asserted, however, is that forethought and planning, along with timely adjustments to methods, will contribute to a more robust research process.

Some general comments about validity and reliability for observational data are presented in this section as food for thought.

Validity is often determined by the theory underlying the analytic lens used by the researcher. For example, the validity of data generated by the RTOP comes from the instrument's foundation in constructivism and national reform documents. High RTOP scores, because of the criteria on which the instrument is built, can be associated with instruction and a classroom environment aligned with reformed science education. A novel means of demonstrating credibility of findings was carried out by Ermeling (42) in an observational study of the effects of a chemistry teacher professional development model. To ensure that changes in instruction could be attributed to the professional development, he identified a "tracer",a clearly defined element that should be observed only as a result of the intervention, and monitored it throughout the study. The tracer he tracked was the teachers' promotion of struggle (facilitation of cognitive dissonance) to expose student misconceptions, because this phenomenon was explicitly taught during the professional development and was not observed until after the teachers learned this strategy.

With respect to observational data, reliability often comes in the form of inter-rater agreement. This can be achieved through multiple observers, coders, and raters (if using a protocol). Yezierski and Herrington (14) used three raters to score observations with the RTOP and negotiated scores that differed by more than 5%. Such negotiations brought to light how challenging it was for one rater to observe all of the interactions along with lesson, student, and teacher dimensions evaluated by the RTOP. Although one video viewed by three researchers yielded primarily convergent scores on items (evidence of the efficacy of training in the protocol resulting in strong inter-rater agreement); more subtle phenomena recognized by one rater could be discussed by the entire rater team during the negotiations. This assured a thorough analysis and contributed to the validity of findings.

Reporting Results of Observational Research

Observational data are transformed by a series of decisions imposed by the research questions, duration, frequency, and focus of the observations, as well as by the analytic methods used by the researchers. It is essential for researchers to carefully describe and explain the rationale for each of these transformations and the resulting limitations of the findings. The limitations properly situate the researcher's claims in a context to help the reader evaluate the transferability of the findings. Transparency and detailed descriptions of methods are not the only means to producing high quality reports of observational research. Reports of how findings can be triangulated among multiple data sources such as artifacts also strengthen the quality of the findings. For example, Treagust, Chittleborough, and Mamiala (43) investigated the relationships between teaching, scientific, mental, and expressed models in organic chemistry and triangulated findings from three sources: classroom observations, interviews with students, and student questionnaires. Descriptions of how findings can be triangulated by multiple sources strengthen their validity and are essential in research reports.

Unlike quantitative data that can be reduced to descriptive statistics, observational data sets including videos and lengthy transcripts are impossible to report in total in a manuscript. However, there are a host of approaches that enable the researcher to provide adequate evidence to support findings from studies including observational data. First, if observational data are analyzed through coding or other similar qualitative technique, the code descriptions along with excerpts from the transcripts which illustrate the codes should be provided. In addition, agreement among raters and how discrepancies were resolved should be described. Second, if observational data are analyzed using a protocol, raw scores or data from the protocols along with reliability estimates generated from multiple raters or repeated ratings should be provided. Overall, the challenge in the report is to balance the richness and complexity of observational data with publication constraints (e.g., word limits) that enable the researcher to make valid and reliable claims about the phenomena observed.

Acknowledgments

I wish to extend my sincerest thanks to Destinee Johnson (Winthrop University Class of 2014) who contributed significantly to the section titled, *"Observation Protocols for Chemistry Teaching and Learning"* through her literature review and synthesis during her work as a Summer Undergraduate Researcher in Chemistry and Biochemistry at Miami University (NSF REU Award 1007875) in 2013.

References

1. Mead, M.; Metraux, R. *The Study of Culture at a Distance*; University of Chicago Press: Chicago, IL, 1953.
2. Glesne, C. *Becoming Qualitative Researchers*; Addison Wesley Longman: Boston, MA, 1998.
3. Mayer, D. *Educ. Eval. Policy Analysis* **1999**, *21*, 29–45.
4. Patton, M . Q. *Qualitative Evaluation and Research Methods*; Sage Publications, Inc.: Newbury Park, CA, 2002.
5. Phelps, A. J. *J. Chem. Educ.* **1994**, *71*, 191–194.
6. Bauer, C. F. Ethical Treatment of the Human Participants in Chemistry Education Research. In *Tools of Chemistry Education Research*; Bunce, D. M., Cole, R. S., Eds.; ACS Symposium Series 1166; American Chemical Society: Washington, DC, 2014; Chapter 15.
7. Policy and Guidance, 2013. U.S. Department of Health and Human Services Office for Human Research Protections. http://www.hhs.gov/ohrp/policy/ (accessed August 1, 2013).
8. Emerson, R. M; Fretz, R. I.; Shaw, L. L. *Writing Ethnographic Fieldnotes*; University of Chicago Press: Chicago, IL. 1995.
9. Harrison, A. G.; de Jong, O. *J. Res. Sci. Teach.* **2005**, *42*, 1135–1159.
10. Bulte, A. M. W.; Westbroek, H. B.; de Jong, O.; Pilot, A. *Int. J. Sci. Educ.* **2006**, *28*, 1063–1086.

11. Wu, H.-K.; Krajcik, J. S.; Soloway, E. *J. Res. Sci. Teach.* **2001**, *38*, 821–842.
12. Högström, P.; Ottander, C.; Benckert, S. *Res. Sci. Educ.* **2010**, *40*, 505–523.
13. van Driel, J. H.; de Jong, O.; Verloop, N. *Sci. Educ.* **2002**, *86*, 572–590.
14. Yezierski, E. J.; Herrington, D. G. *Chem. Educ. Res. Prac.* **2011**, *12*, 344–354.
15. Jewitt, C. *Int. J. Soc. Res. Methodol.* **2011**, *14*, 171–178.
16. Bond-Robinson, J.; Rodriques, R. A. B. *J. Chem. Educ.* **2006**, *83*, 313–323.
17. Derry, S. J. Guidelines for Video Research in Education, 2007. http://drdc.uchicago.edu/what/video-research-guidelines.pdf (accessed November 29, 2013).
18. Wu, H.-K. *Sci. Educ.* **2003**, *87*, 868–891.
19. Talanquer, V. Using Qualitative Analysis Software To Facilitate Qualitative Data Analysis. In *Tools of Chemistry Education Research*; Bunce, D. M., Cole, R. S., Eds.; ACS Symposium Series 1166; American Chemical Society: Washington, DC, 2014; Chapter 5.
20. Bleicher, R.; Tobin, K.; McRobbie, C. *Res. Sci. Educ.* **2003**, *33*, 319–339.
21. Cole, R. S.; Becker, N.; Stanford, C. Discourse Analysis as a Tool To Examine Teaching and Learning in the Classroom. In *Tools of Chemistry Education Research*; Bunce, D. M., Cole, R. S., Eds.; ACS Symposium Series 1166; American Chemical Society: Washington, DC, 2014; Chapter 4.
22. Becker, N.; Rasmussen, C.; Sweeney, G.; Wawro, M.; Towns, M.; Cole, R. S. *Chem. Educ. Res. Prac.* **2013**, *14*, 81–94.
23. Pianta, R. C.; La Paro, K.; Hamre, B. K. *Classroom Assessment Scoring System (CLASS);* Brookes Publishing: Baltimore, MD, 2008.
24. Marshall, J. C.; Smart, J.; Horton, R. M. *Int. J. Sci. Math. Educ.* **2009**, *8*, 299–321.
25. Burry-Stock, J. A.; Oxford, R. L. *J. Personnel Eval. Educ.* **1994**, *8*, 267–297.
26. Danielson, C. *The Handbook for Enhancing Professional Practice*; Association for Supervision and Curriculum Development (ASCD): Alexandria, VA, 2003.
27. Weiss, I.; Pasley, J.; Smith, P.; Banilower, E.; Heck, D. *Looking Inside the Classroom: A Study of K-12 Mathematics and Science Education in the United States*; Horizon Research: Chapel Hill, NC, 2003.
28. Connor, C. M.; Morrison, F. J.; Fishman, B. J.; Ponitz, C. C.; Glasney, S.; Underwood, P. S.; Schatschneider *Ed. Res.* **2009**, *38*, 85–99.
29. Minner, D.; DeLisi, J. *Inquiring into Science Instruction Observation Protocol (ISIOP)*; Educational Development Center, Inc.: Boston, MA, 2012.
30. Wainwright, C. L.; Flick, L.; Morrell, P. *J. Math. Sci.* **2003**, *3*, 21–46.
31. Sawada, D.; Piburn, M.; Judson, E.; Turley, J.; Falconer, K.; Benford, R.; Bloom, I. *School Sci. Math.* **2002**, *102*, 245–253.
32. Dringenberg, E.; Wertz, R. E.; Purzer, S.; Strobel, J. *Development of the Science and Engineering Classroom Learning Observation Protocol*; American Society for Engineering Education: Washington, DC, 2012.
33. Sampson, V. *Sci. Teach.* **2004**, *71*, 30–33.
34. Adams, P. E.; Krockover, G. H. *J. Res. Sci. Teach.* **1999**, *36*, 955–971.

35. Beerer, K.; Bodzin, A. Science Teacher Inquiry Rubric (STIR), 2003. http://www.lehigh.edu/~amb4/stir/stir.pdf (accessed June 11, 2013).

36. Hora, M. T.; Oleson, A.; Ferrare, J. J. Teaching Dimensions Observation Protocol (TDOP), 2013. http://tdop.wceruw.org/Document/TDOP-Users-Guide.pdf (accessed November 30, 2013).

37. Walkington, C.; Arora, P.; Ihorn, S.; Gordon, J.; Walker, M.; Abraham, L.; Marder, M. UTeach Observation Protocol (UTOP), 2009. http://cwalkington.com/UTOP_Paper_2011.pdf (accessed June 11, 2013).

38. *Science for All Americans. Benchmarks for Science Literacy*; American Association for the Advancement of Science: Washington, DC, 1993.

39. *National Science Education Standards*; National Research Council, National Academy Press: Washington, DC, 1996.

40. Christian, B. N. Improving Outcomes at Science Museums: Blending Formal and Informal Environments to Evaluate a Chemical and Physical Change Exhibit. M.S. Thesis, Miami University, Oxford, OH, 2012.

41. Falk, J. H.; Dierking, L. D. *Learning from Museums: Visitor Experiences and the Making of Meaning*; AltaMira Press: Lanham, MD, 2000.

42. Ermeling, B. A. *Teach. Teacher Educ.* **2010**, *26*, 377–388.

43. Treagust, D. F.; Chittleborough, G. D.; Mamiala, T. L. *Res. Sci. Educ.* **2004**, *34*, 1–20.

Chapter 3

Using Interviews in CER Projects: Options, Considerations, and Limitations

Deborah G. Herrington[*,1] and Patrick L. Daubenmire[2]

[1]Department of Chemistry, Grand Valley State University, Allendale, Michigan 49401, United States
[2]Department of Chemistry and Biochemistry, Loyola University Chicago, Chicago, Illinois 60660, United States
*E-mail: herringd@gvsu.edu

Interviews can be a powerful chemistry education research tool. Different from an assessment score or Likert-scale survey number, interviews can provide the researcher with a way to examine and describe what we cannot see, aspects such as feelings, thoughts, or explanations of thinking or behavior. Most people have no doubt seen countless interviews on TV news and talk shows. These sessions might convey interviewing as a spontaneous, easy, and straightforward process. However, using interviews as a meaningful research tool requires considerable thought, preparation, and practice. Embedded within a chemistry education research context, this chapter provides a general introduction to the use of interviews as a research tool including how to plan, conduct, and analyze interviews. It highlights important considerations for designing and conducting fruitful interviews, shares examples of different ways in which interviews have been used effectively in chemistry education research, and provides additional references should the reader want to delve more deeply into particular topics.

Introduction

Improving teaching and learning in chemistry requires an understanding of what students know and the nature of their difficulties with the content. Well-constructed interviews can provide chemistry education researchers with a rich data set that affords a glimpse into students' thought processes. Furthermore, they can help researchers understand what other factors might play a role in students' varying levels of success in chemistry. Consider a comparison of different student assessment methods. A multiple choice assessment is quick to grade and can quickly indicate that a student does not understand the material, but it is less likely to ascertain the student's particular difficulty with the material. On the other hand, an open ended question on a test may take longer to grade, but it can better detect the specific problem a student has with the content. Lastly, an oral final exam, which allows for follow-up questions to probe more deeply into students' understanding of a topic, takes the most time but will allow for the best identification of specific content issues. Interviews are most like this last form of assessment. They may take longer to conduct and analyze, but the wealth of information obtained from even just a few student responses can potentially help improve instruction for the whole class.

Interviews are most useful in answering why and how questions. For example, a multiple choice test or a survey could be used to identify gains in student achievement or attitudes as a result of a particular intervention. Interviews, on the other hand, can help determine *why* these gains are observed or maybe how students are applying elements of a particular intervention in solving problems. There are many good books and papers that provide in depth information regarding interview methods, several of which we have cited in this chapter. The goal of this chapter, however, is to provide an overview of important considerations in using interviews as research tools specifically for chemistry education research (CER), particularly for those new to the use of interviews. Thus, in this chapter we aim to provide examples of two common, but somewhat different types of interviews that have been used successfully in CER studies, highlight many practical considerations for planning for an interview, conducting an interview, and analyzing interview data, and point readers to additional resources should they want to delve into any of the topics in more detail.

Types of Interviews

There are several different types of interviews that can be used for data collection in CER. In this chapter we will focus primarily on the two types of interviews most commonly used in CER and thus probably most valuable for those new to using interviews as a tool for CER: open-ended and think-aloud. Table I provides examples of two CER studies that used interviews as their primary data collection tool. One used a structured open-ended interview format while the other used a think-aloud interview protocol. Throughout the chapter we will refer to these two examples, along with other CER references, to discuss the important aspects to be considered in designing, conducting, and analyzing interviews. Of particular note, while there are numerous studies in science education research, and more specifically in CER, that use interviews as data collection methods, few of them actually provide the interview protocols used in obtaining the data. In this chapter we have chosen to use as examples only studies for which interview protocols were readily available.

Table I. Examples of the Use of Structured Open-Ended and Think-Aloud Interviews in CER Studies

Type	Structured, Open-Ended	Think-Aloud
Title	Exploring Conceptual Integration in Student Thinking: Evidence from a Case Study (1)	"It Gets Me to the Product": How Students Propose Organic Mechanisms (2)
Research Question	To what extent do students achieve "conceptual integration" of the science they are learning about in school, in particular across related topics in chemistry and physics?	What strategies do students enrolled in a first-semester graduate level organic chemistry course use to solve mechanistic problems?
Rationale for Type of Interview	In examining how students integrate concepts across subjects, students need to be asked in some depth about two potentially related areas of science, requiring more in-depth qualitative approaches such as interviews. A structured interview was chosen as it provided a means to collect data about student thinking over a range of topics within a realistic span of time.	In order to make explicit students' organic chemistry problem solving strategies, the researchers chose a think-aloud protocol, asking participants to describe their thoughts while solving a series of organic mechanism problems.

Open-Ended Interviews

Most people are probably familiar with open-ended types of interviews where the interviewer asks a question and the interviewee responds to the question. Although on the surface this may appear to be a fairly easy tool to use, the quality of the data obtained through interviews is highly dependent upon the interviewer and the structure of the interview. Thus, preparing for interviews is very important, and choosing an interview format that is well aligned with the research question(s) and theoretical framework is an essential first step. There are several different types of open-ended interviews, which may have slightly different names depending on the author (3–5), but the main distinguishing feature among them is the level of structure to the interview protocol. Figure 1 depicts the different types of open-ended interviews on a continuum. It should be noted, however, that it is often the case that interview protocols fall into more than one of these categories with some sections that are more structured and some that are less structured.

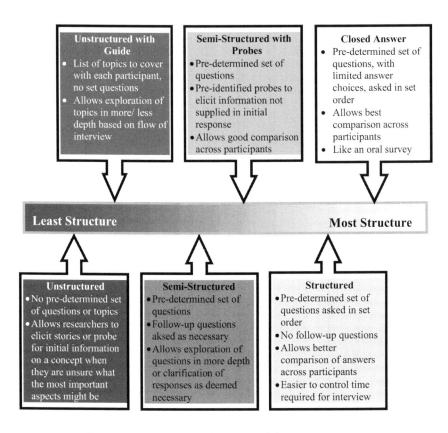

Unstructured with Guide	Semi-Structured with Probes	Closed Answer
• List of topics to cover with each participant, no set questions • Allows exploration of topics in more/ less depth based on flow of interview	• Pre-determined set of questions • Pre-identified probes to elicit information not supplied in initial response • Allows good comparison across participants	• Pre-determined set of questions, with limited answer choices, asked in set order • Allows best comparison across participants • Like an oral survey

Least Structure   **Most Structure**

Unstructured	Semi-Structured	Structured
• No pre-determined set of questions or topics • Allows researchers to elicit stories or probe for initial information on a concept when they are unsure what the most important aspects might be	• Pre-determined set of questions • Follow-up questions aksed as necessary • Allows exploration of questions in more depth or clarification of responses as deemed necessary	• Pre-determined set of questions asked in set order • No follow-up questions • Allows better comparison of answers across participants • Easier to control time required for interview

Figure 1. Continuum of Open-Ended Interview Types.

Unstructured Interviews

The least structured type of interview is often called an unstructured or conversational interview. This type of interview is most commonly used in ethnographic studies where people are interested in stories or in cases where the researcher does not know enough about a phenomenon to ask specific questions. In the latter case, the researcher can use information from these interviews to design a more structured interview protocol (*3, 5*). In this type of interview, there is no set of pre-determined interview questions. The interviewer has freedom to pursue any line of questioning that he/she believes will provide interesting and relevant data. In some cases the researcher will have a list of topics to cover during the interview, but the topics are not covered in any particular order nor are there any set questions about these topics.

Unstructured interviews are not very common in CER studies as many CER studies have specific research questions they are interested in answering or a specific chemistry topic they want to focus on; however, some researchers may chose to use unstructured interviews to gather information that can help them develop more structured interview protocols to use in their data collection. One interesting use of an unstructured interview in CER is a recent study designed to identify effective instructional strategies for assisting a visually impaired student in understanding gas laws (6). The researchers used tutoring sessions as an initial data source. These sessions were essentially unstructured interviews where the questions were posed by the visually impaired student (the participant). As the tutor (the researcher) responded to those questions, he gained an understanding of the difficulties the student encountered in learning the content.

It is important to note that unstructured interviews require the most skill on the part of the interviewer and make it more difficult to make comparisons across participants. Though those new to the use of interviews as a research tool may view unstructured interviews as the easiest to conduct as there is seemingly little, if any, upfront preparation, as discussed later in this chapter, constructing good interview questions requires careful consideration. In unstructured interviews, the interviewer must instinctively, as the interview progresses, determine what questions would best elicit the desired information in a way that does not bias or lead the participant. Furthermore, the questions asked in unstructured interviews will be different for each participant which limits the researcher's ability to compare across participants.

Structured Interviews

More structured types of interviews (semi-structured – structured) are much more commonly used in CER studies. These types of interviews have a predetermined set of questions to ask participants and are particularly useful in investigating specific research questions or topics. In a truly structured interview protocol, participants are all asked exactly the same question in the same order. These types of interview protocols are typically best for studies that want to cover several topics in a limited amount of time, where the ability to easily compare responses between participants is important, and where researchers are more interested in participants' knowledge and/or experience as opposed to their opinions or feelings or problem solving strategies.

The open-ended study highlighted in Table I provides a good example of a study that uses a structured interview protocol (1). As the goal of this study was to investigate "conceptual integration" across particular related topics in physics and chemistry, the authors wanted to be able to cover a wide range of topics in a reasonable amount of time as well as be able to compare answers across participants. Carefully choosing the questions that each student was asked prior to the study allowed the researchers to work their way through several topics in about 45 min – 1 hour. Asking all the participants the same question in the same order made it easier for the researchers to compare responses between participants. Moreover, as this study focused primarily on evaluating aspects of students' knowledge, the use of follow-up questions to dig more deeply into student responses is arguably not as important here as in studies that are interested in opinions, feelings, etc. Thus, a more structured interview aligns well with this study's research questions and design. The questions in one small segment of the interview were presented as follows:

- Do you know what the composition [make-up] of an atom of sodium would be?

    o   (Can you tell me about the structure [arrangement of parts] of the sodium atom?)

- Do you think that a single sodium atom could fall apart? (Could the outer electron fall out of the atom?) (Why?/Why not?)

- What do you think holds the protons together in atomic nuclei?

Another good example of the use of a protocol that is more on the structured end of the continuum is a study by Cacciatore and Sevian (7) that looked at whether changing a single laboratory experiment could improve students' general chemistry performance. In this paper, the authors report both their interview questions along with the rationale behind each question. For example:

"Question: Which labs were most helpful in learning the lecture material? Why?"

"Rationale: Assesses student's beliefs about the connections between lecture and laboratory portions of the chemistry course."

Semi-Structured Interviews

True semi-structured interviews, on the other hand, typically have a pre-determined set of questions, but the order and the exact wording of the questions may differ somewhat depending on the participant and his/her responses. Also, semi-structured interview protocols typically involve asking follow-up questions to dig deeper into particular topics of interest or seek clarification to a participant's response. Semi-structured interviews are best used in studies that look to examine one or two focused topics in more depth, for studies that are more focused on identifying important similarities and differences and less focused on being able to directly compare responses to particular questions between participants, and for studies that are more focused on students' ideas, opinions, beliefs, etc. as opposed to their knowledge about a topic.

For example, Cole and Todd (8) used a semi-structured interview protocol to examine the effects of web-based multimedia homework on student learning in general chemistry. For this study the authors collected several forms of quantitative data (homework, laboratory, and exam scores; standardized test scores (ACT and Math Placement scores); and a version of the Group Assessment of Logical Thinking) to evaluate the impact on knowledge gains. Interviews were used in conjunction with quantitative data to ascertain students' opinions about the value of the on-line homework. Although this chapter focuses on the use of interviews as a CER tool, the use of both quantitative and qualitative data to obtain a more complete picture is common in many CER studies. For the interview portion of this study, the interview protocol consisted of a set of 16 questions or topics that were covered in each of the interviews; however, the order and exact wording of the questions was not necessarily the same for each participant. In the interview protocol for this study (found in the paper's supplemental online material), the authors state that interviewers did not use the protocol verbatim, but each interview covered all of the topics listed in the protocol. The types of questions included in this interview protocol vary from straight forward background questions ("Why are you taking chemistry 103?"), to fairly specific opinion questions with pre-identified probes ("Which course tools have helped you learn the most - lecture, discussion, lab, homework, tutorials, demos, videos, animations, online quizzes, group work?"), to more general topics about which the authors wished to elicit student opinions ("Use the response to [question]13 to probe more about the use of videos, animations, etc. on how technology impacts the student's learning").

A study by Howard and co-workers (9) that examined college students' understanding of atmospheric ozone formation also used semi-structured interviews as a research tool. This study, however, used more of a "semi-structured with probes" protocol. The researchers asked each student a common a set of questions and had a related sub-set of probing questions for each primary question that were used as needed. Given that this study was again more interested in student knowledge (understanding of atmospheric ozone formation), an interview protocol towards the more structured end of the spectrum is reasonable. In this interview protocol, students were presented with a series of figures, problems, or situations related to atmospheric ozone formation about which students were then asked a series of questions. Some of these questions had additional follow-up questions that were used depending on the participant's response. For example, in the first interview question students were presented with a figure that represents the main cyclic tropospheric ozone formation components. They were then asked, "Can you explain what is happening in the Figure?" Depending on the participant's response, the follow-up probe ("Can you describe what NO, NO_2, and $HO^{\bullet}$ are, and their significance to our atmosphere?") may also have been used.

Although this section of the chapter has tried to clearly delineate the differences between unstructured, semi-structured, and structured interview protocols, it is important to remember that as illustrated in Figure 1, these interview types really represent a continuum. Furthermore, it is possible to combine different types of interviews within one interview protocol; some sections of the protocol being more structured and some being more semi-structured or even unstructured. In general, a more structured protocol is advised for people new to interviewing. This allows the researcher to spend time developing questions that will elicit sought after information and avoid using questions that could bias or lead the participant to particular answers (see section on writing interview questions). A more structured protocol is also useful if more than one researcher will be conducting interviews as it helps to minimize variation between interviewers (3).

Focus Group Interviews

Most of the time interviews are thought of as a one-on-one conversation between the interviewer and the interviewee; however, an alternative to this format is the focus group interview. Focus group interviews are also used commonly in educational research. Although originally developed for social psychological research purposes, focus group interviews became heavily associated with consumer market researchers in the 1950s, and have only more recently, 1980s, began to grow in use in academic social research (*10*). One common use of focus group interviews for academic social research is for program evaluations, such as in the assessment of an undergraduate chemistry program (*11*).

A focus group interview is essentially a discussion between a small group of individuals about a particular topic that is facilitated by a moderator (the interviewer). In many ways focus group interviews are similar to one-on-one open ended interviews and thus many of the same considerations for "regular" open-ended interviews can be applied to focus group interviews. For example, focus group interviews can be unstructured (aimed at gathering data to explore a new domain by encouraging a wide variety of viewpoints on a topic) or more structured in format; the use of interview questions that do not bias or lead the participants is crucial to obtaining meaningful information; and good alignment of research questions, design, and interview format is important. Yet, there are several considerations and benefits unique to focus group interviews that are highlighted in this section (for readers looking for more information about the use of focus group interviews see (*12*)).

In general, focus groups should be small, about 4-8 people, to allow for everyone to have a chance to speak. Although general considerations for participant selection, discussed later in this chapter, should also be applied for focus group interviews, the quality of data obtained from a focus group interview is largely dependent upon how comfortable the participants feel discussing ideas and offering alternative opinions. Thus, it is important to choose participants who both have experiences that allows them to contribute to the conversation and are similar enough that they will feel comfortable expressing their opinions. An important role of the interviewer in focus groups is to ensure that everyone has the opportunity to speak and that a minimal level of threat is maintained, so the freedom to share opposing or even counter viewpoints is respected and may be freely offered. Moreover, in focus group interviews it is more difficult to maintain the confidentiality of a participant's responses as a result of the other participants in the room. Therefore, in choosing to use focus group interviews, one must also carefully consider the sensitivity of the issues being discussed.

Despite some of the additional considerations for the use of focus group interviews, this format also provides a number of potential benefits with respect to collecting rich data. This format may allow participants who may be more reluctant in an individual interview to feel more willing and encouraged to respond. Hearing others ideas, too, may prompt one's own thinking and add to the multiple perspectives desired from this type of interview. For example, focus group interviews were used in the CER study by Stojanovska et al. (*13*) to examine misconceptions students held about the particulate nature of matter. In particular, the researchers noted that having students interact and exchange ideas during focus group interviews provided valuable information about misconceptions and source material for follow-up questions.

Think-Aloud Interviews

Think-aloud interviews differ from open-ended interviews in that they ask participants to articulate their thoughts during a specific task or when solving a particular problem. Bowen (*14*) describes this method essentially as way to listen to learners. It is less focused on broader feelings or perceptions of a course or an instructional approach and more focused on a participant's reasoning during pre-selected tasks. Think-aloud protocols are often selected when a researcher wants to know how or why a participant is using knowledge, processes, algorithms, or heuristics to solve problems or complete tasks. These verbal data help researchers make inferences about what information is focused on in a problem and what processes are selected in solving the problem.

When using think-aloud interviews, the selected problem itself becomes the source material for the interview "questions" and the participant usually does most of the talking, describing his/her thoughts and reasoning during the solving of the problem. Other prompts or follow-up questions may be used after the problem is solved. These follow-up questions, though, ask participants questions from a slightly different vantage point. During the problem solving, participants are likely providing descriptions out of introspection, what they are actually thinking during problem solving. Descriptions provided after solving the problem come more from retrospection, reflecting on the problem solution and possibly why they solved the problem the way they did (*14*, *15*). Both question pathways may provide relevant information for the research framework, but are different avenues of knowledge access by the participants and should be treated as such when analyzing these data. Ultimately, think-aloud protocols can serve as a way to uncover knowledge and mental models participants may be activating around a particular term, concept, or type of problem (*10*).

In the study by Bhattacharayya and Bodner (*2*), highlighted in Table I, the researchers asked graduate students to think-aloud while solving complex organic synthesis mechanisms. The ultimate goal of this research was to determine how prior experiences (such as, undergraduate courses) may or may not have prepared students for novel organic chemistry problems they encounter in their own graduate research. The think-aloud interviews allowed researchers to see how students used the curved-arrow or electron-pushing conventions typically taught in undergraduate organic courses. Participants described their reasoning concerning what the arrows actually designated and why they used a particular set of steps in the solution. Other CER studies have used think-aloud protocols to investigate students' thinking when solving algorithmic and conceptual chemistry problems (*16*), to assess chemistry teachers' understanding of chemical equilibrium (*17*), and to describe students' use and connections made among representations of matter (*18*).

Developing the Interview Protocol

Identifying the Desired Information before Starting the Interview Process

Interview participants provide valuable gifts: their time and insights. So, it is important that the information the researcher wants to obtain from the interview is framed prior to starting the interview process. This should be shaped by the theoretical framework, research questions, and hypotheses of the study, unless the investigation is purely exploratory, and even then, the topic of interest should be carefully considered so that questions can optimize collection of the desired data. In their "Seven Stages of an Interview Inquiry," Kvale and Brinkman (10) call this stage: "thematizing" - identifying the why and what before the how. For a structured interview this means identifying important topics to discuss during the interview so that an appropriate set of questions can be developed for the interview protocol. For example, in exploring how students integrate their scientific knowledge, Taber (1) (Table I) made a choice based on the research question. In order to determine how students integrated knowledge across chemistry and physics content areas, Taber reasoned that, "collecting data about thinking over a range of topics was more important than being able to spend time approaching particular topics from a range of perspectives." As a result, he used prior research to identify a series of topics that students would likely be able to integrate across chemistry and physics, and developed questions for an interview protocol based on those topics.

Within a think-aloud protocol, preparing for interviews could mean considering the tasks or problems to be used during the interviews, what the variables are, and why anticipated answers to those problems and students' descriptions of their reasoning would be relevant to answering the research questions (14). For instance, Bhattacharayya and Bodner (2) situated their research within a phenomenographic framework. They chose a think-aloud interview because it would give a voice to the participants by making the students' underlying thought processes explicit, thus providing the researchers an opportunity to uncover and to interpret student strategies. Though the researchers anticipated "multiple voices," they hypothesized that there would be a finite number of differing approaches which would allow them to characterize a finite set of problem solving strategies.

In addition to determining the type of information desired from an interview, researchers must also carefully consider the interview format. For instance, if a researcher is interested in students' perceptions of learning with a particular instructional approach, a focus group interview might seem to be a good format for obtaining multiple and varied viewpoints about these perceptions. In using this format, though, confidentiality has to be protected as much as possible. Comments shared by individual participants should not be shared with the professors who are using the instructional approach being studied. Instead, comments need to compiled and not be attributable to individual students in the class. Ethical considerations must be simultaneously at the forefront in the thematizing stage as well as throughout interview inquiries. To assist the researcher with this, guidelines for working with human subjects have been established and should be followed throughout the research project. Additional descriptions and guidelines for conducting research with human subjects are described in this volume by Bauer (*19*).

How To Construct Good Interview Questions

The questions asked in an interview ultimately depend on the focus of the study, but there are some guidelines to keep in mind when writing interview questions to help ensure that they furnish valuable data. There are several different types of questions that can be asked. Although there are no doubt multiple ways to classify types of interview questions, one useful classification scheme is provided by Patton (*3*). Patton suggests that all interview questions can be classified into one of six different categories which have been summarized in Table II. Table II also highlights the type of information that each type of question is meant to elicit and provides an example of how each type of question has been used in a CER or science education research study.

Although there is no set way to order interview questions, in general, most interviews start out with Background/Demographic type questions and then move to other types of questions. Both Patton (*3*) and Merriam (*5*) suggest that a good progression is to first ask participants to describe a situation (Experience/Behavior or Knowledge questions) and then follow-up with questions about how they feel about the situation (Feeling or Opinion/Value questions). In our experience, this provides a good model to follow. For example, in studying the Target Inquiry professional development program and its impact on teachers understanding of inquiry instruction, Herrington and co-workers (*20*) first asked high school teachers to use a set of cards to construct their model of inquiry-based instruction (Experience/Behavior) and then asked them to explain and justify their model (Opinion/Value).

Another key consideration is how to word (or phrase) the questions. First, it is important that the questions are clear to the person being interviewed. Using terms or disciplinary jargon should be avoided. Second, if participants feel uncomfortable answering a question it is unlikely that they will provide useful data. If asking a somewhat controversial or personal question, posing it in a way that takes the direct focus off of the participant may be helpful. Some examples include asking a hypothetical question, playing Devil's advocate (Some people might say…), or asking about an ideal situation (*5*). For example, asking students what they thought about the feedback they received on assignments in their chemistry class may not yield completely honest answers if students are concerned that their responses might get back to their instructor. However, rewording the question and asking "If you were teaching this course, what kinds of feedback would you want to give students on their assignments?" may allow students to voice opinions about things they felt were lacking from the feedback without worrying about giving a negative response.

Table II. Types of Qualitative Interviews Questions

Question Type	Type of Information Provided	Example in CER
Background/ Demographic	Used to identify characteristics of a person or program (age, chemistry courses taken, major, years of teaching experience, number of students, etc.).	*How many faculty teach the lecture portion of this course? Are they all tenured or tenure track? (how many tenured, tenure-track, and/or contract) (21)*
Experience/ Behavior	Used to elicit information about experience or behaviors that would be visible if the interviewer were present as an observer.	*How do you study for this course? Describe a typical week. (8)*
Opinion/Value	Used to try and understand what a person thinks or their rationale for a certain action/decision. Used commonly as probes during or after think-aloud interview protocols (Can you explain why you chose to use that method?)	*Which labs were most helpful in learning the lecture material? Why ?(7)*
Feeling	Used to identify participant's emotions.	*How do you feel about science subjects at your school? (22)*
Knowledge	Used to determine a person's factual knowledge.	*I dissolve lead sulfate in water to form lead ions and sulfate ions. What will happen if I add solid lead sulfate to this? It is at equilibrium initially. (23)*
Sensory	Used to identify sensory inputs (sight, sound, taste, smell, and touch) experiences in a situation.	*Which tactile representations of images did you find to be helpful / not helpful and why? (6)*

Finally, there are some things that researchers should take care to avoid when crafting interview questions. Asking multiple questions at once is problematic. For example, "How would you assess your learning and effort in CHM 100?" is a poor question because learning and effort are two separate things. If the researcher is interested students' assessments of both learning and effort, each of these should be asked about separately. This separation will also facilitate analysis of the interview data. In general, yes or no questions should also be avoided as they do not yield the rich, descriptive data desired from a qualitative interview. For example, if investigating the structural features of molecules that students focus on when making predictions about chemical reactions, consider the following ways of asking the same question:

(1) Are there any structural features of the molecule that you looked at in deciding what type of reaction would occur?

(2) What specific structure features of the molecule did you look at in deciding what type of reaction would occur?

The first question students can answer with a yes or no. Of course this could be followed up with "which ones?" but this can lead to a back and forth that is more like an inquisition than an interview. The second question, on the other hand, prompts students to identify the structural features without the need for a follow-up. There are, however, some cases where yes or no questions can be appropriate and useful. For example, in the Taber study (*1*) highlighted in Table I, several yes or no questions were used in the interview protocol. In some cases a yes or no question was used to determine whether a student was familiar with a particular phenomenon , such as a balloon sticking to a wall after being rubbed on a sweater, before asking them follow-up questions about that phenomenon. In others, a yes or no question was asked first to determine whether a student thought a particular thing was possible (e.g., *Do you think a single sodium atom can fall apart?*), and then it was followed up with a why/why not to elicit the student's rationale.

Perhaps the most important and most difficult pitfall to avoid is using leading questions. At first glance a question such as "What did you like about CHM 100?" may seem like a good open-ended question; yet, it carries with it the implicit assumption that the class was good. Another way to approach this is to say, "Tell me what you thought about CHM 100." This invites the participant to discuss both positives and negatives of the course.

One final caution is using why questions. Although a well-placed "why do you think that" can sometimes provide valuable insights, why questions can also sometimes hinder the collection of meaningful data (*3*). When asking a participant why they answered a question in a particular way, the interviewee may feel that their answer was somehow inappropriate or inadequate. Simply rephrasing the question as, "Can you tell me more about your thought process in answering that question?" may be more inviting. Furthermore, in some cases there are many reasons "why" a person might chose to do something that could include personal choice, level of understanding of the topic or content, desire to please the interviewer, etc. A participant may not be able to distinguish among these reasons and clearly articulate which one explains his/her answer. In this case, if a researcher is particularly interested in one thing – for example, a feature of the question that prompted a particular response – then more useful data may be obtained from rewording the question to ask, "Can you tell me what features of the question resulted in you choosing that problem solving method?" as opposed to asking, "Why did you solve the problem that way?"

How To Construct Good Tasks for Think-Aloud Protocols

Considerations for think-aloud interviews are somewhat different from constructing good open-ended questions as participants are describing their thought processes as they complete pre-designed tasks. In think-aloud protocols, the goal is to have the participant describe what he/she is thinking while completing the task without any interruptions or prompting questions. This makes the development of the tasks very important. Though the specific tasks will differ based on the research question, there are several things to consider when choosing appropriate tasks. First, the tasks have to be problems that participants cannot solve automatically, avoiding situations where a participant may be able to get the correct answer without actually being able to describe how they got there (*15*). For example, students may be able to draw a Lewis structure for CO_2 from memory without the need to think about how to draw it. On the other hand, tasks should not be too unusual. If a participant does not know where to start, he/she will probably not be able to provide useful interview data. For example, in Bhattacharyya and Bodner's study of organic mechanisms (*2*), they used problems that required 2-4 step mechanisms, many of which were found in standard undergraduate organic chemistry textbooks. Using resources such as textbooks to generate a pool of items for selecting appropriate tasks for developing think-aloud protocols is a great strategy (*14*).

The time required for participants to complete the tasks is also an important consideration. In a study looking at concept learning versus problem solving, Nakhleh and Mitchell (*16*) asked students to solve one conceptual and one algorithmic gas law problem from the exam their recent exam as well as a pair of stoichiometry problems using the think-aloud method. Each interview took approximately 50 minutes. This illustrates that it is important to recognize that think-aloud protocols are typically limited to just a few problems because thinking aloud while solving problems generally requires more time than just completing the task alone. Session that are too time intensive could lead to participant fatigue which in turn can affect the quality of the data obtained.

Another consideration in developing think-aloud tasks is what resources will be provided to participants (periodic table, calculator, textbooks, molecular models, etc.) (*14*). For example, in the Bhattacharyya and Bodner study (*2*), they provided their participants with two different comprehensive organic textbooks and a set of molecular model kits to try and eliminate content knowledge as a confounding variable in their study of organic problem solving.

Piloting the Interview Protocol

The importance of piloting the interview protocol cannot be stressed enough. One way to pilot an interview protocol is to practice with someone who is familiar with using interviews as a tool in educational research as they will likely be able to provide valuable feedback about the interview questions themselves and give some practice dealing with many different issues that could be encountered in the interviews (e.g., the reluctant participant, the participant who has difficulty answering the question you have asked, etc.). However, it is also important to conduct a few pilot interviews with the target population for the study. This is the fastest way to figure out which questions are confusing to the participants, which questions elicit unanticipated answers, and which questions do not provide meaningful data and thus need to be reworded or eliminated completely from the protocol. Furthermore, participant responses from pilot interviews may suggest questions that are missing from the protocol that should be included or probes for certain questions that could be used with participants that tend to be less forthcoming with information. For example, in the study of the Target Inquiry program, after the first year of the program teachers were asked to describe any changes to their teaching over the past year. Several teachers also mentioned changes they had noticed in their students as a result of changes to their teaching. Thus, a follow-up probe was added to the interview protocol (*Have you noticed any changes in your students? If yes, can you describe those changes?*) as a prompt for teachers who did not volunteer information about students in their initial response. Additionally, for participants who were less forthcoming with information, additional follow-up prompts (*Prompts: Motivation? Retention? Understanding? Frustration?*) corresponding to the things teachers most frequently mentioned in relation to changes in their students were included. In McClary and Talanquer's study of student models of acid and base strength (*24*), piloting their interview protocol indicated that asking students to justify all of the acid strength ranking tasks took too long and resulted in cognitive overload. Thus, they modified their protocol so that students only justified the three most complex ranking tasks.

The use of a think-aloud protocol is somewhat different, but it is still important to pilot the interview protocol. Doing so will help determine whether the tasks are performing as expected. Moreover, when using a think-aloud protocol, it is important to practice the think-aloud procedure with each of the participants as this process is often unfamiliar to them. In the study by Nakhleh and Mitchell (*16*), the researcher trained the students by first demonstrating the think-aloud method himself as he completed a practice problem and then had each participant complete their own practice problem using the think-aloud method. More often, however, researchers provide participants with instructions regarding the think-aloud procedure and then give them a practice problem that allows them to try using the think-aloud method.

Selecting Participants

Using interviews for data collection allows the researcher to investigate selected issues or concepts in great depth, but this is only possible with participants who provide information-rich cases to study. Unlike quantitative methods which rely on random sampling to provide the most robust generalizations of statistical comparisons to the larger population, obtaining quality data from interviews requires more purposeful sampling. Although most often interview participants are volunteers, choosing appropriate volunteers is of utmost importance. For example, in Taber's study looking at conceptual integration across topics in chemistry and physics (*1*), he chose students working at an advanced level, who had shown interest in science, had studied both chemistry and physics at a college level, and had been academically successful because, "These are students where we might expect significant evidence of conceptual integration, and who should cope with the challenge of a broad-based interview of around an hour's duration."

Often in CER, researchers may find themselves recruiting participants from a convenient source such as a particular section of a general chemistry lab or lecture. This is known as convenience sampling and for many research studies may be perfectly appropriate. Other times it is important to ensure recruitment of participants with adequate variation across a variable of interest (e.g., low, medium, and high performing students), known as maximum variation sampling. An example of this can be found in the study by Cole and Todd (8). In this study they used the Group Assessment of Logical Thinking (GALT) as a pre-test measure because it has been shown to correlate well with performance in general chemistry. Students in their study were divided into four groups based on their GALT scores (high or low) and homework type (online or textbook). The researchers then randomly selected six students from each of the four groups to participate in interviews. Ensuring participation from each group provided researchers with the opportunity to determine whether a particular homework type was more favored by a particular group of students.

Another great example of purposeful sampling methods for conducting interviews can be found in the study by Bruck, Towns, and Bretz (25). The aim of this study was to identify the goals, strategies, and assessments used by faculty members involved in the development and implementation of laboratory curricula at American Chemical Society (ACS)-approved institutions. In particular, they were interested in investigating the relationships between faculty goals and (1) institution type, (2) course level taught, and (3) whether they had received National Science Foundation Course, Curriculum, and Laboratory Improvement (NSF-CCLI) funding to improve laboratory instruction. Thus the researchers purposefully identified faculty who had received NSF-CCLI funding and those who had not and then used stratified random sampling across institution type and course level taught to select faculty to invite to participate in interviews. These are just a few different strategies for purposeful sampling. Patton (3) describes in detail 15 different purposeful sampling strategies.

Related to choosing participants for interviews is the issue of sample size. Unlike quantitative methods where certain sample sizes are required to provide adequate power to detect significant changes or differences, there is no set required number of interviews for a study. In an ideal situation where timelines and resources are plentiful, Lincoln and Guba (26) recommend that the sample size be dictated by saturation. Data are collected until no new information is gleaned from sample units. In practice, though, this rarely occurs and thus the decisions about sample size are largely tied to the goals of the study along with the time and resources available to the researchers. In some cases, in-depth information from a small number of people (even N=1) can be very valuable. In Taber's study on conceptual integration (1), four students were interviewed, but in order to illustrate the value of a broad research protocol in obtaining meaningful data, in his paper he chose to describe the findings from just one of those interviews in detail as a case study. In other cases, where researchers are looking to identify patterns or variations across a phenomenon, then it is often necessary to interview a larger number of participants in less depth. For example, by interviewing 14 participants (25% of the class) Bhattacharyya and Bodner were able to identify patterns in the use of curved arrow notation that were consistent across several of the participants (2).

Conducting the Interview

Developing a Rapport

Getting good data from interview participants often depends largely on the interviewer's ability to develop a rapport and make the participants feel at ease. Two very important things to consider in relation to this are (i) the location of the interview and (ii) who will conduct the interview. It is important that interviews are conducted in a neutral location where it is possible to ensure that other people will not be able to hear the interview or walk in during the interview. This suggests that holding interviews in a faculty office is not typically a good choice, especially when interviewing students, as this could set up a power dynamic that could make students feel uncomfortable.

To develop a rapport with participants it is important to remain respectful and sensitive to the participant while at the same time remaining neutral and non-judgmental. This is often difficult as interviewers, like all researchers, have their own biases. However, it is critical that the participants feel that they can share honest responses with the interviewer without being judged. This is also difficult to do if there is a power dynamic, either real or perceived, between the interviewer and interviewee, such as that between a professor and a student. For CER studies that involve interviewing students, one option, although not always possible, is to have students conduct the interviews as students typically feel most comfortable talking to other students whom they do not view as being more knowledgeable than they are and are less likely to judge them if they do not know "the" answer to a question. Another good strategy, again if possible, is to develop a rapport with participants before the interview. This importance of rapport has been underscored in the work that we (the chapter authors) have done with teachers. Working with teachers in an environment where they have perceived that their ideas and input are valued and where they have been treated as colleagues has allowed us to gain their confidence and trust. Such trust has resulted in fruitful interviews.

Paying Attention during the Interview

Paying attention during an interview is not only an act of respect to the participant, but can also help an interviewer maintain control of the interview. If an interviewer pays close attention to participant responses, he/she is better prepared to redirect the participant who is not answering the question asked or who may be meandering in his/her responses. Time for both the interviewer and interviewee is a precious commodity and thus it is imperative to use that time to get as much useful data as possible. If a participant gets off topic, the interviewer should find an opportunity to interject and redirect the participant. A couple of examples of how a researcher might do this are:

- I would like to focus back on the difficulty you described having with equilibrium problems.
- That is very important, but I am most interested in how you actually solved the equilibrium problems. Could you tell me specifically about the approach you took to problem #5?

This will also help address participant fatigue as it will prevent the interview from becoming too long. Interviews lasting less than an hour for secondary and post-secondary students are common and typically do not result in participant fatigue. If interviews need to be longer, then appropriate breaks for the participant when fatigue is observed can be beneficial to data collection. Finally, paying attention provides opportunities for the interviewer to probe more deeply into responses, particularly if the responses appear superficial.

Giving Participants Appropriate Feedback and Support

Something that will help build a rapport with participants and ensure continued collection of useful data throughout the interview is being sure to give participants appropriate feedback and support along the way. Remaining neutral is important but that does not preclude the interviewer from letting participants know that their contributions are valued or providing them with encouragement and feedback. For example, telling a participant that their honest feedback about the homework in CHM 100 is appreciated because it will be valuable in helping improve the course, can make a participant feel valued and encourage them to provide additional details. Moreover, simple phrases like the following are great ways to encourage participants to keep them going if they appear to be tiring.

- A number of students struggle with that question.
- I understand how that can be challenging.
- That is great. That is exactly what we are looking for.
- Okay, I just have a couple more questions for you.

Patton provides numerous other examples of ways to rephrase questions, transition from one topic to another, or to give participants supportive feedback that researchers may want to consider, in particular if they are planning to conduct longer interviews (*3*).

Recording Interview Data

Recording the interview is ideal as it frees the interviewer from taking copious notes and allows him/ her to pay attention to the participant responses. Recording, though, brings with it ethical considerations concerning the ability to maintain confidentiality of participants (see also reference (*19*) in this volume). This is especially true when deciding whether to use audio only or audio and video recordings. Video recordings have the benefit of capturing participants' expressions and mannerisms, the non-verbal cues, during an interview. Such features may inform aspects of the research. Video recording devices, though, can be more difficult to set up (angles, positioning, lighting, etc.) and participants may be more tentative with both video and audio recording over audio only. Technology advances (e.g. mobile devices with applications), though, have facilitated both forms of recordings and lessened the intrusion of recording during the interview process.

Ultimately, because interviews are about gathering information from human beings, the comfort level of the participant is important. If participants are hesitant about having their comments, expressions, and/or actions recorded, the data can be skewed or biased. Frequently though, any slight hesitation or nervousness participants may have with being recorded fades if the interview has a comfortable flow. This is one reason why starting with more demographic/knowledge types of questions and developing a rapport with participants is so important. However, it is still important to give participants the option of having the recording stopped at any time. If the recording device continues to make a participant uncomfortable, then an interviewer may want to turn it off completely and just take notes.

Regardless of the use of recording devices to capture an interview, a researcher may still want to take notes during an interview. The first reason being, in some cases the audio recording may fail and the notes might be the only data source from an interview. More importantly, however, is to record initial impressions, indicate a comment for follow-up, or underscore a phrase or term that, within the moment, appeared relevant.

Interviews for the purpose of data collection are often single shot opportunities. There is rarely a chance for multiple trials with the same sample. Even if a researcher is able for some reason to conduct an interview again with the same participant, the questions have already been asked, and so, the participant's responses, even during an immediate redo of the interview, may change somewhat because the participant has already heard the interview questions. Therefore, it is critical, prior to conducting interviews, to test the devices to make sure they are functioning appropriately and placed properly to capture varying volume levels. This testing step can be combined with the steps for piloting the interview protocol that were described earlier. Finally, establishing a consistent template for saving recorded data is crucial. The form of the recording (e.g. file type), what was recorded, from whom, when, how the files are named and stored, paying particular attention to maintaining confidentiality, are small but critical details when using interviews as a research tool.

Media for Conducting Interviews

Most interviewers would probably prefer to conduct interviews face-to-face as it is easier to build a rapport with participants and it provides researchers the benefit of non-verbal cues that can signal the asking of follow-up questions. However, distance and incompatible schedules between researcher and participants can make conducting face-to-face interviews difficult and sometimes more costly. Thus, researchers have looked to other formats for conducting interviews including: (1) phone; (2) email; and (3) video-chat. Given the increased use of these virtual forms of communication, alternate interview formats are likely to become more common. All of these methods provide the benefit of giving the researcher access to participants over a greater geographic area for a relatively small cost as compared to conducting face-to-face interviews with the same participants, but each of these methods also has several other inherent pros and cons. The following sections discuss the use of each of these methods for conducting interviews with some of the most notable pros and cons for each of these methods summarized in Table III.

Phone Interviews

Phone interviews have become very popular with market research where typically very structured interview protocols are employed, but generally have been considered less desirable for less structured qualitative interviews because of the absence of non-verbal cues to help direct the interview. More recently, however, phone interviews have been used in qualitative studies (*27*, *28*). In Irvine's comparison of phone and face-to-face interviews (*28*) she found that (i) on average participants in phone interviews talked for a shorter amount of time and provided less detail and elaboration than participants in face-to-face interviews, and (ii) the interviewer did a larger portion of the talking in phone interviews. Additionally, phone interviews, like face-to-face interviews, require recording and transcribing, thus it is important that you have a good quality audio recorder that can capture your phone conversation. While these are certainly limitations of phone interviews, Holt (*27*) notes several advantages to phone interviews in addition to increased access to participants. Phone interviews provide an added sense of anonymity, which can result in participants being more open about sensitive or personal topics. Furthermore, she notes that phone interviews can serve to eliminate perceived power differences between the researcher and participant that can make it difficult to develop a rapport with the participant.

Email Interviews

Unlike face-to-face, phone, or video-chat interviews, email interviews lack verbal cues that can help the researcher assess the comfort of the participant and reliability of the responses. Additionally, email interviews are asynchronous. The advantage of this is that participants can take their time and think about their answers. This generally results in more thoughtful answers as well as fewer fragmented sentences than synchronous methods (*29*, *30*). Other advantages include (1) the ease of scheduling as there is no need to find a time that works for both researcher and participant, rather a set of questions can be sent out to several participants at once, and (2) eliminating the need for transcription of the data as it is already in text form. On the other hand, the time it takes for data collection may be extended as it depends on how quickly the participants respond to the initial questions and the follow-up questions. Several researchers have also found that participant attrition is higher with email interviews than with other synchronous forms of interviews as participants can drop out at multiple points (after the initial invitation, after the initial set of questions, or after any set of follow-up questions) (*30*, *31*). Finally, there are also some concerns with email interviews about the reliability of the data as it is not possible to verify the identity of the person who is actually providing the information (*29*).

Table III. Pros and Cons of Using Media for Conducting Interviews

Media Type	Phone	Email	Video-chat
Cost	Relatively inexpensive	Relatively inexpensive providing participants have computers	Relatively inexpensive providing participants have computers, webcams, and software
Participant access/ Scheduling	Can access participants at great geographical distance without travel but still have to have compatible scheduled times	Little difficulty as researchers can send out questions to multiple participants at the same time and participants can respond when convenient	Can access participants at great geographical distance without travel but still have to have compatible scheduled times
Verbal and non-verbal cues	Access to verbal cues but not non-verbal cues	No access to verbal or non-verbal cues	Access to both verbal and non-verbal cues
Data processing	Need to have a reliable means of recording audio and interviews need to be transcribed	Data is already in typed format, no need to transcribe	Need to have a reliable means of recording video and audio and interviews need to be transcribed

Video Chat

The increased use of video conferencing software, such as Skype, provides another option for conducting interviews that carries with it similar advantages of other methods while still providing the researcher with the valuable verbal and non-verbal cues. Hanna (*32*) also suggests that for some participants, being able to take part in the interview from the comfort of their own homes may be advantageous. If the participant is in a comfortable environment, then he/she is likely to be more open and honest with responses.

Like face-to-face and phone interviews, Skype interviews still need to be recorded and transcribed. A quick internet search will provide a list of software, some that are free access and some that have fees, that can be downloaded and used to record both the video and audio portions of a Skype call. The ability to capture video data in the interview can provide an additional data source for researchers to analyze.

Other Media for Interviews

There are some types of interviews that do not lend themselves well to formats other than face-to-face. For example, in think-aloud interviews, in addition to collecting verbal data regarding participants thought processes, researchers also tend to observe what participants are writing and collect any artifacts they construct. These verbal and observational data are still best collected simultaneously using face-to-face methods. Nonetheless, there are some forms of technology and programs that can be used to capture students' drawing and monitor their progress as they work through problems.

For example, in a study of students understanding of enzyme-substrate interactions Linenberger and Bretz (*33*) report the use of the Livescribe digital pen to capture the audio of students' explanations overlaid upon digital images of what they have drawn. This technology is finding more use in CER studies as it overcomes several data analysis difficulties. In particular, Linenberger and Bretz reported that even with videotaped interviews that included audio and copies of student drawings, they often had difficulty interpreting student drawings given that students made several markings on the same drawing. The Livescribe pen, however, ties the audio to the specific pen marks students make, thus eliminating this analysis challenge. Another excellent description of the use of technology to collect and analyze data in CER studies is also provided in this volume by Cooper, Underwood, Bryfczynski, and Klymkowsky (*34*).

Analyzing the Interview Data

Transcription

If interviews were recorded, then the first step in analyzing the data is usually transcribing the audio portion of the recordings. Good transcription, which includes line numbers in the document for referencing, takes time, and if using a professional service, can be costly. Industry Production Standards suggest that one hour of interview recording takes about four hours to transcribe (*35*). Merriam (*36*) describes two options for transcribing: verbatim or interview logs. Verbatim transcription captures every word, utterance, and sound from the interview. Interview logs are a process for capturing the main points. Though interview logs may be a more affordable alternative to verbatim transcription, the ways to measure their reliability is not entirely clear.

Ideally transcription is done without bias. However, unbiased transcription is often not possible. For example, what does it mean to transcribe an interview verbatim? Was the sound a laugh or a sigh? Is there any way to capture voice intonation or inflections? What about grammar and punctuation? Does punctuation change the interview data? For instance, consider the same audio recording transcribed in the following two ways:

- "Organic reaction mechanisms are tough, you know. I really think so,"
- "Organic reaction mechanisms are tough. You know I really think so"

Would those two transcripts lead to slightly different interpretations of data? Such considerations are important in moving to data interpretation (*37*). Even though transcription itself may be a slight first level of data interpretation, it remains the best preparation of recorded interviews for analysis, especially when done with as little bias as possible. Though this process is time consuming, accurate transcriptions can facilitate rich analysis of the interview data.

Importance of a Theoretical Framework

Maxwell (*38*) defines a theoretical framework as "the system of concepts, assumptions, expectation, beliefs, and theories that supports and informs your research." The driving question(s) of a research study and how that question is asked is a reflection of the theoretical framework or theoretical orientation behind the study. Different theoretical perspectives allow for researchers to look at the same situation or same data and ask different questions or focus on different elements of the data. Merriam (*5*) gives the example of an educator, sociologist, and psychologist looking at the same classroom. The educator may ask questions about instructional strategies, the sociologist about social interaction patterns, and a psychologist about motivation. In analyzing interview data, the theoretical framework provides the lens through which the researcher views the data. For example, in the Bhattacharyya and Bodner study highlighted in Table I (*2*), the researchers had phenomenography as their theoretical orientation. The authors chose this theoretical framework as they were looking to identify and classify the different strategies students used for solving complex organic mechanistic problems. The phenomenographic perspective presumes that people experience the same phenomena differently; however, the number of different ways people experience a given phenomenon is finite.

A more detailed theoretical framework was employed in a study that looks at how students and faculty connect levels of representation (macroscopic, particulate, and symbolic) (*18*). The authors use a "levels of complexity" framework which allows them to classify participants explanations of phenomena as emergent (macro-level properties resulting from a particulate level mechanism) or submergent (imposing the properties of the macro-system on the particulate). Thus, in the analysis of the interview data, the researchers were looking for how participants were making connections between the two levels. They provided the following examples of emergent and submergent explanations.

Emergent: *"Let's say the gas is comprised of particles; the particles collide with the walls; the collision with the walls creates pressure. [...] The moment I looked at the diagrams, I immediately thought 'particulate model'. According to this approach, it should be divided to the smallest particles that are still relevant to the problem. In this case, the fact that we have H_2 is not relevant. The molecular structure of the gas does not change at all. Therefore we can simplify H_2 to be a sphere, a particle, does not matter what. Then I asked myself: which particulate theory is relevant to the problem? We can use a theory of motion, a basic mechanistic theory (John, faculty – Theoretical Chemistry)."*

Submergent: In asking a student to describe the how the distribution of gas particles would change if the temperature were lowered, the student responded *"OK, I'll go with (b) [gas particles are concentrated in the* *middle of the tank], since the product PV should decrease, because you lowered the temperature, and P remains the same.*" The researchers explain that this is submergent reasoning as "*in his answer, he first considers the ideal gas equation, and gets to the (incorrect) conclusion that the volume of the gas should decrease. This in turn leads him to impose this conclusion on the submicro representation, and consider the particles as concentrated in a smaller volume.*"

In general, in qualitative research the theoretical framework should guide the research question, the data collection methods, and the data analysis methods. Although a thorough discussion of theoretical frameworks is beyond the scope of this chapter, Patton (*3*) provides detailed descriptions of a number of different theoretical orientations along with the types of overarching questions that characterize each perspective and Merriam (*5*) provides a clear and concise explanation of how to identify a theoretical framework. Bodner and Orgill (*39*) also provide good descriptions of theoretical frameworks in CER.

Qualitative Coding of Interview Data

The pages of transcription data for interviews might appear daunting, and the actual number of pages per hour of interview varies somewhat with the interviewee. Though, with a theoretical framework guiding the conceptualization of coding, some key steps can facilitate this process. Essentially, coding helps construct and verify patterns and trends within the interview data. Approaches to coding can allow the codes to emerge directly from the data, as with a grounded theory approach, or be framed more by an analytic framework in which concepts for codes may be pre-established (*40*). For example, consider a researcher interested in how students approach drawing Lewis structures. Since protocols exist for drawing Lewis structures, categories for coding students' solutions can be preconceived (e.g., the use of a connect-the-dots approach of one atom's valence electrons to another atom's valence electrons, or an electron summation and redistribution across bonding atoms approach).

Qualitative research experts may have some variations in their described approaches to coding analysis (*5, 36, 40, 41*), but one approach that captures many common elements is:

1. Go to the data. Get a sense of the whole. Read all transcriptions carefully. Jot down ideas as they come to mind.
2. Pick one transcript and go through it. Do not think about the substance of its information, but focus on its underlying meaning. Take notes on your thoughts. Do this for a few transcripts.
3. Between steps one and two, a list of topics may be emerging, with the possibility of beginning to cluster topics together. At this point, software or application-assisted analysis can facilitate the next steps (again, see Chapter 5: Using Qualitative Analysis Software to Facilitate Qualitative Data Analysis, (Talanquer) in this volume).

4. Go back to the data. Try out the clusters as a preliminary organizing scheme. Consider how well the clusters hold together. Tag transcription statement that relate to initial codes. Pay attention for possible new categories and/or codes to emerge.
5. Go back to the data and tagged statements. Find the most descriptive wording for these topics and turn them into categories. Look for ways to reduce the list of categories by grouping topics that relate to one another.
6. Go back to the data. Assess how well the categories are holding. Make a final decision on each category and organize these as codes.
7. Go back to the data. Analyze with the developed coding scheme.

In this volume, Talanquer (*42*) provides a more detailed description of the coding process with particular considerations in CER and the use of qualitative analysis software.

Triangulation

Just like organic chemists use multiple methods (NMR, IR, GC-MS) to determine the identity of a compound, interview data is often most powerful when it is used in concert with other data to help address aspects of the how and why within the research project. In educational research, this is often referred to as triangulation. Triangulation of interview data with other data sources supports robust data interpretation.

Although good planning for interviews (good questions, timing, setting, developing a rapport, etc.) attempts to ensure quality data are obtained, there are other elements (fatigue, a poor mood, or even an ulterior motive in a respondent) that may compromise the data (*36*). Checking accuracy with other collected data as much as possible is critical. For example, in reaching their conclusion that organic chemistry students can provide correct answers to mechanism problems despite lacking an understanding of the chemical concepts behind their responses, Bhattacharyya and Bonder (*2*) compared responses to think-aloud interviews with students' actual solutions to organic problems and course grades.

Additionally, the combined use of different research tools described in this book and in chapters in its companion volume (*43, 44*) can provide means for triangulation of interview data. Consider, for example, students being interviewed about their experience with a particular instructional approach used in their chemistry class. The use of classroom observations in conjunction with interviews can provide opportunities to check for data alignment (Yezierski provides an excellent review of the use of classroom protocols in this volume (*45*)). Alternatively, in using a think-aloud protocol during the solving of selected problems, a researcher may also choose to use an eye-tracking device as measure of how participants read the text of the problem or examine any structures provided (in this volume Havanki and VandenPlas discuss the different ways eye tracking technology can be used in CER (*46*)).

Using interviews within a research design must involve an acknowledgement that interpretation is part of nearly every stage of the process and additional data sources can improve validity and reliability of the study findings. Alone, interview data do not always provide enough evidence for making robust conclusions. These data, though, within a set of convergent measures can make powerful contributions to the research story.

Summary

The interview is a tool that provides a valuable means for researchers investigating "how" and "why" questions and allows access to the "unseen," namely participants' thoughts, beliefs, and feelings. When guided by a clear theoretical framework and a well designed protocol, interviews can provide rich data about participants' experiences, knowledge, and practices. With careful planning at all stages of development, implementation, and analysis, interview data can act as valuable data sources CER studies.

References

1. Taber, K. S. Exploring conceptual integration in student thinking: Evidence from a case study. *Int. J. Sci. Educ.* **2008**, *30*, 1915–1943.
2. Bhattacharyya, G.; Bodner, G. M. "It gets me to the product": How students propose organic mechanisms. *J. Chem. Educ.* **2005**, *82*, 1402–1407.
3. Patton, M. Q. *Qualitative Research & Education Methods*, 3rd ed.; Sage Publications, Inc.: Thousand Oaks, CA, 2002.
4. McMillan, J. H.; Schumacher, S. *Research in Education: Evidence-Based Inquiry*, 7th ed.; Pearson: Upper Saddle River, NJ, 2010.
5. Merriam, S. B. *Qualitative Research: A Guide to Design and Implementation*; Jossey-Bass: San Francisco, CA, 2009.
6. Harshman, J.; Bretz, S. L.; Yezierski, E. Seeing chemistry through the eyes of the blind: A case study examining multiple gas law representations. *J. Chem. Educ.* **2013**, *90*, 710–716.
7. Cacciatore, K. L.; Sevian, H. Incrementally approaching an inquiry lab curriculum: Can changing a single laboratory experiment improve student performance in general chemistry? *J. Chem. Educ.* **2009**, *86*, 498–505.
8. Cole, R. S.; Todd, J. B. Effects of Web-based multimedia homework with immediate rich feedback on student learning in general chemistry. *J. Chem. Educ.* **2003**, *80*, 1338–1343.
9. Howard, K. E.; Brown, S. A.; Chung, S. H.; Jobson, B. T.; VanReken, T. M. College students' understanding of atmospheric ozone formation. *Chem. Educ. Res. Pract.* **2013**, *14*, 51–61.
10. Kvale, S.; Brinkmann, S. *InterViews: Learning the Craft of Qualitative Research Interviewing*; Sage Publications, Inc.: Los Angeles, 2009.
11. Dreisbach, J. H.; Hogan, T. P. Focus groups and exit interviews are components of chemistry department program assessment. *J. Chem. Educ.* **1998**, *75*.
12. Liamputtong, P. *Focus Group Methodology: Principle and Practice*; Sage Publications, Inc.: Thousand Oaks, CA, 2001.

13. Stojanovska, M. I.; Šoptrajanov, B. T.; Petruševski, V. M. Addressing misconceptions about the particulate nature of matter among secondary-school and high-school students in the Republic of Macedonia. *Creative Educ.* **2012**, *03*, 619–631.

14. Bowen, C. B. Think-aloud methods in chemistry education: Understanding student thinking. *J. Chem. Educ.* **1994**, *71*, 184–190.

15. van Someren, M. W.; Barnard, Y. F.; Sandberg, J. A. C. *The Think Aloud Method : A Practical Guide To Modelling Cognitive Processes*; Academic Press: London, 1994.

16. Nakhleh, M. B.; Mitchell, R. C. Concept learning versus problem solving. *J. Chem. Educ.* **1993**, *70*, 190–192.

17. Cheung, D. Using think-aloud protocols to investigate secondary school chemistry teachers' misconceptions about chemical equilibrium. *Chem. Educ. Res. Pract.* **2009**, *10*, 97–108.

18. Rappoport, L. T.; Ashkenazi, G. Connecting levels of representation: Emergent versus submergent perspective. *Int. J. Sci. Educ.* **2008**, *30*, 1585–1603.

19. Bauer, C. F. Ethical Treatment of the Human Participants in Chemistry Education Research. In *Tools of Chemistry Education Research*; Bunce, D. M., Cole, R. S., Eds.; ACS Symposium Series 1166; American Chemical Society: Washington, DC, 2014; Chapter 15.

20. Herrington, D. G.; Yezierski, E. J.; Luxford, K.; Luxford, C. Target inquiry: Changing teachers' beliefs about inquiry instruction and their classroom practice. *Chem. Educ. Res. Pract.* **2011**, *12*, 74–84.

21. Bruck, L. B. Faculty Perspectives of the Chemistry Undergraduate Laboratory: Goals, Limitations, and Assessments. M.S. Thesis, Purdue University, 2009.

22. Raved, L.; Assaraf, O. B. Z. Attitudes towards science learning among 10th-grade students: A qualitative look. *Int. J. Sci. Educ.* **2011**, *33*, 1219–1243.

23. Tyson, L.; Treagust, D. F.; Bucat, R. B. The complexity of teaching and learning chemical equilibrium. *J. Chem. Educ.* **1999**, *76*, 554–558.

24. McClary, L.; Talanquer, V. College chemistry students' mental models of acids and acid strength. *J. Res. Sci. Teach.* **2011**, *48*, 396–413.

25. Bruck, L. B.; Towns, M.; Bretz, S. L. Faculty perspectives of undergraduate chemistry laboratory: Goals and obstacles to success. *J. Chem. Educ.* **2010**, *87*, 1416–1422.

26. Lincoln, Y. S.; Guba, E. G. *Naturalistic Inquiry;* Sage Publications, Inc.: Thousand Oaks, CA, 1985.

27. Holt, A. Using the telephone for narrative interviewing: A research note. *J. Qual. Res.* **2010**, *20*, 113–121.

28. Irvine, A. Duration, dominance and depth in telephone and face-to-face interviews: A comparative exploration. *Int. J. Qual. Meth.* **2011**, *10*, 202–220.

29. James, N.; Busher, H. *Online Interviewing;* Sage Publications, Inc.: London, 2009.

30. Meho, L. I. E-mail interviewing in qualitative research: A methodological discussion. *J. Am. Soc. Inf. Sci. Technol.* **2006**, *57*, 1284–1295.
31. Kazmer, M. M.; Xie, B. Qualitative interviewing in Internet studies: Playing with the media, playing with the method. *Inform. Commun. Soc.* **2008**, *11*, 257–278.
32. Hanna, P. Using Internet technologies (such as Skype) as a research medium: A research note. *J. Qual. Res.* **2012**, *12*, 239–242.
33. Linenberger, K. J.; Bretz, S. L. A novel technology to investigate students' understandings of enzyme representations. *J. Col. Sci. Teach.* **2012**, *42*, 45–49.
34. Cooper, M. M.; Underwood, S. M.; Bryfczynski, S. P.; Klymkowsky, M. W. A Short History of the Use of Technology To Model and Analyze Student Data for Teaching and Research. In *Tools of Chemistry Education Research*; Bunce, D. M., Cole, R. S., Eds.; ACS Symposium Series 1166; American Chemical Society: Washington, DC, 2014; Chapter 12.
35. *Industry Production Standards Guide*; Executive Suite Association and Association of Business Support Services International, Inc., 1998.
36. Merriam, S. B. *Qualitative Research and Case Study Applications in Education*; Jossey-Bass Publishers: San Francisco, 2001.
37. Poland, B. D. Transcription Quality. In *Inside interviewing: New Lenses, New Concerns*; Holstein, J. A., Gubrium, J. F., Eds.; Sage Publications, Inc.: Thousand Oaks, CA, 2003; pp 267–287.
38. Maxwell, J. A. *Qualitative Research Design: An Interactive Approach*, 2nd ed.; Sage Publications, Inc.: Thousand Oaks, CA, 2005.
39. Bodner, G. M.; Orgill, M. *Theoretical Frameworks for Research in Chemistry/Science Education*; Pearson Prentice Hall: Upper Saddle River, NJ, 2007.
40. Creswell, J. W. *Research Design : Qualitative, Quantitative, and Mixed Methods Approaches*; Sage Publications, Inc.: Los Angeles, 2009.
41. Miles, M. B.; Huberman, A. M. *Qualitative Data Analysis: An Expanded Sourcebook*; Sage Publications, Inc.: Thousand Oaks, CA, 1994.
42. Talanquer, V. Using Qualitative Analysis Software To Facilitate Qualitative Data Analysis. In *Tools of Chemistry Education Research*; Bunce, D. M., Cole, R. S., Eds.; ACS Symposium Series 1166; American Chemical Society: Washington, DC, 2014; Chapter 5.
43. Bretz, S. L. Qualitative Research Designs in Chemistry Education Research. In *Nuts and Bolts of Chemical Education Research*; Bunce, D. M., Cole, R. S., Eds.; ACS Symposium Series 976; American Chemical Society: Washington, DC, 2008; Chapter 7.
44. Towns, M. H. Mixed Methods Designs in Chemical Education Research. In *Nuts and Bolts of Chemical Education Research*; Bunce, D. M., Cole, R. S., Eds.; ACS Symposium Series 976; American Chemical Society: Washington, DC, 2008; Chapter 9.
45. Yezierski, E. J. Observation as a Tool for Investigating Chemistry Teaching and Learning. In *Tools of Chemistry Education Research*; Bunce, D. M., Cole, R. S., Eds.; ACS Symposium Series 1166; American Chemical Society: Washington, DC, 2014; Chapter 2.

46. Havanki, K. L; VandenPlas, J. R. Eye Tracking Methodology for Chemistry Education Research. In *Tools of Chemistry Education Research*; Bunce, D. M., Cole, R. S., Eds.; ACS Symposium Series 1166; American Chemical Society: Washington, DC, 2014; Chapter 11.

Chapter 4

Discourse Analysis as a Tool To Examine Teaching and Learning in the Classroom

Renée S. Cole,*,1 Nicole Becker,2 and Courtney Stanford1

1Department of Chemistry, University of Iowa,
Iowa City, Iowa 52242, United States
2Department of Chemistry, Michigan State University,
East Lansing, Michigan 48824, United States
*E-mail: renee-cole@uiowa.edu

Discourse analysis can be a powerful tool for exploring issues of teaching and learning in chemistry contexts. This chapter presents an overview of classroom discourse analysis, including discussion of potential analytical frameworks and considerations for collecting and analyzing data. Examples of how discourse analysis has been used in chemistry education research studies of high school and undergraduate chemistry courses are provided to illustrate the types of questions and analyses that have been used to examine teaching and learning in the classroom.

Introduction

There exists an extensive body of research that suggests that collaborative learning environments improve student learning in a variety of contexts (1, 2). Engaging students in collaborative activity has been highlighted for its ability to not only support students' conceptual development, but also to support students' ability to engage in scientific practices such as argumentation. Various approaches to creating collaborative learning environments have emerged in response to calls for a stronger focus on collaborative and active-learning environments (1, 3). Peer Led Team Learning (PLTL) and Process Oriented Guided Inquiry Learning (POGIL) (1, 3, 4) are two commonly used approaches for which there is a growing body of research evidence. PLTL has been shown to improve students' performance in both general and organic chemistry contexts at the undergraduate level (5–8). Similarly, participation in POGIL learning environments has been correlated with improved course performance and retention (3, 9–11).

The recent National Research Council (NRC) report on the status and contributions of Discipline-Based Education Research (1) highlights that while there is significant evidence that collaborative learning environments may benefit students in terms of individual learning outcomes, there are many questions related to collaborative activity that remain under-explored. For instance, what approaches to facilitating small group interactions are most effective? How do small group dynamics influence the ways in which students reason about chemistry concepts? What types of curricular materials and scaffolds are most effective in supporting learning? There is clearly a need for additional research that examines how specific factors impact student learning in collaborative environments. In order to identify and measure the efficacy of instructional approaches, researchers need robust strategies for characterizing learning environments.

In addition to methods such as interviews and classroom observation protocols, which are addressed in this volume by Herrington and Daubenmire (12) and Yezierski (13), discourse analysis provides another qualitative research tool that researchers can use to investigate teaching and learning in chemistry. Analysis of classroom discourse is one route to providing a deeper understanding of how opportunities for learning arise from classroom interactions (14). Analyses of classroom discourse may be used alone or in conjunction with other research methods to gain a more complete understanding of how particular aspects of learning environments support or constrain student learning.

In this chapter, we present an overview of approaches to classroom (and course related) discourse analysis. By discussing examples of discourse analysis studies from the field of chemistry education research, we highlight the breadth of questions about teaching and learning that may be addressed by this approach. We also discuss key methodological issues related to the analysis of classroom discourse and ways in which the chemistry education research community has addressed them.

What Do We Mean When We Say Classroom Discourse Analysis?

Discourse is a widely used term that broadly centers on the language and representational practices used in communicative acts between individuals or within groups. Our interpretation of classroom discourse is commensurate with Gee (15) who differentiates between language in use (referred to as "discourse" with a lower case "d"), and "Discourse" (with a capital "D"). The latter also encompasses factors such as the attitudes, beliefs, values, ways of thinking, and socially constructed identities of the participants that frame the ways that they use language and engage with other participants (15). In part, these factors originate in participants' membership in different discursive communities; groups that have characteristic ways of using language and representing phenomena, including the expectations for how individuals should interact with one another and justify their reasoning (14, 16).

The term "discourse analysis" has been used to refer to a variety of approaches to analyzing verbal or written language. While the origins of discourse analysis are in branches of philosophy, sociology, linguistics, and literary theory, work in this area has expanded to a wide variety of disciplines, becoming a prominent area of emphasis for research in anthropology, communication, psychology, and education. The different traditions focus on very different aspects and functions of discourse (*17, 18*), which makes the multi- and interdisciplinary traditions both a strength and a source of confusion as different frameworks are developed and adapted for use in different areas. In education research, sociocultural perspectives of learning provide particularly useful frameworks for exploring the role of collaborative discourse in supporting classroom learning because these frameworks highlight relationships between social and individual processes in the construction of knowledge (*19–22*).

Classroom discourse analysis is a highly interdisciplinary field that is broadly centered on examining the role of spoken language in teaching and learning (*23–26*). Underlying most approaches to analyzing classroom discourse is the idea that the processes of teaching and learning cannot be fully understood without taking into consideration the inherently social nature of human interactions (*16, 23, 27, 28*). Originating in the work of Vygotsky (*20*), who considered social resources such as language and symbolic representations to be crucial mediators of learning processes, sociocultural perspectives on classroom discourse highlight the way in which social factors contribute to the ways in which individuals interact (*23*). From this perspective, learners jointly build ideas and new understandings as they interact with one another and more experienced members of the discourse community (i.e. teachers).

Within STEM fields such a chemistry and physics, the disciplinary discourse of the field is characterized by particular ways of using language and other semiotic resources as tools to predict and explain natural phenomena (*14, 29, 30*). Semiotics refers to the ways of knowing or meaning making in a discipline. Airey and Linder describe semiotic resources as "those parts that make up the complex of representations, tools and activities of a discipline." ((*29*), pg 44)

In classroom chemistry contexts, instructors serve as expert representatives of disciplinary traditions, facilitating student enculturation into the broader discipline. This includes helping students understand disciplinary expectations for what counts as an appropriate justification or reasoning in chemistry contexts. Instructors are also responsible for helping students negotiate expectations for how they should explain their thinking or participate in classroom conversations (*16, 26*).

While some of these expectations (norms) reflect the norms of the discipline, the specific practices that emerge in each classroom are unique since the members of the class establish their own expectations for how students should work together to solve problems and explain their reasoning (*27*). Student success and learning in such environments thus becomes a function of not just individual students' cognitive capabilities or the skill of their instructors, but also of the quality of the educational dialogue that is established in each class (*28*).

Analyses of classroom discourse have the potential to explore many of the social factors that frame classroom learning. For instance, while factors such as social norms are rarely explicit, these factors may be reflected in patterns of interaction (*15*). Similarly, there may be patterns in interaction and reasoning that reflect the ways that the instructor engages with the class and promotes students' reasoning, or the way in which classroom reasoning emerges in response to particular instructional tasks. Close examination of these trends can provide valuable insight into effective instructor facilitation strategies, curricular materials, and the structuring of student groups for better engagement in collaborative learning environments.

Approaches to Analyzing Classroom Discourse in CER Contexts

Discourse analysis provides a means to systematically examine the teaching and learning that occurs in active-learning classrooms. Some emerging areas of interest to the CER community include the ways students develop understanding of chemistry concepts by engaging with peers and instructors; the ways in which students reason and justify their thinking during discussion; and the ways in which students use representations in communicating ideas. Discourse analysis also provides a means of exploring similarities and differences among populations of students, implementations of a specific active learning approach, and different instructional methods.

Discourse analysis is particularly relevant when the research focus is on the social context of learning, particularly the socially interactive aspects of language — the "language-in-use" with an emphasis on the function of the discourse (*15*, *17*). Broadly, a researcher analyzing classroom discourse attempts to "identify what is done and how it is done, that is, to identify the function of the talk not only considering its content, but also by taking it apart to see how it is structured and organized" ((*17*), pg 28). In reading the literature and communicating the results of discourse analysis, it is useful to distinguish between discursive moves and discursive practices. Discursive moves are deliberate acts that have a particular function and intended consequence (*31*), while discursive practices are better understood as the conventions or "rules" for constructing knowledge in a particular community (*32*).

Inductive and Deductive Approaches to Analysis

As with the analysis of any qualitative data, approaches to analyzing the data may be inductive or deductive. Inductive approaches that draw from traditions such as grounded theory and the constant comparative method may be well suited for research areas that are largely exploratory (*33*). In such analyses, identification of analytical categories is guided by the research questions framing the study (*34*).

An example of an inductive approach to examining classroom discourse is seen in the work of Krystyniak and Heikkinen (*35*). The authors analyzed verbal interactions between undergraduate chemistry students and instructors in inquiry versus non-inquiry laboratory activities in order to explore how each type of activity contributed to students' reasoning. Using an inductive method derived from a constant comparative approach, the authors characterized verbal interactions based on the perceived quality of reasoning elicited by the activities, for instance, as they related to higher-order thinking" or "process-skills" versus talk about non-science concepts. The discourse of both the students and the instructors was influenced by the nature of the laboratory activity, with more sophisticated language use observed during inquiry activities.

Given the complexity of the classroom learning environment, many researchers narrow their research focus using a more deductive approach to analyzing classroom discourse. While some researchers may use established methodological approaches to analyze specific aspects of classroom discourse, others may opt to develop coding categories from theoretical constructs or aspects of the interaction the researcher finds most salient. In the following sections we illustrate some of the various deductive approaches to analyzing classroom discourse in CER contexts by discussing several examples from the CER literature. We focus our discussion on studies related to students' engagement in argumentation in collaborative classrooms and the role of the instructor in facilitating classroom discourse.

Analysis of Argumentation in Collaborative Classrooms—Toulmin

Argumentation has been highlighted as a scientific practice that is central to the construction of explanations, models, and theories across STEM disciplines (*36*) and one that supports student learning of science content (*37*, *38*). In order to explore how students engage in argumentation within collaborative learning environments, several researchers have adapted Toulmin's model of argumentation as a methodological framework for examining the ways in which students engage in argumentation within classroom contexts (*39–45*). Toulmin's model provides a way of examining the structure and function of both individual arguments and those that are co-constructed as a classroom of learners interacts (*46*). As such, this model has been used to assess the quality of student argumentation (*40*, *44*), examine evidence used to support reasoning in chemistry contexts (*41*), and explore how the structure of curricular materials and the nature of instructor facilitation contribute to student reasoning (*43*, *45*).

According to Toulmin's model, an argument is comprised of three core components, namely: claim, data, and warrant (*36*). The *claim* of an argument represents a conclusion that is supported with evidence and reasoning. *Data*, or the grounds of the argument, is any evidence presented in support of a claim. While in STEM contexts data may include experimental evidence, it may also include known facts or other information (*42*). The *warrant* of an argument explains how the data lead to the claim. These three components (claim, data, and warrant) comprise the core of an argument, but more complex arguments may also include other components such as backings, qualifiers, and rebuttals. More complete characterizations of Toulmin's model and its use as an analytical framework can be found in the literature (*40*, *42*, *46*).

One example that can be used to illustrate the use of Toulmin's model in the context of chemistry education research is an investigation of an undergraduate general chemistry class using POGIL instructional materials and a PLTL approach to engaging students in small-group discussions. Kulatunga, Moog, and Lewis adapted Toulmin's model of argumentation as a methodological framework for exploring relationships between student participation patterns and the quality of argumentation (*44*) and for exploring the role of different types of question prompts in supporting student argumentation (*45*). The authors examined the nature and frequency of arguments produced by general chemistry students as they worked in small groups on POGIL instructional tasks and documented patterns in individual and co-constructed argumentation. By relating these trends to small group dynamics and the nature of the POGIL question prompts, findings from this work inform an understanding of how the structure of POGIL activities and small group dynamics in a PLTL approach contribute to student engagement in the scientific practice of argumentation.

Strengths of this deductive approach to analysis drawing on Toulmin's model of argumentation include the fact that it provides a structured approach to defining argumentation in collaborative contexts and operationalizing components of arguments as a coding framework (*39*). Examining arguments in this way enabled Kulatunga et al. to explore the sophistication of student reasoning and the way in which patterns of reasoning were influenced by particular instructional tasks. However, there are also limitations to the use of a deductive analytical framework such as Toulmin's model of argumentation. For example, while it can be useful for evaluating the presence and completeness of an argument, it does not determine the correctness of the argument (*37*). Examining the extent to which students provide correct (or incorrect) evidence in support of arguments may in some contexts be useful as it may provide insights into student misconceptions or difficulties they experience in understanding chemistry concepts.

Additionally, there may be forms of arguments presented in the class that may not be captured by Toulmin's model. For example, students may evaluate two or more predictions about observable phenomena. While hypothetico-predictive arguments are highly valued by the scientific community, this type of reasoning may not be easily captured using Toulmin's model since the model focuses more on articulating and comparing predictions about scientific phenomena, rather than coordinating evidence to a claim (*47*).

Alternate Frameworks for Analyzing Argumentation

To address limitations such as these, several alternate lenses for examining classroom argumentation have been used in studies of classroom argumentation (*48*). For instance, Lawson (*47*) developed an analytical approach for examining hypothetico-predictive arguments while Schwarz and colleagues (*49*) developed a framework for analyzing students' written arguments that focuses on the structure and complexity of the evidence and reasoning for an argument. Sampson and Clark (*48*) review a number of analytical approaches for examining classroom argumentation and illustrate opportunities and challenges presented for various approaches by analyzing example arguments using different lenses.

Alternately, an analytical tool related to argumentation that has been used in the context of general chemistry laboratory activities is the Assessment of

Scientific Argumentation in the Classroom (ASAC) observation protocol, which is designed to evaluate the quality of students' arguments (*50*). Compared to Toulmin's model of argumentation, the ASAC serves to analyze argumentation in a more holistic fashion that takes into consideration social, epistemic, and cognitive factors. The ASAC has been used in both a longitudinal study that examines how students' ability to participate in scientific argumentation changes over time and a study comparing argumentation across different instructional approaches (*50*).

Analyzing More than Argumentation—Coordinating Instructor Discursive Moves with Argumentation

While many of the discourse analysis studies in chemistry education research have focused on some form of student argumentation, there are other aspects of classroom learning environments that can be analyzed. Even when analyzing argumentation, classroom discussions often include a significant quantity of dialogue that is not aimed at constructing an argument. Instances where group members interact with one another to organize their collective activity or in which the teacher helps students reframe their thinking may not be directly related to explanations of chemical behavior. However, these interactions may give valuable insight about the nature of interactions in the classroom. To analyze these other aspects of the learning environments, researchers analyze data using different analytical frameworks.

When the goal of the research is to look for relationships between the characteristics of discourse in the classroom and the impact it has on particular outcomes, such as argumentation, it is often useful to analyze the data using multiple analytical frameworks. In an upper-division physical chemistry context, Cole, Stanford, Towns, and Moon (*51*) adapted the Inquiry Oriented Discursive Moves (IODM) framework to examine the role of instructor discursive moves in advancing classroom argumentation in an inquiry-oriented undergraduate physical chemistry class. Originally developed in the context of an undergraduate differential equations class (*52*), the IODM framework represents a characterization of discursive moves used by instructors (or other actors in the classroom community) in order to sustain and advance collaborative inquiry. Among the major types of discursive moves addressed by this framework are revoicing, questioning/requesting, telling, and managing. The revoicing move addresses instances in which the instructor restates or reframes contributions to classroom reasoning made by students. While it most often serves to highlight and affirm a particular idea, thereby advancing the class's reasoning, revoicing may fulfill other roles depending on the context. For instance, revoicing a student's contribution may serve to attribute ownership of an idea to a student or to translate student contributions into a format more consistent with discipline-specific norms. By coordinating discursive moves such as these with classroom argumentation patterns, Cole and colleagues were able to document instructor facilitation practices that were effective in supporting student argumentation in POGIL physical chemistry courses.

Similarly, Kulatunga and Lewis (*43*) coordinated analysis using Toulmin's argumentation scheme with a framework for analyzing the role of teacher discursive moves. The goal was to explore the role of the peer-leader (teacher) in supporting students' attempts at argumentation (*53*). Findings from this research highlight the relationship between different types of instructor verbal behaviors and the quality and frequency of student argumentation. This work provides evidence for how simple scaffolding strategies used by the teachers may help students build stronger scientific arguments.

More Frameworks for Analyzing the Role of the Instructor

A number of studies described in the chemistry education research literature have analyzed classroom discourse with a focus on teachers' discursive practices in order to explore how instructors engage students in classroom discourse and support student engagement and conceptual development (*41, 43, 54–57*). As an example of one approach to exploring how instructors support students' understanding of chemistry content, Stieff and colleagues (*58*) analyzed classroom talk in two high school chemistry classrooms. The goal was to investigate how teacher-student classroom discursive practices contributed to "levels confusion", that is the observation that students mistakenly attribute characteristics of macroscopic phenomena to submicroscopic phenomena (or vice-versa). The authors combined the sociocultural construct of intersubjective agreement—participants in a classroom coming to a state of agreement or sharing a reference framework—with Lidar's (*59*) framework for examining teacher epistemological moves. A common feature of these moves is directing students' attention to what is important in the exchange. By identifying epistemological moves (*59*) that confirmed students' interpretation of descriptive levels or re-oriented students to focus on other levels, this work contributes to the body of knowledge that looks at the role of the instructor in supporting students' sense making with domain-specific representational systems and concepts.

Criswell (*54*) coordinated the instructor's use of discursive moves with the conceptual trajectory of the classroom discussion in order to understand how instructional strategies support student conceptual development. Criswell (*54*) used a system of identifying discursive moves that was developed by Wells (*22*). The discursive moves were characterized as initiation, response, or follow-up. Each move was then further analyzed to identify its function in a specific utterance. Transcripts of teacher interviews and video recordings of classroom interactions were analyzed to describe which approaches were most prevalent and their influence on observed patterns of interaction in the classroom. The results of the analysis provided evidence that the behaviors of the teachers did not always support the teachers' desired outcomes.

In a different analysis of the video data (*55*), Criswell and Rushton presented excerpts of classroom dialogue interpreted through two different theoretical lenses. The analysis used a framework (*60*) that includes classifications of monitoring, selecting, and sequencing ideas as part of sharing ideas within the class to illustrate the ways in which the teacher and students structured the sharing of ideas. This work illustrates ways in which instructors can scaffold the class's discussion from intuitive ideas related to stoichiometric relationships to more sophisticated disciplinary use of stoichiometric relationships.

As illustrated by the examples presented here, studies of classroom discourse may vary considerably in the aspects of discourse on which they focus. All, however, are broadly committed to examining specific features of the classroom learning environment and the way the environment can create opportunities for student learning. In the next section, we discuss some of the methods used to explore classroom discourse and comment on some key issues for conducting discourse analysis.

Methodological Considerations for Discourse Analysis Studies

In this section we present an overview of some of the more common means of collecting data for discourse analysis studies and highlight key considerations that are unique to these types of studies.

Data Collection

In much of the work on classroom discourse in STEM education contexts, discourse is characterized as "language in use" (*14*). The key to using discourse analysis in a study is to capture the interactions among participants while they are engaged in discourse (*14, 15, 17, 18, 61*). Data that give access to participants' use of language and support interpretations of the way language in used are central to discourse analysis studies. For this reason, discourse analysis will typically focus on language use as captured in a video or audio file or in the transcript of a video or audio file. In some studies the analysis will also extend to written artifacts.

Audio and Video Recordings

Of particular importance to data collection in studies of classroom discourse are issues related to deciding which elements of the learning environment should be recorded and what technological tools will be used to collect and archive data. While some researchers may think of audio or video recording as a "true" and unbiased account of classroom interactions, the way in which a phenomenon is filmed can influence what people see and, therefore, influence how they interpret data (*62*). Decisions of who, what, and when to record data reflect the researcher's perceptions of which aspects of interaction are most relevant. Derry (*63*) suggests that in deciding how best to systematically collect and store data for a particular study, the researcher should consider data collection from multiple perspectives (including the perspectives of researcher, instructional designer, director, technologist, etc.) in order to better understand how different approaches will impact potential findings and implications of the work.

Audio recordings of classroom interactions provide a rich means of capturing verbal interactions and aspects of prosody—the intonation and rhythm of speech—that can be key to inferring meanings. Stress on different words, intonations (which can differentiate a question from a statement), differences in volume, and length of sounds can all be important aspects of discourse. Pauses, vocal inflections or other speech sounds such as "ums", laughter, etc. may provide indications of confidence or context for interpreting the function of a statement in a particular context.

Depending on the focus of the analysis, it may be preferable to use video recordings as a primary source of data as they enable the researcher to document gestures as well as verbal dialogue. A high quality video with clear sound is the easiest to analyze and has been used in many published studies of classroom discourse analysis. As with prosody of spoken language, gestural cues may help the researcher interpret the meaning of spoken language (*61*) and as such, many analyses combine analysis of verbal components of speech with analysis of non-verbal elements such as eye gaze, gestures, etc. (*64*). As noted before, it must be acknowledged that the researcher's interpretation of video data is influenced by the researchers' bias and theoretical assumptions (*29*).

Selecting the appropriate technology for audio or video data collection is crucial to obtaining clear audio that appropriately represents interactions and from which individual student voices can be identified. A consideration that is unique to settings involving interaction of multiple participants is the fact that classroom settings involve many different actors. If distinguishing between different members of the classroom community is important to subsequent analysis, it is critical to obtain clear audio or video from which individual student voices can be identified. This is sometimes difficult in a classroom setting and using the microphone on video cameras may not provide sufficiently clear audio. Positioning remote microphones close to the participants may make it easier to obtain high-quality audio recordings.

Further complicating data collection is the fact that in active-learning classrooms, participants are seldom stationary. The instructor may move about the classroom and students may reorganize themselves for different instructional tasks. A stationary video camera may make it difficult to adequately record interactions between members of the class. Researchers may choose to address this issue by using multiple cameras in order to more easily follow both the instructor and students (and their written work if this will be central to later analysis). Depending on the context of the classroom interaction, it may be necessary to have a researcher actively control and adjust the camera over the course of the class period. Having an individual operate the camera provides a greater likelihood that the camera can be focused on the participants as needed.

As with any data collection involving audio or video data, it is important to check the quality of both frequently. When capturing video and audio of multiple students, it may be difficult to predict whether audio recordings will produce high quality data unless they are tested with the class in place for the various types of interactions. We recommend testing video and audio quality at the beginning and end of each data collection session. If there are problems with the equipment, it is better to lose some data for a single class period than to lose all the data for an extended period of time. Derry (*65*) provides additional guidance in addressing challenges for collecting and using video recordings to examine teaching and learning in classroom environments. Many aspects of collecting video data to document classroom activity are also discussed by Yezierski (*13*) in this volume, in her chapter on classroom observations.

Additional Sources of Data

While audio and video data are commonly used as primary sources of data, other sources may be valuable for making sense of classroom dialogue and triangulating emerging themes. For example, a record of the written artifacts produced as the class interacts may help in interpreting references to equations, diagrams, or figures referenced in verbal exchanges. In situations where teacher and students use white boards or other media to share completed work during whole class discussions, it may be possible to capture such written information using video recordings or electronic files of smartboard data. When students share written work with one another during small group interactions, researchers may consider collecting student written work at the end of the class period in order to help interpret audio or video data.

However, collecting written work after the class may make it challenging to coordinate written work with verbal discourse. In interview contexts, several researchers have used digital pens such as the LiveScribe Smart Pen to capture audio and written data during semi-structured interviews (*66*). Weibel et al. (*67*) also described how the pens may be used to collect more complete field notes in qualitative studies. The technology may be promising for coordinating written work with verbal interactions in the context of discourse analysis.

Field notes may also be valuable for keeping a record of non-verbal exchanges and production of written artifacts as well as documenting classroom dynamics. Christian and Talanquer (*68*) characterized modes of reasoning, content focus, and levels of cognitive processing used by students while solving organic chemistry problems in study groups. The authors used observational notes in order to provide a record of written work generated during the study sessions and to facilitate interpretation of the audio recordings of small group talk. Through this approach, they were able to develop insights into what students actually do when reviewing content on their own.

Transcription and Its Role in Data Analysis

When working with audio data in the context of one-on-one interviews, data analysis often begins with listening to and transcribing verbal discourse from the recordings, reviewing transcripts, and selecting a focus for subsequent analysis. Some aspects of transcription are described in Herrington and Daubenmire's chapter on using interviews in chemistry education research (*12*). Transcribing audio or video data from classroom discourse, however, presents a unique set of challenges compared to transcribing audio of one-on-one interviews. For instance, concurrent overlapping speech may be difficult to interpret and it may be challenging to appropriately attribute statements when there are many actors in the evolving classroom discourse. Furthermore, transcribing classroom discourse may be considerably more time consuming than transcribing one-on-one interviews. A single fifty-minute class period can easily generate a 25-30 page transcript.

Some researchers may opt to analyze audio or video recordings directly, which can by facilitated by using computer assisted qualitative data analysis packages. Talanquer discusses how these packages can facilitate analysis in a chapter in this volume (*69*). In some cases, avoiding transcription may allow the researchers to maximize time and resources. This route may be especially useful when research is intended to evaluate an instructional approach or provide timely feedback to stakeholders about the impact of a learning environment (*70*).

For other research designs, transcripts of verbal interactions with time stamps and clear speaker attributions may aid analysis considerably. As such, researchers may choose to create transcripts for some or all audio or video data. For instance, Kulatunga, Moog, and Lewis (*44*) transcribed 12 weeks of small group interaction in a peer led team learning (PLTL) general chemistry class. This provided a complete record of student interactions over the course of the semester, which was then analyzed for patterns in students' reasoning. In contrast, Stieff and colleagues (*58*) conducted an initial review of video prior to beginning transcription in order to select only episodes of classroom interaction that involved classroom discussion of multiple representational levels.

In many cases, annotating aspects of non-verbal interaction in transcriptions of audio or video data may help support the researchers' interpretation of verbal exchanges. One common approach for handling non-verbal interactions involves the designation of transcription conventions for encoding relevant features of non-verbal interaction within the text of the transcript. For example, in order to support their interpretation of participants' references to different levels of representation, Stieff and colleagues annotated pauses, pitch, emphasis, and other cues in transcripts of verbal dialogue using a variety of transcription conventions.

While researchers may designate their own transcription conventions, there are a number of established convention systems that researchers may want to consider. For example, the Jeffersonian Notation System, commonly used in the field of conversation analysis (*18*, *71*) uses text formatting options such as underline, italics, and commas to denote cues including speech emphasis, non-verbal activity, and change in intonation (*72*). Criswell (*54*) includes an appendix identifying the transcription conventions used in the excerpts of data of classroom discourse in his study of how the conceptual change strategies favored by teachers during instruction influenced their discursive practices.

Depending on the focus of the study (and time or personnel constraints), researchers may also opt to use a professional transcription service. However, in such cases, it will still be necessary to carefully check the transcripts for accuracy, particularly in regards to technical language. It is important be clear about what needs to be transcribed – some services will provide a "clean" transcript rather than one with all the "ums" and false starts. As we noted earlier, it is also possible to document voice inflections and other aspects of discourse that aren't encoded in the words themselves but which provide additional data. These vocalizations may provide information that is critical for analyses, such as students' confidence in their statements.

It is important to keep in mind that the choices made in the process of transcription are theory-laden and play a large role in constraining analysis and interpretation of the data (73). The extent to which researchers choose to incorporate non-verbal cues certainly depends on how the data is to be analyzed. For example, some coordinate analysis of transcripts of classroom discourse with an analysis of other data in order to provide a more complete account of the classroom context. The complexities of the transcription process have been addressed by many sources, but Bird (74) provides a perspective that many readers may find engaging. Her first-hand account of learning "to love transcription" includes a discussion of transcribing different types of data, ranging from one-on-one interviews to complex classroom interactions, and she includes references to additional resources that may help the reader further explore affordances and constraints of various approaches to transcription.

Unit of Analysis

An important step in discourse analysis is deciding on the appropriate unit of analysis for the study. The unit of analysis establishes the grain size that will be analyzed for meaning in the discourse and can consist of single words, phrases, sentences, paragraphs, or even longer passages depending on the analytical focus (15, 17). Often, units of analysis are defined according to tone units, speech patterns such as changes in vocal inflection that may indicate the end of a thought. Common units of analysis based on tone units include utterances (a stream of speech that can separated based on an idea or function) and stanzas (a group of utterances with a common theme or perspective). Alternately, units of analysis may be defined based on conversational turns.

As an example of how different units of analysis may be used to facilitate analysis of classroom interaction, consider Criswell and Rushton's (55) use of multiple units of analysis in an investigation of patterns of classroom talk in a high school chemistry class engaged in inquiry-oriented problem solving activities. The authors parsed transcripts of classroom dialogue into conversational turns, which the authors defined, following Edwards (75), as a consecutive utterance by a single speaker without a pause longer than 3 seconds. This unit of analysis was used to examine the function of individual speakers' contributions to classroom discourse. At a larger grain size, data were segmented into task segments that marked shifts in the focus of the problem solving activity within the class (e.g. between posing the problem and generating the solution). Coordinating analyses of these units of analysis enabled the authors to explore relationships between micro-level features of classroom talk (i.e. the function of individual conversational turns) and the way in which students' ideas were explored in the context of specific problem-solving activities.

It is important to note that what is considered an appropriate unit of analysis may vary considerably across different disciplinary contexts and research traditions. Thus it is critical that researchers describe and justify how the relevant unit of analysis is defined and determined for a particular study.

Reliability and Validity

As with any qualitative research, researchers must contend with issues of reliability and validity of the analysis. Gee (*14*) highlights three key factors contributing to the validity of discourse analysis: convergence, agreement, and coverage.

Convergence arises when analyses of different data sets (or the same data set with different analytical approaches) lead to similar conclusions. As we have discussed, many researchers address this aspect of validity by analyzing multiple sources of data in order to triangulate emerging themes.

Establishing agreement involves verifying interpretations of the data with other researchers who have expertise within the discipline under study (as well as those who may not). One common route to establishing the agreement with respect to applications of an analytical framework is to examine agreement among multiple independent analysts. Establishing inter-rater agreement also contributes to clarification of coding criteria and insights into the interpretation of data. This is generally done in one of two ways. First, if the study is conducted by a team of researchers, having multiple researchers code the transcripts followed by a session to resolve any differences allows for greater confidence in the reliability of the coding process. If there are significant discrepancies among coders, it may indicate that analytical categories are not sufficiently well defined or do not fit the data well. In the study by Becker et al. (*41*), agreement was approached by negotiating reliable applications of a coding framework as a team, and by confirming applications of the coding scheme as well as interpretations of the data with all members of the team. The discussion of coding using components of Toulmin's model of argumentation in this case, led to insights about how to recognize portions of arguments and differentiate between evidence and reasoning.

As an alternate approach, a single researcher can code all the data with a second researcher coding a subsample of the data. Generally, the second coder will code 10-20% of the transcripts to check for reliability. When using this procedure, it is recommended that both coders analyze a set of data and resolve any differences in coding before more extensive coding is completed. As an example of this type of approach, Krystyniak and Heikkinen (*35*), used an analytical approach focused on identifying the nature and frequency of interactions between instructor and students. They reported frequency of agreement between two raters and a discussion of the development of their coding scheme as support for the validity of their approach. They also discussed examples of problematic passages of discourse and the way in which the researchers resolved their coding discrepancies.

The idea of coverage in establishing the validity of an analytical approach pertains to the transferability of the analytical approach to new data sets or contexts (*5*). An analytical approach that has substantial coverage enables predictions to be made about what might happen in related situations. For instance, Criswell and Rushton (*55*) addressed this aspect of validity by using the same analytical approach in five different classroom contexts.

Limitations

As with any qualitative study, a limitation of most studies involving discourse analysis is that they focus on a small number of students in a single class. This means that researchers must guard against over generalizing about the impact of particular discursive moves or the transferability of development of student understanding of chemistry in the specified learning environment. To address issues of representativeness, studies are more robust if they look at data from multiple classroom contexts, which may involve different instructors, institutions, or instructional approaches.

Another limitation of classroom discourse analysis is that a single analytical lens may not provide an adequate characterization of complex and dynamic classroom learning environments. As we have discussed, there are many social factors that frame classroom learning. Coordinating multiple frameworks can provide more complete understanding of how the different social factors interact. This coordination was illustrated in the previously described studies by Stieff (58) and Criswell (54, 55). Further analysis, where the coded data becomes the new data source to be interpreted and analyzed, is often required to answer the question being investigated. For example, when using Toulmin analysis, the argumentation logs become the data set for further interpretation. Kulatunga and colleagues (44) characterized the complexity of the coded arguments and correlated the occurrences to individually and co-constructed arguments. In Becker et al.'s study (41) of the development of socio-chemical norms in a physical chemistry class, the first phase was identifying arguments. The subsequent stage of analysis focused on identifying patterns of argumentation. Examining these patterns in the context of the broader classroom interactions allowed the authors to make inferences about the social norms that framed classroom learning.

Conclusions

Documenting particular discursive moves and connecting these to targeted learning outcomes provides a mechanism to characterize classroom (and course related) learning environments. The chemistry education research studies used as examples describe a variety of ways to characterize discursive practices and draw conclusions about the nature of teaching and learning chemistry in the context under study. These conclusions provide substantial insights for improving classroom practice.

Various frameworks for analyzing classroom discourse make it possible to explore how specific aspects of the learning environment support student learning, including interactions between teacher and students and the structure of curricular activities. We have also discussed ways in which certain aspects of classroom discourse analysis may present unique challenges for researchers. For instance collecting data in a dynamic classroom environment may require considerable planning, and transcribing classroom discourse for subsequent analysis may require significant time and resources. Nonetheless, the examples presented here highlight that there is much to be learned about the teaching and learning of chemistry through discourse analysis.

Discourse analysis is particularly well suited to address some of the recommendations of the National Research Council committee regarding areas of research that should be addressed by the discipline-based education research community (*1*). The areas best suited to discourse analysis are:

- *Research is needed to explore similarities and differences among different groups of students.*
- *Research is needed in a wider variety of undergraduate course settings.*
- *DBER should measure a wider range of outcomes and should explore relationships among different types of outcomes.*
- *The emphasis of research on instructional strategies should shift to examine more nuanced aspects of instruction.* ((*1*), pg 9.12).

Discourse analysis provides a lens to explore the similarities and differences among different student populations and how they experience different learning environments. As shown in the examples presented here, a closer examination of how meaning is built through interactions in high school, introductory, and upper-level chemistry courses allows chemistry education researchers to better inform the design and implementation of more effective learning environments. Discourse analysis can also be paired with other measures of student outcomes to provide a more robust understanding of student knowledge, skills, and attitudes and how these factors are impacted by the learning environments experienced by students.

Resources

Researchers interested in learning about discourse analysis in more depth are directed to the following resources:

Cazden, C. *Classroom Discourse: The Language of Teaching and Learning*, 2nd ed.; Heinemann: Portsmouth, NH, 2001.

Gee, J. P. *An Introduction to Discourse Analysis: Theory and Method*, 2nd ed.; Routledge: London, 2005.

Gee, J. P. *How To Do Discourse Analysis: A Toolkit*; Routledge: Florence, KY, 2010.

Lemke, J. L. Analyzing Verbal Data: Principles, Methods, And Problems. In *Second International Handbook of Science Education*; Fraser, B., Tobin, K., McRobbie, C., Eds.; Kluwer Academic: London, 2012.

Rymes, B. *Classroom Discourse analysis: A Tool for Critical Reflection*; Hampton Press: Cresskill, NJ, 2009.

Wood, L. A.; Kroger, R. O. *Doing Discourse Analysis: Methods for Studying Action in Talk and Text*; Sage Publications, Inc.: Thousand Oaks, CA, 2000.

References

1. *Discipline-Based Education Research: Understanding and Improving Learning in Undergraduate Science and Engineering*; National Research Council: Washington, DC, 2012.

2. Olson, S.; Riordan, D. G. *Report to the President. Engage to Excel: Producing One Million Additional College Graduates with Degrees in Science, Technology, Engineering, and Mathematics*; President's Council of Advisors on Science and Technology: Washington, DC, 2012.

3. Eberlein, T.; Kampmeier, J.; Minderhout, V.; Moog, R. S.; Platt, T.; Varma-Nelson, P.; White, H. B. Pedagogies of engagement in science: A comparison of PBL, POGIL, and PLTL. *Biochem. Mol. Biol. Educ.* **2008**, *36*, 262–273.

4. Towns, M.; Kraft, A. *Review and Synthesis of Research in Chemical Education from 2000-2010*; White Paper for The National Academies National Research Council Board of Science Education; 2011.

5. Drane, D.; Smith, H. D.; Light, G.; Pinto, L.; Swarat, S. The gateway science workshop program: Enhancing student performance and retention in the sciences through peer-facilitated discussion. *J. Sci. Educ. Technol.* **2005**, *14* (3), 337–352.

6. Streitwieser, B.; Light, G. When undergraduates teach undergraduates: Conceptions of and approaches to teaching in a peer led team learning intervention in the STEM disciplines: Results of a two year study. *Int. J. Teach. Learn. Higher Educ.* **2010**, *22* (3), 346–356.

7. Tien, L. T.; Roth, V.; Kampmeier, J. A. Implementation of a peer-led team learning instructional approach in an undergraduate organic chemistry course. *J. Res. Sci. Teach.* **2002**, *39* (7), 606–632.

8. Varma-Nelson, P. *Peer-Led Team Learning: Evaluation, Dissemination, and Institutionalization of a College Level Initiative*; Springer: New York, 2008; Vol. 16.

9. Straumanis, A.; Simons, E. A. A Multi-Institutional Assessment of the Use of POGIL in Organic Chemistry. In *Process Oriented Guided Inquiry Learning*; Moog, R. S., Spencer, J. N., Eds.; Oxford University Press: New York, 2008; pp 226–239.

10. Bailey, C. P.; Minderhout, V.; Loertscher, J. Learning transferable skills in large lecture halls: Implementing a POGIL approach in biochemistry. *Biochem. Mol. Biol. Educ.* **2012**, *40* (1), 1–7.

11. Hein, S. M. Positive impacts using POGIL in organic chemistry. *J. Chem. Educ.* **2012**, *89* (7), 860–864.

12. Herrington, D. G.; Daubenmire, P. L. Using Interviews in CER Projects: Options, Considerations, and Limitations. In *Tools of Chemistry Education Research*; Bunce, D. M., Cole, R. S., Eds.; ACS Symposium Series 1166; American Chemical Society: Washington, DC, 2014; Chapter 3.

13. Yezierski, E. J. Observation as a Tool for Investigating Chemistry Teaching and Learning. In *Tools of Chemistry Education Research*; Bunce, D. M., Cole, R. S., Eds.; ACS Symposium Series 1166; American Chemical Society: Washington, DC, 2014; Chapter 2.

14. Gee, J. P.; Green, J. L. Discourse analysis, learning, and social practice: A methodological study. *Rev. Res. Educ.* **1998**, *23*, 119–169.

15. Gee, J. P. *An Introduction to Discourse Analysis: Theory and Method*, 2nd ed.; Routledge: London, 2005.

16. Cobb, P.; Yackel, E. Constructivist, emergent, and sociocultural perspectives in the context of developmental research. *Educ. Psychol.* **1996**, *31*, 175–190.

17. Wood, L. A.; Kroger, R. O. *Doing Discourse Analysis: Methods for Studying Action in Talk and Text*; Sage Publications, Inc.: Thousand Oaks, CA, 2000.
18. Wooffitt, R. *Conversation Analysis and Discourse Analysis: A Comparative and Critical Introduction*; Sage Publications, Inc.: London, 2005.
19. Leach, J.; Scott, P. Individual and sociocultural views of learning in science education. *Sci. Educ.* **2003**, *12* (1), 91–113.
20. Vygotsky, L. S. *Mind in Society: The Development of Higher Physiological Processes*; Harvard University Press: Cambridge, MA, 1978.
21. Lave, J. *Cognition in Practice: Mind, Mathematics, and Culture in Everyday Life*; Cambridge University Press: Cambridge, U.K., 1988.
22. Wells, G. *Dialogic Inquiry: Toward a Sociocultural Practice and Theory of Education*; Cambridge University Press: Port Chester, NY, 1999.
23. Lemke, J. L. Analyzing Verbal Data: Principles, Methods, and Problems. In *Second International Handbook of Science Education*; Fraser, B., Tobin, K., McRobbie, C. J., Eds.; Springer: New York, 2012; pp 1471–1484.
24. Rymes, B. *Classroom Discourse Analysis: A Tool for Critical Reflection*; Hampton Press: New York, 2009.
25. Cazden, C. B. *The Language of Teaching and Learning*; Harvard Education Publishing Group: Cambridge, MA, 2001.
26. Lave, J.; Wenger, E. *Situated Learning: Legitimate Peripheral Participation*. Cambridge University Press: Cambridge, U.K., 1991.
27. Cobb, P.; Stephan, M.; McClain, K.; Gravemeijer, K. Participating in classroom mathematical practices. *J. Learn. Sci.* **2001**, *10*, 113–164.
28. Mercer, N. Sociocultural discourse analysis: Analysing classroom talk as a social mode of thinking. *J. Appl. Linguist.* **2007**, *1* (2), 137–168.
29. Airey, J.; Linder, C. A disciplinary discourse perspective on university science learning: Achieving fluency in a critical constellation of modes. *J. Res. Sci. Teach.* **2009**, *46* (1), 27–49.
30. Lemke, J. L. *Talking Science: Language, Learning, and Values*; Ablex Publishing Corporation: Norwood, NJ, 1990.
31. Krussel, L.; Edwards, B.; Springer, G. T. The teacher's discourse moves: A framework for analyzing discourse in mathematics classrooms. *School Sci. Math.* **2004**, *104* (7), 307–312.
32. Kelly, G.; Chen, C.; Crawford, T. Methodological considerations for studying science-in-the-making in educational settings. *Res. Sci. Educ.* **1998**, *28* (1), 23–49.
33. Corbin, J.; Strauss, A. *Basics of Qualitative Research: Techniques and Procedures for Developing Grounded Theory*; Sage Publications, Inc.: Thousand Oaks, CA, 2008.
34. Patton, M. Q. *Qualitative Research*; Wiley Online Library, 2005.
35. Krystyniak, R. A.; Heikkinen, H. W. Analysis of verbal interactions during an extended open-inquiry general chemistry laboratory investigation. *J.Res. Sci. Teach.* **2007**, *44*, 1160–1186.
36. Toulmin, S. E. The Uses of Argument; Cambridge University Press: Cambridge, U.K. 1958.
37. Driver, R.; Newton, P.; Osborne, J. Establishing the norms of scientific argumentation in classrooms. *Sci. Educ.* **2000**, *84* (3), 287–312.

38. Nussbaum, E. M.; Sinatra, G. M. Argument and conceptual engagement. *Contemp. Educ. Psychol.* **2003**, *28* (3), 384–395.

39. Erduran, S. Methodological Foundations in the Study of Argumentation in Science Classrooms. In *Argumentation in Science Education: Recent Developments and Future Directions*; Erduran, S., Jimenez-Aleixandre, M. P., Eds.; Springer Academic Publishers: Dordrecht, The Netherlands, 2007; pp 47–69.

40. Erduran, S.; Simon, S.; Osborne, J. TAPping into argumentation: Developments in the application of Toulmin's argument pattern for studying science discourse. *Sci. Educ.* **2004**, *88*, 915–933.

41. Becker, N.; Rasmussen, C.; Sweeney, G.; Wawro, M.; Towns, M.; Cole, R. S. Reasoning using particulate nature of matter: An example of a sociochemical norm in a university-level physical chemistry class. *Chem. Educ. Res. Pract.* **2013**, *14* (1), 81.

42. Cole, R.; Becker, N.; Towns, M.; Sweeney, G.; Wawro, M.; Rasmussen, C. Adapting a methodology from mathematics education research to chemistry education research: Documenting collective activity. *Int. J. Sci. Math. Educ.* **2012**, *10* (1), 193–211.

43. Kulatunga, U.; Lewis, J. E. Exploration of peer leader verbal behaviors as they intervene with small groups in college general chemistry. *Chem. Educ. Res. Pract.* **2013**, *14*, 576–588.

44. Kulatunga, U.; Moog, R. S.; Lewis, J. E. Argumentation and participation patterns in general chemistry peer-led sessions. *J. Res. Sci. Teach.* **2013**, *50* (10), 1207–1231.

45. Kulatunga, U.; Moog, R. S.; Lewis, J. E. Use of Toulmin's argumentation scheme for student discourse to gain insight about guided inquiry activities in college chemistry. *J. Coll. Sci. Teach.* **2014**, *43* (5), 78–86.

46. Rasmussen, C.; Stephan, M. A Methodology for Documenting Collective Activity. In *Handbook of Innovative Design Research in Science, Technology, Engineering, Mathematics (STEM) Education*; Kelly, A. E., Lesh, R. A., Baek, J. Y., Eds.; Taylor and Francis: New York, 2008; pp 195–215.

47. Lawson, A. The nature and development of hypothetico-predictive argumentation with implications for science teaching. *Int. J. Sci. Educ.* **2003**, *25* (11), 1387–1408.

48. Sampson, V.; Clark, D. B. Assessment of the ways students generate arguments in science education: Current perspectives and recommendations for future directions. *Sci. Educ.* **2008**, *92* (3), 447–472.

49. Schwarz, B. B.; Neuman, Y.; Gil, J.; Ilya, M. Construction of collective and individual knowledge in argumentative activity. *J. Learn. Sci.* **2003**, *12* (2), 219–256.

50. Sampson, V.; Enderle, P.; Walker, J. The Development and Validation of the Assessment of Scientific Argumentation in the Classroom (ASAC) Observation Protocol: A Tool for Evaluating How Students Participate in Scientific Argumentation. In *Perspectives on Scientific Argumentation*; Khine, M. S., Ed.; Springer: Dordrecht, The Netherlands, 2012; pp 235–264.

51. Kwon, O. N.; Ju, M. K.; Rasmussen, C.; Marrongelle, K.; Park, J. H.; Cho, K. Y.; Park, J. S. Utilization of revoicing based on learners' thinking in an inquiry-oriented differential equations class. *SNU J. Educ. Res.* **2008**, *17*, 111–134.

52. Stanford, C.; Cole, R. S.; Towns, M.; Moon, A. How Instructors Influence the Development of Scientific Arguments in POGIL Classrooms. In *Biennial Conference on Chemical Education*, 2014.

53. Gillies, R. M. The effects of communication training on teachers' and students' verbal behaviours during cooperative learning. *Int. J. Educ. Res.* **2004**, *41* (3), 257–279.

54. Criswell, B. A. Reducing the degrees of freedom in chemistry classroom conversations. *Chem. Educ. Res. Pract.* **2012**, *13* (1), 17–29.

55. Criswell, B. A.; Rushton, G. T. Conceptual change, productive practices, and themata: supporting chemistry classroom talk. *J. Chem. Educ.* **2012**, *89* (10), 1236–1242.

56. Warfa, A.-R. M.; Roehrig, G. H.; Schneider, J. L.; Nyachwaya, J. Role of teacher-initiated discourses in students' development of representational fluency in chemistry: A case study. *J. Chem. Educ.* **2014**, *91* (6), 784–792.

57. Hernández, G. E.; Criswell, B. A.; Kirk, N. J.; Sauder, D. G.; Rushton, G. T. Pushing for particulate level models of adiabatic and isothermal processes in upper-level chemistry courses: A qualitative study. *Chem. Educ. Res. Pract.* **2014**, *15*, 354–365.

58. Stieff, M.; Ryu, M.; Yip, J. C. Speaking across levels – Generating and addressing levels confusion in discourse. *Chem. Educ. Res. Pract.* **2013**, *14*, 376–389.

59. Lidar, M.; Lundqvist, E.; Östman, L. Teaching and learning in the science classroom: The interplay between teachers' epistemological moves and students' practical epistemology. *Sci. Educ.* **2006**, *90* (1), 148–163.

60. Stein, M. K.; Engle, R. A.; Smith, M. S.; Hughes, E. K. Orchestrating productive mathematical discussions: Five practices for helping teachers move beyond show and tell. *Math. Think. Learn.* **2008**, *10* (4), 313–340.

61. Gee, J. P. *How To Do Discourse Analysis: A Toolkit*; Routledge: Florence, KY, 2010.

62. Miller, K.; Zhou, X. Learning from Classroom Video: What Makes It Compelling and What Makes It Hard. In *Video Research in the Learning Sciences*; Goldman, R., Pea, R. D., Barron, B., Derry, S. J., Eds.; Lawrence Earlbaum Associates: Mahwah, NJ, 2007; pp 321–334.

63. Derry, S. J. Video Research in Classroom and Teacher Learning (Standardize That!). In *Video Research in the Learning Sciences*; Goldman, R., Pea, R. D., Barron, B., Derry, S. J., Eds.; Lawrence Erlbaum Associates: Mahwah, NJ, 2007; pp 305–320.

64. Alibali, M. W.; Nathan, M. J. Teachers' Gestures As a Means of Scaffolding Students' Understanding: Evidence from an Early Algebra Lesson. In *Video Research in the Learning Sciences*; Goldman, R., Pea, R. D., Barron, B., Derry, S. J., Eds.; Lawrence Earlbaum Associates: Mahwah, NJ, 2007; pp 349–365.

65. Derry, S. J.; Pea, R. D.; Barron, B.; Engle, R. A.; Erickson, F.; Goldman, R.; Hall, R.; Koschmann, T.; Lemke, J. L.; Sherin, M. G. Conducting video research in the learning sciences: Guidance on selection, analysis, technology, and ethics. *J. Learn. Sci.* **2010**, *19* (1), 3–53.

66. Linenberger, K. J.; Bretz, S. L. A Novel Technology to Investigate Students' Understandings of Enzyme Representations. *J. Coll. Sci. Teach.* **2012**, *42* (1), 45–49.

67. Weibel, N.; Fouse, A.; Emmenegger, C.; Friedman, W.; Hutchins, E.; Hollan, J. Digital Pen and Paper Practices in Observational Research. In *Proceedings of the SIGCHI Conference on Human Factors in Computing Systems*; ACM: Austin, TX, 2012; pp 1331−1340.

68. Christian, K.; Talanquer, V. Modes of reasoning in self-initiated study groups in chemistry. *Chem. Educ. Res. Pract.* **2012**, *13* (3), 286–295.

69. Talanquer, V. Using Qualitative Analysis Software To Facilitate Qualitative Data Analysis. In *Tools of Chemistry Education Research*; Bunce, D. M., Cole, R. S., Eds.; ACS Symposium Series 1166; American Chemical Society: Washington, DC, 2014; Chapter 5.

70. Neal, J. W.; Neal, Z. P.; VanDyke, E.; Kornbluh, M. Expediting the analysis of qualitative data in evaluation: A procedure for the rapid identification of themes from audio recordings (RITA). *Am. J. Eval.* **2014**, DOI: 10.1177/1098214014536601.

71. Psathas, G.; Anderson, T. The 'Practices' of Transcription in Conversation Analysis. In *Semiotica*; De Gruyter: Berlin, 1990; Vol. 78, p 75.

72. Liddicoat, A. J. An Introduction to Conversation Analysis, 2nd ed.; Continuum International Publishing Group: London, 2011.

73. Lapadat, J. C.; Lindsay, A. C. Transcription in research and practice: From standardization of technique to interpretive positionings. *Qual. Inq.* **1999**, *5* (1), 64–86.

74. Bird, C. M. How I stopped dreading and learned to love transcription. *Qual. Inq.* **2005**, *11* (2), 226–248.

75. Edwards, J. A. The Transcription of Discourse. In *The Handbook of Discourse Analysis*; Schiffrin, D., Tannen, D., Hamilton, H. E., Eds.; Blackwell: Oxford, 2003; p 321.

Chapter 5

Using Qualitative Analysis Software To Facilitate Qualitative Data Analysis

Vicente Talanquer*

Department of Chemistry and Biochemistry, University of Arizona, Tucson, Arizona 85721, United States
*E-mail: vicente@u.arizona.edu

Technological advances in the last twenty years have led to the development of powerful software for qualitative data analysis. Computer assisted qualitative data analysis (CAQDAS) packages facilitate managing multiple tasks in qualitative research, from organizing data sources based on relevant characteristics, segmenting and categorizing data according to themes, searching for and retrieving information, to building visual representations that more easily elicit significant patterns in the data. This chapter presents an introduction to the use of CAQDAS packages as tools that can greatly support qualitative research activities in chemistry education.

Introduction

The answers to many relevant research questions in chemistry education, such as how students approach the drawing of Lewis structures (*1*) or how the nature of laboratory activities influence students' conversations (*2*), can be obtained through investigations that generate in-depth information about the knowledge, beliefs, reasoning, attitudes, or behaviors of study participants (*3, 4*). This information may be gathered using different qualitative research techniques such as classroom observations, field notes, individual interviews, focus groups, open response questionnaires, written reports, and journals; some of these and other qualitative research strategies are discussed in depth in the chapters by Cole, Becker, and Stanford (5), Yezierski (6), and Herrington and Daubenmire (7) in this volume. The data collected using these strategies, such as text, images, and diagrams, is not numerical in nature (qualitative data) and tends to be very lengthy, requiring intensive and repeated examination in the search for answers to research questions (8, 9). This analysis can be done manually or with the help of computer software commonly known as *computer assisted qualitative data analysis* (CAQDAS) packages that simplify the tasks of organizing, exploring, integrating, and reflecting on the information (*10–12*). The central goal of this chapter is to provide an overview of useful strategies to take advantage of these types of technological resources to support and facilitate qualitative research in chemistry education. The chapter is directed to researchers who have not yet used qualitative data analysis software to support their work.

CAQDAS

Advances in analytical software in the last twenty years have led to the development of powerful tools for qualitative data management. These types of resources facilitate the implementation of core tasks in most common qualitative research designs, from case studies to discourse analysis to grounded theory (*3–9*). Commercially available CAQDAS packages such as ATLAS.ti (*13*), Dedoose (*14*), MAXQDA (*15*), and NVivo (*16*) allow users to import different types of data (e.g., text, images, audio, and video) and categorize them according to user-defined attributes or descriptors (e.g., gender, educational level, or course grade of study participants). These data can then be segmented and organized into categories using codes created by the user either before or during the analysis. Most CAQDAS packages also include tools for selective retrieval of text or coded data, powerful systems for writing and reviewing comments and memos linked to documents, audio or video files, text segments, or specific codes, as well as diverse avenues for data representation that facilitate the identification of conceptual themes or patterns (*10*, *11*).

The use of qualitative software for data analysis has several advantages. First, it frees researchers from tedious managing tasks, allowing them to focus on the data and their reactions to it. It provides dynamic and simultaneous access to different components of the data analysis, from excerpts to codes to annotations to demographic data. Additionally, it more quickly directs researchers' attention to themes and relationships emerging from the analysis. In many ways, the software allows users to get closer to the data and further explore its intricacies. Nevertheless, CAQDAS packages do not carry out core qualitative analysis tasks for the researcher. Common qualitative software does not, for example, transcribe audio or video files, define relevant attributes for data sources, identify meaningful segments in a text, or build and apply codes in an independent and automatic fashion. Nor do these types of resources independently detect patterns in the data or identify overarching themes. Researchers have full control, and responsibility, over the analytical and interpretative processes. Thus, the quality of the results generated by using qualitative software is only as good as that of the research design and methods of data analysis. As is the case with packages for quantitative analysis (e.g., SPSS), the quality of the output strongly depends on the choices made by the researcher about what data to collect, how to collect the data, and what strategies to apply to analyze the data.

Differences among common CAQDAS packages are subtle and likely to become less explicit as the underlying technology advances in upcoming years. Commercially available products tend to offer the same core tools and functions, and decisions about what package to use may thus be difficult, particularly for novice users. Researchers should look for the resource that best fits their research goals and methodological approaches. Detailed comparative analysis involving the most widely used CAQDAS packages can be found in the literature (10, 11). In this chapter, I will illustrate the application of core strategies in computer assisted qualitative analysis using the on-line software Dedoose when presenting specific examples. This choice is somewhat arbitrary and does not imply a personal endorsement or recommendation of any particular type of software. Dedoose is a web-based application that is easily accessible via the Internet and includes the major analytical tools present in modern CAQDAS packages. It has a simple user interface and has been designed to facilitate collaboration by geographically dispersed researchers who may work with Mac or PC systems (14). Comparisons between common CAQDAS packages (ATLAS.ti, Dedoose, MAXQDA, NVivo) will be presented in those areas in which differences are judged to be more substantive.

Analyzing Qualitative Data

The analysis of qualitative data involves a variety of tasks: copying, transcribing, and organizing data; segmenting and coding transcripts, images, audio, and video; stepping back from the data, seeking to identify overarching themes and patterns; building and applying coding schemes at different levels (e.g., descriptive, interpretative); identifying quantitative patterns in coded segments across different participants (3–6). CAQDAS packages allow researchers to perform most of these tasks in more systematic and efficient manners. The following sections summarize fundamental strategies in the analysis of qualitative data using these types of computer applications.

Handling and Organizing Data

Qualitative research studies generate vast amounts of textual and non-textual data of different types, from background information about the study participants to primary data collected in the field to supporting or secondary data that may be relevant for the analysis. CAQDAS packages allow researchers to import all of these different resources, integrate them into a single project, and build links or cross-references between them. Common applications will directly import documents in plain text (.txt), rich text format (.rtf), and Microsoft Word (.doc or .docx) formats, and include tools to edit such texts within the software. In most cases, such documents may contain tables, embedded images, and rich objects, but format specifications may vary among different CAQDAS packages (*10, 11*). Most of these applications also allow audio and video files to be directly incorporated into the software project. However, existing software packages do not have the capability to transcribe these files into written documents.

One of the key elements of CAQDAS packages is that data sources can be organized according to known characteristics. This enables researchers to easily search for, identify, and narrow their focus on subsets of data, which facilitates comparisons. More importantly, this organization lays the foundation on which the dataset can be analyzed in search of themes, patterns, and relationships across data sources. Existing CAQDAS packages organize known data characteristics differently, and use various terms to refer to them, such as families (ATLAS.ti), attributes (MAXQDA, NVivo), or descriptors (Dedoose). All of these applications allow researchers to assign these characteristics within the software or upload spreadsheets that contain such information. A critical step in the data analysis process is to carefully reflect on the attributes or descriptors that are relevant in a study. These decisions should be based on the questions researchers want to ask, the ideas they want to test, and the parts of the data set that they want to isolate for comparative purposes. These ideas will be illustrated with a concrete example.

Recently, we completed a study focused on the characterization of the ideas (assumptions) and reasoning strategies (heuristics) used by college general chemistry students when asked to judge the relative thermodynamic likelihood of different chemical processes (*17*). This research project was based on data collected using individual interviews in which participants were asked to think out loud when solving five different problems. Verbatim transcripts of the recorded interviews were our primary data and, for purposes of analysis, we decided to build different text files containing the answer of a single student to a single question. This decision stemmed from our interest in eventually comparing not only the assumptions and heuristics used by different students, but also the assumptions and heuristics applied across different questions. As shown in Figure 1, when the different text files were uploaded into the Dedoose CAQDAS package, each of the files was assigned a descriptor that indicated the student to which the file belonged (e.g., S1, S2, …) and the question to which it corresponded (e.g., Q1, Q2, …). Additionally, we built descriptors to indicate the gender of each participant and their final grade in the general chemistry class in which they were enrolled at the time of the interview. These descriptors were introduced because we were also interested in exploring the association between the types of assumptions and heuristics that students applied and these individual attributes.

In most software applications, attributes can be applied to individual documents or to sets of documents. Although it is easier and more efficient to organize data sources using attributes or descriptors early in the analytical process, CAQDAS packages allow researchers to assign known characteristics at any stage during the analysis.

	Selected	Linked Media	Memos	Student ▲	Question	Gender	Final Grade
Columns & Filters	☐	1	0	S1	Q1	Male	C
	☐	1	0	S1	Q2	Male	C
	☐	1	0	S1	Q3	Male	C
	☐	1	0	S1	Q4	Male	C
	☐	1	0	S1	Q5	Male	C
	☐	1	0	S10	Q2	Female	B
	☐	1	0	S10	Q1	Female	B
	☐	1	0	S10	Q5	Female	B
	☐	1	0	S10	Q3	Female	B
	☐	1	0	S10	Q4	Female	B
	☐	1	0	S11	Q4	Female	B
	☐	1	0	S11	Q1	Female	B
	☐	1	0	S11	Q2	Female	B

Figure 1. Dedoose screen capture showing the set of descriptors (student, question, gender, final grade) linked to different data sources (media) in our research project about students' assumptions and heuristics. Copyright 2013 Dedoose Version 4.5, web application for managing, analyzing, and presenting qualitative and mixed method research data (2013). Los Angeles, CA: SocioCultural Research Consultants, LLC (www.dedoose.com).

Coding of Data

Several qualitative research approaches or designs rely on coding to conceptually organize data (3–9). During this process, relevant segments of the available data are identified and classified as instances or exemplars of major ideas or themes. Codes define conceptual categories that are expected to run both within and across different data sources, and may help provide meaningful responses to research questions. These codes could be predefined, arising from the theoretical framework or research questions, or they can emerge from the data during the analysis. Most coding involves the combination of these deductive and inductive approaches. In general, coding is an iterative process in which the researcher moves back and forth between careful reading of the data and critical reflection of its meaning. As a researcher dives in and steps back from the data, codes can be generated at different levels. Initially, codes may be very specific, describing significant ideas, events, or actions detected in the data. As the analysis progresses, descriptive codes may be organized or collapsed into higher order categories that are more interpretative in nature (8, 9).

CAQDAS packages facilitate the coding process at many levels. These applications allow the researcher to easily build a coding system either from the outset or in an ongoing manner during data analysis. Labels and descriptions of individual codes can be modified, and codes can be expanded, deleted, or rearranged when needed. Segments of text, audio, or video in any given file can be identified and highlighted to create excerpts, or audio and video fragments, that can be linked to one or multiple codes by simple dragging and dropping movements. Consider the example presented in Figure 2, where a segment (called an excerpt within the Dedoose software) in an interview transcript in our Assumptions and Heuristics project has been highlighted, and linked to two relevant codes in our study (labeled "One-Reason Decision Making" and "Easiness"). This figure also shows the hierarchy of codes built and applied to categorize different assumptions and heuristics that emerged from our data analysis.

Chemistry educators engaged in qualitative research often use CAQDAS packages to facilitate the coding of interview transcripts or written responses to open questions. The use of this type of software can help chemistry education researchers more easily identify trends in students' alternative conceptions (18), explore teachers' instructional goals and strategies (19, 20), collect exploratory data to guide the development of quantitative research instruments (21), identify educational approaches that support learning (22), or elicit students' beliefs and perspective about diverse educational issues (23).

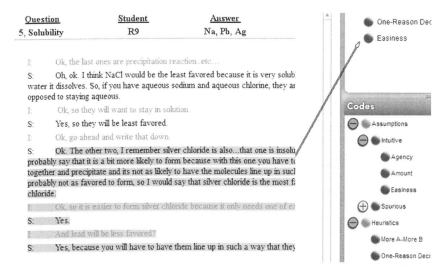

Figure 2. Dedoose screen capture showing a highlighted excerpt in an interview transcript and the two codes associated to this excerpt. The overall coding scheme used in this analysis is shown in the lower right panel. Copyright 2013 Dedoose Version 4.5, web application for managing, analyzing, and presenting qualitative and mixed method research data (2013). Los Angeles, CA: SocioCultural Research Consultants, LLC (www.dedoose.com).

The default structure of the coding scheme in common CAQDAS packages tends to be hierarchical (except in ATLAS.ti). This facilitates the creation of embedded sets of subcodes that can be used to categorize segments or excerpts at different levels. For example, in the case illustrated in Figure 2, the major code "Assumptions" was subdivided into the subcodes "Intuitive" and Spurious," which in turn included various subcategories. These software programs provide the user with the freedom to use as much or as little hierarchy as needed in the organization of the coding scheme. Some software packages, such as MAXQDA and Dedoose, use different colors to more clearly indicate the status of a code within a hierarchy. Most applications allow the easy reshuffling of codes in and out of different categories. Codes can be easily renamed, grouped, merged, or reallocated within different classes. The fluid and flexible manipulation of the coding scheme within these CAQDAS packages greatly facilitates qualitative data analysis since any change automatically propagates throughout all coded segments in every document within the software project. In general, changes that involve code reorganization, merging, and streamlining are easier to implement than changes that require separating data originally linked to a single code into two or more new categories. This task demands recoding of excerpts by the researcher who needs to make decisions about how the data should be reassigned.

Retrieval of Coded Data

When performing qualitative data analysis, it is important to step back, review the progress made, reflect on results, group similar data, identify gaps or inconsistencies, and re-strategize if needed. The ability to easily search for and retrieve coded data is thus critical and constitutes another great advantage of using CAQDAS packages. In general, these applications facilitate the completion of the following tasks:

- View all codes linked to different segments in a given document;
- Retrieve all coded data associated with an individual code or a set of codes (see Figure 3);
- Recode retrieved segments;
- Dynamically view different code frequencies as the analysis proceeds;
- Export and retrieve information in the form of reports that can be printed or saved in common formats (e.g., MS Excel or Word).

Although commercially available software packages offer these different functionalities, they differ greatly in the steps that users must follow to implement them. Some packages, such as ATLAS.ti and MAXQDA, include interactive margin views that visually display all codes linked to different segments in a single text. In other cases, such as Dedoose, these codes are displayed in a separate panel when selecting a segment (see Figure 3). In most cases, access to all coded segments linked to a given code across data sources is provided through a code index or a code panel. Different strategies may need to be applied to filter data in order to view all segments that are simultaneously linked to two or more codes. Despite these differences, common CAQDAS packages allow researchers to analyze their data in multiple manners and generate diverse reports that more easily uncover meaningful trends. For example, the analysis of excerpts linked to both the code "Easiness" and the code "One-Reason Decision Making" in our Assumptions and Heuristics project (Figure 3) revealed the existence of a potential relationship between these two types of reasoning. Reports may also serve to identify potential problems in the analysis. Reviewing all of the segments coded under a certain category can help researchers decide whether a code actually captures a targeted theme or whether all segments belong to the same group. The software will not automatically resolve any potential issues, but will allow researchers to systematically create opportunities to pause, reflect, and make decisions to improve data analysis.

Figure 3. Screen capture of the code window in Dedoose. Selection of a code, such as "Easiness," shows all excerpts coded in that category. By making these set active, subsets of excerpts linked to a second code, such as "One-Reason Decision Making," can be identified. Copyright 2013 Dedoose Version 4.5, web application for managing, analyzing, and presenting qualitative and mixed method research data (2013). Los Angeles, CA: SocioCultural Research Consultants, LLC (www.dedoose.com).

Annotating Data

Qualitative data analysis demands researchers to constantly discuss their ideas and reflect on their actions and interpretations. Researchers must ensure that they record their thoughts, insights, questions, and concerns as they delve into the data. When done manually, this writing process may become messy and inconsistent as researchers deal with different data sources and analytical components, and use a variety of annotating strategies and procedures. CAQDAS packages facilitate these tasks by providing tools that researchers can use to build notes, comments, memos, or entire research journals that can be linked to one or more pieces of the software project, from single documents, to segments within a document, to individual codes in the coding scheme. These writing elements can be built purposely as research journals or spontaneously as researchers interact with the data; they can be managed in systematic ways, searched and retrieved when needed, and exported to be printed or saved as external documents.

Using a laptop or tablet computer, qualitative software can be used to directly write observational logs or field notes to be integrated as data sources into the research project. CAQDAS packages can also be useful in keeping research diaries or journals that record day-to-day project activities. Commercially available applications have been designed to facilitate the spontaneous construction of analytic memos to aid during the coding and interpretative processes. These memos can be used for multiple purposes, such as making comments on specific aspects of the data, posing questions for further reflection, recording tentative interpretations, or sharing thoughts with other researchers who may have access to the data. The ability to link these memos to other objects (documents, data segments, codes) allows researchers to integrate their thoughts with other components of the project.

Visualizing Data

CAQDAS packages can also be used to create visual representations such as maps, diagrams, and graphs that aid analysis by more readily revealing trends and relationships, and helping researchers in the development of interpretations. These visual representations are usually linked to the underlying data used in their construction, allowing researchers to easily move back and forth between the data and the graphic model.

Software applications such as ATLAS.ti, MAXQDA, and NVivo include a variety of mapping tools to build networks (ATLAS.ti), maps (MAXQDA), or models (NVivo). These tools include: icons to represent codes, memos, documents, or theoretical concepts; connectors to link these objects; and linking labels to define the nature of the connections. These resources can be used to build visual constructs that resemble interactive concept maps that elicit relevant themes and relationships emerging from the data analysis. A good example of the use of these visual and conceptual aids in chemistry education research can be found in the study by Del Carlo and Bodner on students' perceptions of academic dishonesty (24). As shown in this work, these types of representations are very useful in creating visual structures of complex systems and phenomena. Nevertheless, it is important to keep in mind that these maps are built with input from the researcher and should thus be interpreted with caution.

The on-line software Dedoose does not include mapping functionalities but is designed to support the graphic visualization of data. For example, the application can be used to visualize code frequencies as a function of selected descriptors, or the degree of co-occurrence of two or more codes in defined sets of data (other CAQDAS packages also include this latter functionality). The resulting graphs are interactive and the associated qualitative data can be explored by clicking on different portions of a graph. All these results can be exported to external documents or printed if needed. Figure 4 illustrates some of the graphical representations that provided insights into trends and relationships in our Assumptions and Heuristics project. These graphs revealed, for example, the overreliance of our study participants on the assumption of "Easiness" in making predictions about the thermodynamic likelihood of chemical reactions (*17*). Similarly, they helped us identify the correlation between the application of this assumption and the use of a "One-Reason Decision Making" heuristic. The software also allowed us to explore how these patterns of reasoning were affected by the nature of the question posed, the gender of the participants, and their final grade in the general chemistry course.

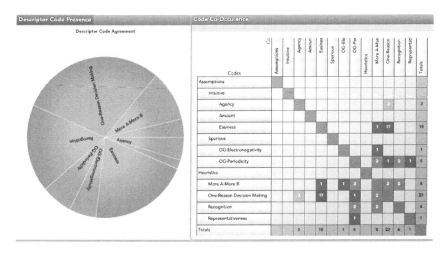

Figure 4. Examples of the types of graphic representations generated by Dedoose based on code frequencies. The pie chart on the left shows the relative frequency of all codes; the graph to the right displays the degree of co-occurrence of different codes. Copyright 2013 Dedoose Version 4.5, web application for managing, analyzing, and presenting qualitative and mixed method research data (2013). Los Angeles, CA: SocioCultural Research Consultants, LLC (www.dedoose.com).

Other Functionalities

Common CAQDAS packages include other tools and functionalities not described in previous sections that can greatly facilitate data management and analysis. For example, these applications have basic editing tools that allow researchers to color, underline, embolden, or italicize words or phrases to make them more salient. Most of these computer programs also incorporate search engines that can be used to track specific words or run word frequency checks. Additionally, many systems allow the creation of hyperlinks between different points in the dataset, which can be used to link specific parts of different documents to track a line of reasoning, compare and contrast approaches to solve a given problem, or associate consistent ideas. All of these tools are particularly useful when performing text or discourse analysis (*8, 9*). Nevertheless, researchers should be cautious when using search tools to avoid building coding systems that are based on the mere frequency of specific words.

More recent innovations in CAQDAS packages include tools to capture data from web pages, social media, and tweets that can be directly incorporated into the software projects. The number of different text, audio, and video formats that can be directly imported into the applications is steadily increasing. Systems like Dedoose, which is an on-line application, allow multiple researchers in different locations to simultaneously work and collaborate on the same project. Traditional stand-alone software packages, such as NVivo, are quickly incorporating similar capabilities, allowing multiple users to access the same files when working with the same computer or with a set of interlinked devices. Access to the same data pool by several researchers facilitates the comparison of coding done by multiple users, and the calculation of inter-rater reliability coefficients (25) within the CAQDAS package. For example, ATLAS.ti includes functionalities for calculating Cohen's kappa and Krippendorff's alpha, while Dedoose facilitates the calculation of percentage of agreement between multiple researchers. Easy access to a software project by multiple people creates unique opportunities to not only carry out collaborative research, but to interactively train researchers in the use of qualitative methods of data analysis.

Final Considerations

The use of software to support qualitative data analysis is often associated with certain "myths" that it is important to dispel (26). On the one hand, some people envision CAQDAS packages as all-powerful tools capable of completing data analysis without much input from the researcher. On the other hand, others judge that the use of these applications will constrain researchers' ability to control the direction and approach to the analysis of the data. As shown in this chapter, none of these scenarios is real. CAQDAS packages are very useful tools for data analysis that allow researchers to keep full control of the analytical process, greatly facilitate multiple managing tasks, but do not independently and automatically decide how to segment, code, or interpret the data. The quality of analysis will not be determined by the sophistication of the software used, but rather by the thoughtfulness and rigor of the analytical procedures employed by the researcher.

Given the multiple tasks that CAQDAS packages can carry out, learning how to use and take advantage of all of their different tools may seem daunting to the beginner. Nevertheless, commercially available applications tend to have intuitive interfaces and multiple resources to support their users (e.g., help manuals and videos). The structure of these software applications allows for open exploration. Thus, I would suggest approaching the initial use of the package in a flexible and fluid manner. One should recognize that not every available functionality needs to be used in each project, and that the ability to easily import, manage, and interact with multiple documents may lead some researchers to over-saturate and overanalyze their data pool. In general, mistakes can be easily corrected by undoing tasks or re-uploading documents and major issues may be controlled and reduced by constantly stepping back from the data to critically reflect on existing results and analytical strategies.

The constant incorporation of technological advances into CAQDAS packages certainly increases their analytical power, but also carries new challenges. For example, accessing and retaining rich multimedia data may raise ethical issues in terms of ownership, privacy, anonymity, and confidentiality. Participation of multiple individuals who can easily access and exchange research data will demand careful research design and monitoring. Researchers should thus carefully reflect on the specific practical, legal, and ethical issues that may emerge from the use of qualitative software. As discussed in Bauer's chapter in this volume (*27*), we, as researchers, are ultimately responsible for ensuring the proper collection, management, and disposal of any data incorporated into our research projects.

References

1. Cooper, M. M.; Grove, N.; Underwood, S. M.; Klymkowsky, M. W. *J. Chem. Educ.* **2010**, *87* (8), 869–874.
2. Xu, H.; Talanquer, V. *J. Chem. Educ.* **2013**, *90* (1), 29–36.
3. Bretz, S. L. Qualitative Research Designs in Chemistry Education Research. In *Nuts and Bolts of Chemical Education Research*; Bunce, D. M., Cole, R. S., Eds.; ACS Symposium Series 976; American Chemical Society: Washington, DC, 2008; Chapter 7.
4. Towns, M. H. Mixed Methods Designs in Chemical Education Research. In *Nuts and Bolts of Chemical Education Research*; Bunce, D. M., Cole, R. S., Eds.; ACS Symposium Series 976; American Chemical Society: Washington, DC, 2008; Chapter 9.
5. Cole, R. S.; Becker, N.; Stanford, C. Discourse Analysis as a Tool To Examine Teaching and Learning in the Classroom. In *Tools of Chemistry Education Research*; Bunce, D. M., Cole, R. S., Eds.; ACS Symposium Series 1166; American Chemical Society: Washington, DC, 2014; Chapter 4.
6. Yezierski, E. J. Observation as a Tool for Investigating Chemistry Teaching and Learning. In *Tools of Chemistry Education Research*; Bunce, D. M., Cole, R. S., Eds.; ACS Symposium Series 1166; American Chemical Society: Washington, DC, 2014; Chapter 2.
7. Herrington, D. G.; Daubenmire, P. L. Using Interviews in CER Projects: Options, Considerations, and Limitations. In *Tools of Chemistry Education Research*; Bunce, D. M., Cole, R. S., Eds.; ACS Symposium Series 1166; American Chemical Society: Washington, DC, 2014; Chapter 3.
8. Miles, M. B.; Huberman, A. M.; Saldaña, J. *Qualitative Data Analysis: A Methods Sourcebook*; Sage Publications, Inc.: Thousand Oaks, CA, 2013.
9. Patton, M . Q. *Qualitative Evaluation and Research Methods*; Sage Publications, Inc.: Thousand Oaks, CA, 2002.
10. Lewins, A.; Silver, C. *Using Software in Qualitative Research: A Step-by-Step Guide*; Sage Publications, Inc.: Thousand Oaks, CA, 2007.
11. CAQDAS Networking Project. http://www.surrey.ac.uk/sociology/research/researchcentres/caqdas/ (accessed August 2013).

12. Silver, C.; Fielding, N. In *The SAGE Handbook of Qualitative Research in Psychology*; Willig, C., Stainton-Rogers, W., Eds.; Sage Publications, Inc.: Thousand Oaks, CA, 2008.

13. ATLAS.ti. http://www.atlasti.com/ (accessed August 2013).

14. Dedoose. http://www.dedoose.com/ (accessed August 2013).

15. MAXQDA. http://www.maxqda.com/ (accessed August 2013).

16. NVivo. http://www.qsrinternational.com/ (accessed August 2013).

17. Maeyer, J.; Talanquer, V. *J. Res. Sci. Teach.* **2013**, *50* (6), 748–767.

18. Rushton, G. T.; Hardy, R. C.; Gwaltney, K. P.; Lewis, S. E. *Chem. Educ. Res. Pract.* **2008**, *9*, 122–130.

19. Bruck, L. B.; Towns, M.; Lowery-Bretz, S. *J. Chem. Educ.* **2010**, *87*, 1416–1424.

20. Bruck, A. D.; Towns, M. *J. Chem. Educ.* **2013**, *90*, 685–693.

21. Harsh, J. A.; Maltese, A. V.; Tai, R. H. *J. Chem. Educ.* **2012**, *89*, 1364–1370.

22. Harshman, J.; Lowery-Bretz, S.; Yezierski, E. *J. Chem. Educ.* **2013**, *90*, 710–716.

23. Russell, C. B.; Weaver, G. C. *Chem. Educ. Res. Pract.* **2011**, *12*, 57–67.

24. Del Carlo, D. I.; Bodner, G. *J. Res. Sci. Teach.* **2004**, *41* (1), 47–64.

25. Stemler, S. E.; Tai, J. In *Best Practices in Quantitative Methods*; Osborne, J. W., Ed.; Sage Publications, Inc.: Thousand Oaks, CA, 2008; pp 29–50.

26. Bong, S. A. *Forum Qual. Soc. Res.* **2002**, *3* (2) Article 10.

27. Bauer, C. F. Ethical Treatment of the Human Participants in Chemistry Education Research. In *Tools of Chemistry Education Research*; Bunce, D. M., Cole, R. S., Eds.; ACS Symposium Series 1166; American Chemical Society: Washington, DC, 2014; Chapter 15.

Analyzing Quantitative
Research Data

Chapter 6

Introduction to the Use of Analysis of Variance in Chemistry Education Research

Thomas C. Pentecost*

Department of Chemistry, Grand Valley State University,
Allendale, Michigan 49401, United States
*E-mail: pentecot@gvsu.edu

Statistical analysis of data and the presence of a statistical difference is the evidence used to justify a claim about the efficacy of some new teaching pedagogy or other instructional intervention. Chemistry education researchers have many, relatively easy to use, software packages that can carry out these analyses. What these software packages cannot do is to determine the type of analysis that is appropriate. The decision about the type of analysis to be done is intimately tied to the design of the study. The final comparison is often comparing the scores of two or more groups on some measure. Exactly how this comparison is done, and the choice of the specific statistical test used, is the focus of this chapter. The chapter begins with a brief review of the t-test and then moves through more complex Analysis of Variance techniques (ANOVA). It is not the goal of this chapter to provide a step-by-step method for doing these analyses; instead the goal is for the reader to come away with an understanding of the fundamental concept behind the ANOVA techniques and the types of designs that lend themselves to the various techniques.

Introduction

The analysis of quantitative data is very common in chemistry education research (CER). Statistical analysis of data and the presence of a statistical difference is the evidence used to justify a claim about the efficacy of some new teaching pedagogy or other instructional intervention. The ability to support a claim is intimately tied to the quality of the study done. Two key components to the quality of a study are the research design and the analysis techniques used. Together these two components make up the methodology of the study. In a previous book in this series, Sanger (1) reviewed the role of inferential statistics in CER. This chapter will build on that work by describing one specific type of analysis method, Analysis of Variance (ANOVA), in more detail. A thorough discussion of all of the various ANOVA techniques is the subject of a semester long graduate course in statistics. Obviously, this chapter cannot take the place of this type of course; instead the goals of this chapter are to provide an introduction to the terminology of ANOVA, illustrate the relationship between the ANOVA technique used and the research design, and provide the interested reader with a list of more detailed references. More simply put, the goal is that when the readers of this chapter meet and discusses their research with a statistical consultant, always advisable, the readers will be able to understand the language used by the statistician. For a more rigorous mathematical treatment of the calculations see the classic text by Howell (2).

Before beginning the discussion of the various ANOVA techniques, it is important to emphasize the relationship between the experimental design and the statistical analysis technique used. This relationship can be illustrated with a chemistry analogy–an enzyme catalyzed process. Just as an enzyme and substrate must come together for a process to occur, in a CER experiment the research design and the analysis method must work together so that a claim can be made. To extend the analogy further, the specific enzyme will only work with certain substrates; just as certain research designs will dictate the statistical analysis used.

We will begin this chapter with a description of the simplest design and method of comparison, one that is familiar to chemists, the t-test. This will allow a review of some terminology and provide a basis for the introduction of more complex designs and ANOVA techniques. It should be pointed out that to use the statistical techniques described below, the data are assumed to be normally distributed. If this is not the case, other approaches may be needed (see the chapter by S. Lewis in this volume). More details describing the requirements of the data that are necessary to use these statistical techniques will be described below.

How Do We Decide If There Is a Difference?

In analytical chemistry, students are introduced to the "Students'-t test" (3). This is often used as a way to compare the means of two sets of measurements. This is actually the simplest experimental design and a good place to start a discussion of statistical methodologies. Consider the following situation: a chemist is comparing the amount of a product produced by two different synthesis methods, A and B. Multiple runs of each method are done and an average percent yield, with its associated standard deviation, is determined for each method. In this scenario the average percent yield is the **dependent variable** (*4*) and the synthesis method is the **independent variable**. Note that in educational research the independent variable is often called the **treatment variable** since this is the variable we manipulate to cause a change in the dependent variable. In this example the independent variable is limited to two options, method A or method B. In a design where there is only one dependent variable and one independent variable with two possible values or conditions, the t-test is the appropriate statistic to use. The t-test statistic, t, can be represented as:

$$t = \frac{\text{difference in means}}{\text{error of measuring the difference}} \qquad (1)$$

Equation 1 illustrates the fundamental form of many statistical test statistics. The observed difference between measurements is compared to the error associated with the measurements. The term in the denominator of equation 1 is associated with the **variance** of the data. Chemists are familiar with the standard deviation, which is the square root of variance. The standard deviation is used in the denominator for the t-test. In the tests used in ANOVA, the actual variance is used.

If the observed difference in means is large or the error is small, the value of the statistic will be large indicating that the difference is likely not due to chance. If on the other hand, the observed difference in means is small or the error is large, it is likely the observed difference is simply due to chance and not a result of a real difference in the two conditions. To determine if the value of the test statistic is significant, it is compared to a critical value. The critical value used depends on the **level of significance** (often referred to as a **α-value**) and the sample size. The α-value is the level of risk of making a **Type I** error you are willing to take when making your claim. Making a Type I error means claiming a difference when, in fact, there is no difference. For example, an α of 0.05 means that there is only a 5% chance the observed difference is the result of chance and not a real difference. If the test statistic is greater than the critical value (obtained from a table for the given value of α), then we can claim a significant difference in the measured means.

An additional point to consider after establishing a statistically significant difference is the magnitude of the difference. The **effect size** is the standardized measure of the magnitude of the observed difference. There several common measures used to calculate the effect size (*2*). Measures of effect size commonly used in CER are Cohen's d, Pearson's correlation coefficient, Eta squared (η^2), and omega squared (ω^2)–sometimes reported as ω. To interpret the magnitude of the effect size, Cohen (*5*) proposed the following guidelines for r and d: 0.10 (small effect); 0.30 (medium effect); and 0.50 (large effect). Guidelines for interpreting ω^2 values are 0.01, 0.6, and 0.14 for small, medium, and large effect sizes (*6*) Although these are widely used guidelines, they are not to be invoked as an absolute indicator of the magnitude of the effect (*2*). No matter which measure of effect size is used, it should be included in any report (*7*). It is important to note that while many statistical software packages will calculate these, it is often necessary to calculate these by hand.

Assumptions About the Data

The t-test and the ANOVA techniques to be described later are all **parametric tests**, which means that the data must meet certain assumptions before the statistical test can be used. The underlying assumptions common to all parametric statistics are (*8*)

- The data are normally distributed.
- The variances of all samples in the data are the same. This is also called *homoscedasticity*.
- The samples or measurements are independent of each other.

Most statistical computer programs provide ways to test for violation of the first two assumptions (*6*). If the data are not normally distributed, but treated as so, then the value of the test statistic calculated will be in error. The result being that the Type I error rate being used is no longer accurate. If the data are not normally distributed, there are several techniques that can be used to transform the data so that parametric statistics can be used (*9*). If the variances are not equal, the chances of claiming a significant difference when there isn't one, increase. Having equal sample sizes mitigates the impact of violations of homoscedascity. Many common statistical tests, such as the t-test and ANOVA techniques have been shown to be **robust statistics**, that is, they are applicable in all but extreme violations of the first two assumptions (*8*). So before doing statistical tests, it is important to establish the "normality" of your data. If the data are not normally distributed, then alternative techniques, such as nonparametric statistics, are required. A review of nonparametric statistics can be found in this volume (*10*) in the chapter by S. Lewis.

The assumption of independent samples often presents a problem in educational research and this highlights the relationship between the experimental design and the analysis methods used. In a CER research project, independent samples means that the data are collected from two separate groups of students. For example, two separate classes of students are given the same test of chemistry knowledge, but are taught with different methods. The responses of the two groups of students are independent. However, if a pre/post test was used, the responses to the pre- and post-test would not be independent because it is the same person responding to each measurement, pre and post. This type of design, where a person is measured more than once is called a **repeated measures** design. Note that repeated measures violates the assumption of independent measurements. Fortunately, there are ways to handle this violation by correcting the denominator of equation 1 which gives us a **paired t-test** or a **repeated measures** ANOVA. It is important to note that simply correcting the statistical tests for this violation is not sufficient. The data also must be collected in such a way to minimize the threats to any claims made as a result of the statistical analysis. Types of designs that minimize these threats have been developed (*11*).

Need for Analysis of Variance Techniques

The t-test is the appropriate test for comparing two groups. However, we often find ourselves in the situation when we want to compare more than two groups' performance on a dependent variable. It would be tempting to simply perform multiple t-tests on the groups, but this is not advisable. Recall that an α value of 0.05 means that there is a 95% probability of not making a Type I error. If you are doing five t-tests, the probability of not making a Type I error is $0.95^5 = 0.77$. This implies the probability of making a Type I error is 1 - 0.77, or 0.33. An α value of 0.33 is an unacceptably high probability of a Type I error,. When more than one comparison needs to be done, multiple t-tests are not appropriate. These situations call for the use of analysis of variance (ANOVA) methods.

Which ANOVA Technique Do We Use?

Before discussing the various ANOVA techniques and the experimental designs they are suited for, it is important to develop a conceptual understanding of the F-ratio, the test statistic common to all ANOVA techniques. In both the t-test and the F-ratio, we are comparing some difference in the data to the error in the measurements. The statistics produced by both are compared to a critical value to determine if the observed difference in the two values is greater than a difference that would arise from experimental error. The distinction lies in values used. Whereas the t-test uses the differences in means, the F-ratio uses mathematical descriptions of the variations in the data (variance). Each type of ANOVA technique partitions the total variance found in the data into components. The simplest partition is into **between-group** and **within-group** variance.

Total Variance = between-group variance + within-group variance   (2)

Between-group variance is related to the spread in the values of the dependent variable when the groups, sometimes called conditions or treatments, are compared. The within-group variance is the spread of values within a single group. Each ANOVA technique will use a specific variance term as the variance of interest (the variance caused by the intervention) and a variance attributed to chance, or random error. The F-ratio is the ratio of the appropriate variances. The larger the variance caused by the treatment, with respect to the variance due to chance, the larger the value of F and the more likely the treatment effect is real, or statistically significant. The general form of the F-ratio is:

$$F = \frac{\text{variance caused by treatment}}{\text{experimental error}} \qquad (3)$$

Roberts and Russo (9) provide an excellent conceptual definition of the F-ratio: *"It is a measure of the extent to which experimental error, not treatment effects, caused the scores to differ. (p37)"* The actual design of the study will determine what values are used to measure the treatment (model) variance and the experimental error.

Independent Measures of a Single Dependent Variable

One-Way ANOVA

The simplest ANOVA consists of an experiment that has two or more levels of a factor (or condition) with each subject being in only one group and only responding to the dependent measure once. Note this is sometimes referred to as a "one-factor ANOVA". In this model the appropriate partition of the total variance is:

$$\text{Total Variance} = \text{Between-group Variance} + \text{Within-group Variance} \tag{4}$$

Because the treatment is occurring in different groups the between-group variance is the appropriate variance for the numerator in equation 3 and within-group variance is used for the experimental error in the denominator. The F ratio then becomes:

$$F = \frac{\text{between-group variance}}{\text{within-group variance}} \tag{5}$$

A large value of F indicates that the variation between groups is larger than the variation within groups, and that group membership does influence the value of the dependent variable. Details of how these values are calculated are beyond the scope of this work. However, there are several excellent texts that illustrate these calculations (9, 12). Generally, these calculations are done using a computer software package, such as SPSS, SAS, or R (13), and the output is in the form of an ANOVA table.

Before looking at specific studies that have used One-Way ANOVA, a simplified example study will be used to illustrate the important points. In this example the investigators wanted to study the effect of three teaching methods on students' quiz scores. The experiment used three separate classes or groups of students, which is necessary so that the requirement of independence is met. ANOVA techniques are most powerful when the number of persons in each group is the same, but most statistical software can make the necessary corrections for unequal group sizes. The data set for this experiment looks like a spread sheet with each row representing a student and a column designating which teaching method the student experienced and a column with the student's quiz score (14). In this design, the between-groups variance is the variance between the three teaching methods and the within-group variance is the variance of the quiz scores within each treatment group. An ANOVA table for the results of this experiment is shown on Figure 1.

The last three columns contain the key information. The variances necessary to calculate F are found in the column labeled "Mean Squares". The last column contains the p value. In this case since $p < 0.05$ (because the α for significance is 0.05), one (or more) of the group's average score on the dependent variable (quiz score) is statistically different from the others. This result should be reported as follows: "There was a significant effect of teaching method on quiz scores ($F(2,12) = 5.12$, $p < 0.05$, $< = 0.60$)" (15). The effect size (ω) of 0.60 resulting from a separate analysis constitutes a large effect. See Field (2) for further discussion and details on how to calculate the effect size. Note that the ANOVA table does not tell us which group or groups are significantly different from the others. It only tells us that there is at least one difference among the groups. To determine which group is different, *post-hoc* tests must be done. There are a variety of *post-hoc* tests that can be run, but they all are essentially t-tests with some corrections. The variety of tests and how to select which to use are described in detail elsewhere (6). When reporting *post-hoc* results, an effect size should be included. It is often this effect size, and not the overall effect size of the general ANOVA, that is of the most interest.

ANOVA

Score

	Sum of Squares	df	Mean Square	F	Sig.
Between Groups	20.133	2	10.067	5.119	.025
Within Groups	23.600	12	1.967		
Total	43.733	14			

Figure 1. Sample ANOVA Table.

Before looking at an example of a study using the One-Way ANOVA, a point should be made about the relationship between the F ratio and the t-test. The preceding paragraph described the use of *post-hoc* t-tests to identify the specific differences present among groups. We have already demonstrated that the use of the F ratio is necessary to prevent an unplanned escalation of the α value. If the independent variable has only two categories, or possible values, then the ANOVA analysis and the independent t-test will produce the same results with regards to statistical significance. The value of the F ratio in this case is simply the value of the t statistic squared. It is not uncommon in the literature to see ANOVA used in cases where an independent t-test would have sufficed. Do not let this confuse you since in this special case (one independent variable with only two categories), the two tests are essentially identical.

Example of a One-Way ANOVA

The One-Way ANOVA is often used as part of a larger study to establish equivalence of intact groups in quasi-experimental designs. In a quasi-experimental design, it is important to establish that the different groups are similar in ability before intervention. An example of this use of One-Way ANOVA is a study by Doymous (16) that compared the effect of using jigsaw methods with the use of animations on college students' understanding of electrochemistry. The design utilized three intact classes: jigsaw, animations, and control. Prior to the intervention in each class, the students responded to two instruments–one a measure of the students' formal thinking ability (TOSR) and the other a measure of students' knowledge of electrochemistry concepts (PNMET). Two separate One-Way ANOVAs were run, one for each instrument. Students' scores on the instruments were used as the dependent variables in each analysis and the treatment they received was the independent variable. The results showed no significant difference between the TOSR scores of the three groups at the $\alpha = 0.05$ level, ($F_{(2,117)} = 1.786$: $p > 0.05$) (*16*). However, there was a significant difference in the analysis of the PNMET scores, ($F_{(2,113)} = 14.336$: $p < 0.05$). Post-hoc tests indicated that the group of students receiving the animation treatment had significantly higher scores on the PNMET. As a result of this, the statistical analysis after treatment had to take this initial difference into account. This required the use of a multivariate technique with more than one dependent variable, an Analysis of Covariance (ANCOVA). This type of analysis is not described in this chapter. It should be noted that a measure of effect size was not reported in this study for this analysis. While one could have been reported for this ANOVA, the purpose of the ANOVA was to establish equivalency of the groups. Later in the study the authors did report an effect size for the ANCOVA analysis.

Two (or more)-Way ANOVA

If we extend our example of the study above to look at the variance in scores of students who were exposed to three teaching techniques to include an additional independent variable, then we have a Two-Way ANOVA. In our data collection we might have used three different teachers. This would present a problem in our earlier analysis since we could not say whether the difference we observed was due to the teaching method or due to the teacher. It should be pointed out that just adding in another factor should only be done if there is a sound theoretical reason for including it. Eye color could be included as a factor in our example but unless there is a theoretical rationale why the students' eye color would impact their performance, it should not be included. In a Two-Way ANOVA, the total variance in the data is portioned into four parts:

$$\begin{matrix} \text{Total} \\ \text{Variance} \end{matrix} = \begin{matrix} \text{Factor A} \\ \text{Between-group} \\ \text{Variance} \\ \text{(ignoring B)} \end{matrix} + \begin{matrix} \text{Factor B} \\ \text{Between-group} \\ \text{Variance} \\ \text{(ignoring A)} \end{matrix} + \begin{matrix} \text{Interaction} \\ \text{between A \& B} \\ \text{Variance} \end{matrix} + \begin{matrix} \text{Residual} \\ \text{Variance} \end{matrix} \quad (6)$$

The first three terms on the right in this equation represent the variance explained by the experiment, which is sometimes called the model variance. The last term is the residual variance or the variance due to error. By portioning the variance in this manner, we can calculate multiple F ratios, each being used to test for the presence of an effect.

When we test if one of the main factors, A or B, is significant we are looking for a **main effect**. The relevant F ratios are:

$$F_A = \frac{\text{Factor A Beween-group Variance}}{\text{Residual Variance}} \qquad F_B = \frac{\text{Factor B Beween-group Variance}}{\text{Residual Variance}} \qquad (7)$$

When calculating the variance due to factor A, the presence of factor B is ignored, and vice versa. There is also a variance due to the **interaction** of the two factors. Interaction variance measures the differential effect of one factor on the other. In our example of teachers and teaching methods, a significant interaction would suggest that the teacher effect is not the same in all instructional methods. It is important to check for the presence of a significant interaction because the presence of an interaction can complicate the interpretation of any significant main effects. The appropriate F ratio for the interaction, labeled AxB, is:

$$F_{AxB} = \frac{\text{Interaction Between A \& B Variance}}{\text{Residual Variance}} \tag{8}$$

The output table for our sample Two-Way ANOVA is given in Figure 2. Interpretation begins by noting that there is no significant interaction between method and teacher ($F(2,6) = 2.443$, $p > 0.05$). There is a significant main effect for method ($F(2,6) = 6.473$, $p < 0.05$, $\omega = 0.45$).

Tests of Between-Subjects Effects

Dependent Variable: Score

Source	Type III Sum of Squares	df	Mean Square	F	Sig.
Corrected Model	35.067[a]	8	4.383	3.035	.097
Intercept	134.349	1	134.349	93.011	.000
Method	18.698	2	9.349	6.473	.032
Teacher	.984	2	.492	.341	.724
Method * Teacher	14.114	4	3.529	2.443	.157
Error	8.667	6	1.444		
Total	224.000	15			
Corrected Total	43.733	14			

a. R Squared = .802 (Adjusted R Squared = .538)

Figure 2. Two-Way ANOVA Output.

The presence of interactions in multifactor ANOVAs puts a limit on the number of factors that should be included in a design. In our example, there were only two factors, so there was only one interaction. In a Three-Way ANOVA there are three factors (A, B, C) and this results in four possible interactions (A*B, B*C, A*C, A*B*C). It is clear that as the number of factors increases so does the number of possible interactions. The best approach is to only include factors that the underlying theory suggests are important.

The presence of interactions and their interpretations is often aided through a graphical presentation of the group means (interaction plot). Sample interactions plots are shown in Figures 3 and 4. If there is no interaction between the factors, then the group means will form a series of parallel lines. If the lines are not parallel, then an interaction is indicated (Figure 4).

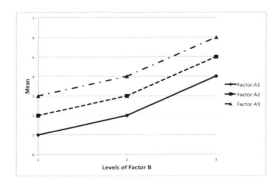

Figure 3. Interaction Plot indicating no interaction.

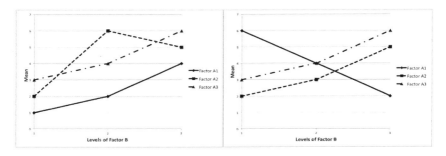

Figure 4. Interaction Plots indicating the presence of an interaction.

Examples of Three-Way ANOVA

A Three-Way ANOVA was used by Lamb (*17*) to investigate the effectiveness of online chemistry modules by high school students. The authors used intact classrooms in an effort to align the research with a normal classroom setting. In the study, a student received either traditional instruction in three areas of chemistry or was instructed with online instructional modules on the same three areas of chemistry. The chemistry knowledge was measured before instruction, during the treatment, and at the end of the experiment. The dependent variable in the ANOVA was the students' gain scores on achievement tests from pre to post. One independent variable was the type of instruction (traditional or online). A second independent variable was teacher experience and the third was student socioeconomic status (SES). The authors found no significant interactions, but the main effect of instruction type was found to be significant (F(1, 350) = 3.94, p = 0.042, Pη^2 = 0.11) (*16, 18*). The socioeconomic status of the student and the teaching experience of the classroom teacher were not found to have a significant impact on the students' gain scores. The authors suggest that the use of the online methodology mitigated the effect of SES and teacher experience on student learning. The lack of a teacher-instructional method interaction indicated that even inexperienced teachers made effective use of the online modules. Similarly, the absence of an interaction between SES and instructional method indicate that all students, regardless of SES status, can benefit from the use of the online modules.

Repeated Measures of a Single Dependent Variable

Repeated-Measures versus Independent-Groups Designs

In many instances in chemistry education research, the study involves participants responding to multiple levels of the treatment variable. This is especially common when there is no control group available. For example, a study could be designed to compare the effectiveness of three feedback techniques on student attitude toward learning chemistry. Instead of having three separate classes, each receiving a specific type of feedback, the experiment could be conducted in one class. In this design, the class would be divided into three groups (A, B, C) and each group would receive a different type of feedback on each test. Since each participant is supplying multiple scores on the dependent variable (attitude), this is a repeated measure design. The use of repeated measures violates an assumption of ANOVA, i.e., independence of measurements. This requires different handling of the data analysis. The partitioning of the variance in a repeated measures design is as follows:

$$\text{Total Variance} = \text{Between-group Variance} + \text{Variance Caused by Treatment} + \text{Residual Variance} \quad (9)$$

If there is more than one independent variable, the variance caused by the treatment term is broken down into a term for between-group variance for each treatment factor as it was in the discussion of the Two-Way ANOVA in Equation 6.

There are two advantages of using repeated measures designs over independent groups designs (*11*). The first is economy. In a repeated measures design the same subjects are used repeatedly. The second advantage is sensitivity. Repeated-measures designs are more sensitive to the treatment effect. This is a result of using the same people where the inherent individual variations, i.e. noise in the data, will be factored out in the analysis. For a One-Way repeated measures ANOVA, the appropriate F ratio is:

$$F = \frac{\text{Variance caused by treatment}}{\text{Residual Variance}} \quad (10)$$

From this expression and the partitioning of variance, there is a gain in sensitivity, because we are able to remove all of the between-group variance from the residual variance.

These advantages come at a price. One issue with a repeated measures design is order effects. Since each participant is receiving each treatment, any differences you see can be due to the intervention or the order in which the interventions were presented. There are ways to account for these effects during the design stage. The Latin Squares design (*11*) is specifically designed to mitigate order

109

effects. The second issue with repeated measures is the necessity for complete data sets. If a participant's score is missing for any items, then it is necessary to remove that person's data completely from the data set. The third, more serious disadvantage of repeated-measures designs is more fundamental to the statistical basis for the ANOVA itself, and that is the requirement of independent samples. This is obviously violated in a repeated measures design, because the scores provided by one person in different treatment conditions are going to be related to each other. To account for this relationship in scores, an additional requirement is placed on the data. For the F-test to remain accurate, the relationships that exist between the scores provided by an individual must be constant across the experimental conditions. This is the **sphericity** assumption and this assumption must be checked before a Repeated Measures ANOVA analysis is performed. When sphericity is violated, corrections must be applied so that a valid F-ratio is produced. Details of these corrections and how to interpret them are beyond the scope of this chapter, but can be found in Field (6).

Examples of One Factor Repeated-Measures ANOVA

A common use of repeated-measures designs in chemistry education is looking at how a measure of interest changes over time. This time period can be the duration of a single class or years. In a study of students' ability to maintain their attention in class, Bunce (19) had students self-report attention lapses, using a clicker, when their attention drifted during class. One of the research questions was to determine the average number of student attention lapses before and after the implementation of a student-centered pedagogy– demonstrations in one course and content clicker questions in another. The dependent variable in this study was the number of student clicker clicks that indicated an attention lapse had occurred. The independent variable was the use of which student-centered pedagogy. The mean number of clicker clicks before and after the student-centered pedagogy was implemented was analyzed with a One-Way Repeated Measures ANOVA. The number of student lapses in attention decreased after the use of a student-centered pedagogy both in the course using clicker questions ($F(1,66) = 8.70$, $p = 0.004$) and the course using demonstrations ($F(1,64) = 4.25$, $p = 0.043$). The implication for teaching is that professors can use student-centered pedagogies as a method for refocusing students' attention during lectures.

In a long term study of the effectiveness of a professional development program for high school chemistry teachers (20), a One-Way Repeated Measures ANOVA was performed. The teachers participated in a 2.5 year program designed to increase their use of inquiry in their classrooms. Classroom observations of the teachers were scored with an instrument, RTOP, to determine the level of inquiry-oriented and standards-based instruction occurring. Observations were done before beginning the program, after key components of the program, and after completion of the program. Since the same instrument (RTOP) was used, the scores represent the repeated-measure of the dependent variable with time as the independent variable. The Repeated Measures ANOVA indicated that the scores on the RTOP were statistically different ($p = 0.001$, $\eta^2 = 0.80$) and follow-up testing indicated that the significant change in the RTOP scores occurred over a two year interval. These results indicated that the change in teaching practices was the result of the accumulation of experiences in the program rather than any one experience.

Mixed Between and Within Designs

The last ANOVA technique that will be discussed is a combination of the previous ones. In this design there is at least one independent (between-groups) factor and at least one repeated measures (within-groups) factor. While there is no limit to the number of between or within factors that can be used, using more than three total factors will result in difficulty interpreting the results.

This type of mixed design, sometimes called a mixed model, is common in psychology and is becoming more common in CER. A point should be made about terminology. The term "mixed models" refers to experimental designs that involve both independent and repeated-measures factors. When the phrase "mixed models" is used to describe an analysis technique, it is referring to a more complex set of procedures than simple ANOVA techniques.

The partitioning of the variance in this design is a combination of the partitioning previously described. The example below shows the partitioning of variance for a One-Way Between and One-Way Within ANOVA. Factor A is the between-group independent variable and Factor B is the repeated measure independent variable.

$$\text{Total Variance} = \text{Between-group Variance} + \text{Within-group Variance} \tag{11}$$

$$\text{Total Variance} = \left\{ \begin{array}{c} \text{Factor A} \\ \text{Between-group} \\ \text{Variance} \\ \text{(ignoring B)} \end{array} + \begin{array}{c} \text{Factor A} \\ \text{Within-group} \\ \text{Variance} \end{array} \right\} + \left\{ \begin{array}{c} \text{Factor B} \\ \text{Between-group} \\ \text{Variance} \\ \text{(ignoring A)} \end{array} + \begin{array}{c} \text{Interaction} \\ \text{Between A \& B} \\ \text{Variance} \end{array} + \begin{array}{c} \text{Residual} \\ \text{Variance} \end{array} \right\} \tag{12}$$

The second term in the first bracket is the error term for the between-group F-ratio and the last term in the second bracket is the error term for the two F-ratios that come from the repeated measures portion. The first term in the second bracket is the main effect of Factor B and the second term in this bracket is the interaction between the two factors, A and B. This partitioning will become even more complicated with the inclusion of a third factor, either an independent or repeated measure factor. The number of interactions and the associated error terms increases dramatically with the number of terms While computer analysis tools can handle this, the interpretation becomes more difficult.

Example of Two-Way Mixed ANOVA

Bunce (21) used a Two-Way Mixed ANOVA design to investigate how the use of either a student response system (SRS) or online quizzes (WebCT) affected students' performance on teacher written exams. The students were classified into three levels of logical thinking ability based on their scores on the Group

Assessment of Logical Thinking (GALT). This classification of logical thinking served as the independent (between-group) factor. The four levels of the within-group factor were: SRS only, WebCT only, both SRS and WebCT, and neither SRS nor WebCT. This was a repeated measures design because there were three exam scores for each student in each within-group level. No interaction between GALT and treatment level was reported. The only significant main effect was for treatment ($F(3.38) = 40.07$, $p = 0.00$, $P\eta^2 = 0.76$). Post-hoc analysis indicated that the students in the WebCT-only group had the highest achievement. Students in the SRS-only group had the lowest achievement with the combined WebCT-SRS and neither WebCT nor SRS groups' achievement in the middle.

Multiple Dependent Variables

All of the techniques considered are applicable to situations when there is only one dependent variable. These ANOVA techniques are part of a class of statistics called **univariate statistics**. Situations with more than one dependent variable fall in to the category of **multivariate statistics**. The technique for this situation is Multiple Analysis of Variance (MANOVA). The focus of this chapter is on the univariate ANOVA since a thorough understanding of univariate is necessary to understand multivariate statistics. Field's text (6) provides a good introduction to MANOVA, while the classic text by Stevens (22) presents MANOVA and other multivariate techniques in rigorous mathematical detail.

Summary

The use of ANOVA in CER is widespread. Unfortunately it is often used without a thorough understanding of the concepts involved. Current statistical software has made an ANOVA analysis relatively easy to perform. Even with easy to use software, the researcher will face a myriad of choices when setting up the analysis. These choices will be easier for the researcher who has a deeper understanding of the concepts and assumptions of the ANOVA technique being used. It has been the goal of this chapter to provide the reader with an introduction to these concepts and assumptions. The interested reader is encouraged to consult the texts included in the references for a more complete understanding of ANOVA. These texts also provide an introduction to more complicated designs, including those that involve a covariate or more than one dependent variable or both. These belong to a class of techniques know as **multivariate** techniques.

In an effort not to overwhelm the reader, some aspects of using ANOVA in CER have not been included in this chapter. One of these is guidance on how to write up the results of the analysis for publication. This is vital to the quality of the work, but often overlooked by both authors and reviewers of manuscripts in CER. One strength of the text by Field (6) is his deliberate efforts to illustrate the proper way of reporting ANOVA results. Another aspect of the reporting process worth mentioning involves checking the assumptions. Too often researchers will report the results of an ANOVA without disclosing whether they checked the validity of the relevant assumptions that must be met for the analysis to be meaningful. It should not be left to the reader to guess that data were normally distributed or if the variances were equal. The tests of the assumptions should be reported in the article.

The last point to make is that CER did not invent ANOVA. These techniques are in wide use in many other areas of study and there are experts in other departments (such as statistics, education, psychology, sociology, and social work) on a college campus available to assist CER researchers when designing their experiments. The earlier you consult an expert about your design and intended analysis, the better.

References

1. Sanger, M. J. Using Inferential Statistics To Answer Quantitative Chemical Education Research Questions. In *Nuts and Bolts of Chemical Education Research*; Bunce, D. M., Cole, R. S., Eds.; ACS Symposium Series 976; American Chemical Society: Washington, DC, 2008; Chapter 8.
2. Howell, D. C. *Statistical Methods for Psychology*; 7th ed.; Cengage Wadsworth: Belmont, CA, 2010.
3. For an interesting description of the history of the t-test and who "student" was, see David Salsburg's book: *The Lady Tasting Tea: How Statistics Revolutionized Science in the Twentieth Century*; Henry Holt and Company, LLC: New York, 2002.
4. For an easy to read and more detailed discussion of the terms introduced here, I recommend Andy Field's book: *Discovering Statistics Using SPSS*; Sage Publications, Inc.: Thousand Oaks, CA, 2005.
5. Cohen, J. *Statistical Power Analysis for the Behavirial Sciences*, 2nd ed.; Academic Press: New York, 1988.
6. Field, A. *Discovering Statistics Using SPSS*, 3rd ed.; Sage Publications, Inc.: Thousand Oaks, CA, 2009.
7. Applebaum, M.; Cooper, H.; Maxwell, S.; Stone, A.; Sher, K. J. *Am. Psychol.* **2008**, *63*, 839–851.
8. Glass, G. V.; Hopkins, K. D. *Statistical Methods in Education and Psychology*, 2 ed.; Allyn and Bacon: Boston, MA, 1984.
9. Roberts, M. J.; Russo, R. *A Student's Guide to Analysis of Variance*; Routledge: London, 1999.
10. Lewis, S. E. An Introduction to Nonparametric Statistics in Chemistry Education Research. In *Tools of Chemistry Education Research*; Bunce, D. M., Cole, R. S., Eds.; ACS Symposium Series 1166; American Chemical Society: Washington, DC, 2014; Chapter 7.
11. Field, A.; Hole, G. *How To Design and Report Experiments*; Sage Publications, Inc.: Thousand Oaks, CA, 2003.
12. Turner, J. R.; Thayer, J. F. *Introduction to Analysis of Variance: Design, Analysis, and Interpretation*; Sage Publications, Inc.: Thousand Oaks, CA, 2001.
13. Tang, H.; Ji, P.; Using the Statistical Progam "R" Instead of SPSS To Analyze Data. In *Tools of Chemistry Education Research*; Bunce, D.

M., Cole, R. S., Eds.; ACS Symposium Series 1166; American Chemical Society: Washington, DC, 2014; Chapter 8.

14. When designing your experiment, it is important to remember that you will most likely have to get your data into a spreadsheet for analysis. If you cannot visualize the spreadsheet or data table that will result from your design, stop and think carefully before going on.

15. When reporting the results of an ANOVA, the value of the F statistic is usually reported with its degrees of freedom numbers in parenthesis, which is related to the number of measurements made for this test. If you are interested, see page 39 of reference 3 for an excellent explanation of the F statistic.

16. Doymus, K.; Karacop, A.; Simsek, U. *Education. Tech. Res. Dev.* **2010**, *58*, 671–691.

17. Lamb, R. L.; Annetta, L. The use of online modules and the effect on student outcomes in a high school chemistry class. *J. Sci. Educ. Technol.* **2012** DOI:10.1007/s10956-012-9417-5.

18. The effect size measure used, Pη-squared, is related to the eta-squared effect size measure described above.

19. Bunce, D. M.; Flens, E. A.; Neiles, K. Y. *J. Chem. Educ.* **2010**, *87*, 1438–1443.

20. Yezierski, E. J.; Herrington, D. G. *CERP* **2011**, *12*, 344–354.

21. Bunce, D. M.; VandenPlas, J. R.; Havanki, K. L. *J. Chem. Educ.* **2006**, *83*, 488–493.

22. Stevens, J. P. *Applied Multivariate Statistics for the Social Sciences*, 2nd ed.; Erlbaum: Hillsdale, NJ, 1996.

Chapter 7

An Introduction to Nonparametric Statistics in Chemistry Education Research

Scott E. Lewis*

University of South Florida, 4202 E. Fowler Ave., CHE205,
Tampa, Florida 33620, United States
*E-mail: slewis@usf.edu

The intent of this chapter is to present an overview of nonparametric statistics, particularly as they are employed in chemistry education research. The nonparametric statistics tests presented are: chi-square, Spearmen's rho, Kendall's tau, logistic regression, the Wilcoxan signed rank test, the Mann-Whitney test and the Kruskal-Wallis test. Each of these is presented with a hypothetical chemistry education research example and followed with a review of how the test has been used in the research literature. This overview is intended for researchers who are performing or considering projects in chemistry education research and are familiar with the general processes of statistical testing and interpretation of results, but are unfamiliar with nonparametric statistical tests.

Introduction

Chemistry education research (CER) is at the intersection of multiple established disciplines and as a result, faculty who undertake research projects in this field have a wide variety of training. It is unlikely a researcher is proficient in the wide range of techniques present in the field. This makes the efforts to share the tools of research in chemistry education research, through both this book and the *Nuts and Bolts of Chemical Education Research* (*1*), uniquely important. The intent of this chapter is to present an overview of nonparametric statistics, particularly as they are employed in chemistry education research. This overview is intended for researchers who are performing or considering projects in chemistry education research and are familiar with the general processes of statistical testing and interpretation of the results, but are unfamiliar with nonparametric statistical tests.

Data Scales

Prior to introducing the statistical tests, an overview of the different scales of data is needed to ground the discussion of the appropriate uses for each test. Stevens (2) proposed a hierarchy of data scales to describe the type of information that could be conveyed with each scale, a classification which remains in widespread use in statistics texts. The first data scale is termed *nominal*, and describes data which represents categories and offers no potential for ranking. In CER, examples of nominal data can include student demographics such as sex or race, or categorizing student qualities or experiences, such as whether students have taken part in a teaching reform.

The next data scale is *ordinal* which also represent categories but with a potential for ranking. One of the important features of ordinal data is that there is no assumption made regarding the distance between the rankings. A common CER example of ordinal data is a set of responses to a Likert-style survey, for example, when respondents are asked to rate an item on the frequency of occurrence using the scale of Never, Occasionally, Sometimes, Usually, or Always. There is a clear ranking of the responses but no assumption can be made regarding the difference between the rankings. For example, the difference between Never and Occasionally is not necessarily the same as the difference between Occasionally and Sometimes. Another example of ordinal data is the ubiquitous letter grade scale used to rate student performance. Depending on how the grades are assigned, it is possible that the difference between a grade of A and B, may not be the same as the difference between a grade of B and C. Instead, classifying this data as ordinal conveys only that students receiving an A were rated higher than those receiving a B.

The *interval* data scale has both a ranking and an assumption that the difference between the rankings is consistent across the scale. Another characteristic of interval data is that the value for zero is arbitrarily defined. A CER example of interval data could be SAT scores on each subject test, which have a range of 200 to 800. The scale could have just as easily have been defined as 0 to 600. As a result of the arbitrary zero, ratio relationships are not consistent among interval data. That is, an SAT subject score of 400 has not demonstrated twice that of a subject score of 200 on any associated metric. The difference between the rankings remains consistent, as the 100 point difference between a 400 and a 300 subject score shares similarity with the difference between a 500 and a 400 subject score; each difference represents a distance of approximately one standard deviation.

The last data scale is *ratio*, which builds upon interval but the zero point has a meaningful definition. A common CER example of ratio data include most test scores, where a zero indicates answering no questions correctly. Ratio level data allows for comparisons of both differences and ratios. In classic test theory, a test score of 60 represents a student who scored twice as many correct as a test score of 30. Differences also remain consistent, as the number of questions needed to go from a score of 30 to a score of 45 is the same as the number of questions needed to go from 60 to 75. Ratio data completes the hierarchy of data scales that progresses from nominal, to ordinal, then interval and finally ratio. The hierarchy is arranged in terms of the amount of information that is conveyed, where interval data convey more information than ordinal data, for example. A statistical test proposed for one data scale, say ordinal, can be employed for any higher level of data scale, in this case interval or ratio data. However, typically, when a statistical test is described for a data scale, it should not be used with any lower-level data. That is, a test prescribed for ordinal data should not be used with nominal level data.

The data scales presented are not without controversy. Velleman and Wilkinson (3) provide a thorough introduction to these concerns. Among some of the critiques, there is concern that some data are not well described by the available scales. Percentages, for example, follow ratio level data but also feature additional information. Also, there is the potential that determining the appropriate data scales depends on the context of the data, while the general description of the data scales provides the impression that data can be assigned a scale independent of context. For example, consider binary data that classifies students as passing or failing a class. One context for this data is to determine how many students may repeat a class and the data can be treated as nominal. Alternatively, in a study investigating student success, a researcher may treat pass as a higher ranked outcome thus making the data ordinal. Finally, there is a tendency to prescribe each statistics tests to a particular data scale, which may prevent researchers from using other statistical tests which could provide evidence of meaningful relationships. The intent of this chapter, then, as an introduction, is to indicate the data scales which are most commonly associated with each statistical test, with the caveat that employing other tests should also be considered. To assist in determining the appropriateness of a test, an effort is made to present the underlying math to each statistical test presented in this chapter.

Nonparametric versus Parametric

Statistical tests are classified based on their reliance on the normality assumption; parametric statistics assume normality and nonparametric do not. The normality assumption is that the data used in the test will follow a normal distribution, which has the appearance of a bell curve. Nominal data cannot follow a normal distribution, as the data does not indicate a ranking. Ordinal data also does not follow a normal distribution, as there is no indication about the consistency of differences among the rankings. A histogram of ordinal data may at first appear normal, but since no assumption on the distance between ordinal data can be made, the distance between categories cannot be assumed to follow any particular pattern. As a result, the normality assumption on nominal or ordinal level data would not be satisfied and nonparametric statistics are recommended.

Interval and ratio data scales can follow a normal distribution, but the normality distribution should be examined. Initial tests involve examining the skewness and kurtosis values of the distribution in the context of the standard error of skewness and kurtosis. Follow-on tests can include a visual inspection of a frequency plot, particularly to ensure the data do not have multiple modes. While many parametric tests are robust to violations of normality (4), not all tests are robust; and in these cases nonparametric statistics can offer a suitable alternative. It is worth noting, though, that many parametric tests such as independent sample t-tests, ANOVA, and multiple regression, rely on a normality assumption only on the dependent variable. In an independent sample t-test, for example, which compares two groups, the independent variable group identification is nominal and therefore cannot follow a normal distribution. Table I indicates the relationship between common parametric tests and their nonparametric alternatives.

Table I. Mapping Nonparametric Tests to Parametric Counterparts

Type of Relationship	Parametric Test	Counterpart Nonparametric Test
Measures of Association between Variables	Pearson Correlation	Chi-square (χ^2) test
		Spearman's rho (ρ)
		Kendall's tau (τ)
	Multiple Regression	Logistic regression
Pre/Post Comparison of a Repeated Measure	Paired t-test	Wilcoxon Signed Rank test
Comparison of Two Independent Groups	Independent Samples t-test	Mann-Whitney test
Comparison of More than Two Independent Groups	Analysis of Variance (ANOVA)	Kruskal-Wallis test

Nonparametric tests can result in determinations of statistical significance in much the same way parametric tests do. The determination of statistical significance involves establishing a limit for Type I error rate termed the α-value prior to analysis. The analysis results in an observed p-value that is a measure of the probability that the data would arise if the null hypothesis is assumed to be true. The p-value is compared to the α-value, and if it is below the α-value, the null hypothesis is rejected. In many nonparametric tests, the value for p can be calculated directly by hand or by using statistical software. The recommendation is to calculate statistical significance by hand for small sample sizes, as many software programs make automatic corrections for continuity or in the case of ties (5). When performing calculations by hand the statistical significance can be determined using tables of p-values. All tests presented in this chapter have an applicable table presented in Leach (6), with logistic regression being the only exception. For larger samples, where the corrections employed are more applicable, the use of statistical software is recommended and possibly necessary.

Some statistics software packages do not include all the tests here, and the researcher is recommended to check the availability of the desired tests prior to making a decision on purchasing.

Nonparametric Statistical Tests

Measures of Association

Correlations are one of the most useful techniques in education research as they allow a quick examination of association, or absences thereof, between two variables. The term correlation often refers to the Pearson Product Moment Correlation, which is a parametric statistic, relying on a normality assumption for each of the variables examined. Multiple nonparametric measures of association exist, depending on the data scales used, and are reviewed below.

Chi-Square (χ^2) Test

The χ^2 test examines associations between data that are at the nominal data scale. Nominal is the lowest data scale, and thus this test can be employed in any of the other data scales, though interpretation can become problematic when the number of data points possible becomes larger (e.g., a nominal scale may have three categories, but a ratio scale may have hundreds of possible data values). To demonstrate the chi-square test, a contingency table must be created. In a contingency table, variables are listed as the heading for either the columns or rows and the frequency of each cross-tabulation among the variables is reported. Table II uses fictional data to demonstrate a contingency table between gender and passing a class.

Table II. Example Contingency Table

	Fail	Pass	Total
Male	24	16	40
Female	20	40	60
Total	44	56	100

In this example, a χ^2 test can determine if there is a relationship between the gender of the student and the chance for passing a class. To do this, the test determines an expected value for each cell, which assumes that there is no relationship between the variables. The expected value for male students who fail would be: the percentage of students who are male (0.40) multiplied by the percent of students who failed (0.44) multiplied by the total number of students (100). This calculation provides a value of 17.6. Table III includes the data from Table II, coupled with the expected value for each cell.

Table III. Example Contingency Table with Expected Values

		Fail	*Pass*	*Total*
Male	Observed	24	16	*40*
	Expected	17.6	22.4	
Female	Observed	20	40	*60*
	Expected	26.4	33.6	
Total		*44*	*56*	*100*

The χ^2 test then examines the difference between the observed value and the expected value for each cell to determine if a relationship is present. Because the expected value assumes no relationship between the variables, any differences between observed and expected are evidence of a relationship. χ^2 is calculated using the formula (6):

$$\chi^2 = \sum \frac{(E-O)^2}{E}$$

In this example, the χ^2 value equals 6.926. The degrees of freedom (df) for a contingency table, not counting the totals, is equal to:

$$df = (rows - 1) * (columns - 1)$$

In this example, df = 1. Using both the χ^2 value and df value can indicate a p-value and ultimately aid in the decision regarding the null hypothesis. The p-value for this example was found to be 0.008, which may lead to rejection of the null hypothesis based on the threshold of Type I error (the α value) the researcher decides is appropriate. The incorporation of expected values in Table III offers insight into the relationship that is observed, which the χ^2 value alone cannot provide. In Table III, we see that female students were more likely to pass the course than expected, and male students were more likely to fail the course.

The χ^2 test is common in CER projects as illustrated by three recent examples in the research literature. Gron et al. (7) converted the ratio data of students' recorded percent error into ordinal data by categorizing results as <1%, 1 – 2 % and so on. Data were collected over a three year period, during which a new assessment technique was implemented. The placement of precision categories was analyzed against the year the data was collected using a χ^2 test to determine if students' precision differed across the years. Two separate χ^2 tests were conducted, comparing the third year to the first year and comparing the third year to the second year, providing evidence that students' reported experimental precision improved in the later years. In this example, the χ^2 test was used to determine the association of an independent variable, year, on an outcome variable, reported precision.

Chase et al. (8) administered two attitude-based surveys as part of their evaluation of Process Oriented Guided Inquiry Learning. The surveys were administered voluntarily. A χ^2 test was performed between survey participation and proficiency, with the intent to determine if students who took the survey differed from those who did not in terms of their course performance. The

χ^2 test revealed no statistically significant relationship and was thus used to support the ruling out of a potential spurious explanation. Similarly, Bergin et al. (9) examined the relationship between demographic variables and students' self-report of whether they leave comments on an online instructor evaluation site. A separate χ^2 test was run comparing each demographic variable against students' status as a contributor. The results also showed no statistically significant relationship; there was no evidence of an association between each demographic variable and whether or not a student contributed to the online evaluation site.

Using the lack of statistical significance to support a conclusion of no relationship between variables can be problematic, as the error level would be represented by the power of the statistical test, not the α value cut-off that was used with the p-value. Reporting the effect size of the relationship in addition to the results of the χ^2 test can strengthen the argument that no relationship exists. One measure for effect size for a χ^2 test is Cohen's w (10) where w is defined as:

$$w = \sqrt{\sum \frac{(P_0 - P_E)^2}{P_E}}$$

and P_0 is the proportion of observed in each cell (number of observed divided by total observed) and P_E is the proportion of expected in each cell. This formula can be simplified to:

$$w = \sqrt{\frac{1}{N} * \chi^2}$$

where N is the total sample size. The effect size can be described qualitatively as small when **w** is approximately 0.1, medium at 0.3 and large at 0.5. By reporting the effect size along with χ^2 the strength of the relationship can be described as well as the statistical significance. In making a case that two variables are not related, finding a small effect size would be supportive. On the other hand, consider if the χ^2 test resulted in non significance, but a large effect size was observed, which is possible, particularly with a small sample size. Rather than concluding no relationship exists (which contradicts the observed effect size) a researcher can conclude that the sample size was insufficient to find statistical significance.

Spearman's Rho (ρ)

The Spearman's ρ test is designed to examine the relationship between variables within a sample that are at the ordinal or higher data scale. Spearman's ρ is conceptually closer to the conventional Pearson's Correlation Coefficient than χ^2 in that the value for Spearman's ρ indicates the extent of the relationship with a range of -1.0 to 1.0, where a negative sign indicates that large values on one variable correspond to the smaller values on the other variable.

As is common in tests designed for ordinal data there is a reliance on ranking data. As an example, consider the hypothetical determinination of a correlation between student grades (ordinal data on an A through F scale) in Chemistry versus

Math SAT scores (interval data). Fictional data is presented in Table IV. Each student is ranked on each variable relative to their peers. By convention, the lowest score is assigned the rank of 1. Once the rankings are made, the difference in rankings is calculated for each student.

Table IV. Spearman's ρ Example Data

Student	1	2	3	4	5
Chemistry Grade	B	A	C	D	F
Math SAT Score	560	720	610	490	580
Chemistry Ranking	4	5	3	2	1
Math SAT Ranking	2	5	4	1	3
Chem Rank – Math Rank	2	0	-1	1	-2
Difference Squared	4	0	1	1	4

Spearman's ρ is calculated using the formula (6)

$$\rho = 1 - \frac{6D}{n(n^2 - 1)}$$

Where D is the sum of the differences squared, in this example D = 10, and n is the number of observations in the sample, in this example 5. This leads to a ρ value of 0.5 observed for this sample, which can be interpreted similarly to a conventional correlation value. This correlation is below the critical value found in the relevant table (11), indicating insufficient evidence to reject the null hypothesis, which states there is no relation between the variables.

One consideration for calculating Spearman's ρ is the treatment of ties in the ranking. In the event of a tie, the average ranking would be assigned. For example, if two cases tied with the lowest score, they would each be assigned a ranking of 1.5 to average the rankings of 1 and 2 that the two lowest cases would hold. If the number of ties becomes large, particularly if the sample size is large and the number of possible data values are small (e.g., 50 students ranked by their performance on a scale of five letter grades), the distribution will need to be adjusted (6). One possible alternative when ties are common is to employ tests that rely on a contingency table, such as χ^2.

In CER projects, Spearman's ρ can be used in place of the conventional correlation when the data is ordinal or when the normality assumption is not met. Christain and Yeziersky (12), in the development of the chemical and physical change assessment, used Spearman's ρ to show that true-false items on the assessment were correlated with other items. Along with other considerations, the results of the Spearman's ρ test supported the justification for the removal of the true-false items. Hilbing and Barke (13) examined the relationship between attitude toward chemistry and attitude toward chemistry education, with each variable having the possible data values of negative, indifferent or positive. The results from a Spearman's ρ test was used to support the claim that a positive chemistry education experience can lead to a positive attitude toward chemistry. Obenland et al. (14) used a Spearman's ρ to determine the relationship between performance on the Chemistry Concept Reasoning Test and student responses on two Likert-style survey questions regarding perceptions of the usefulness of Socratic dialogue. The observed ρ values led the authors to conclude that a weak correlation existed between the perceived usefulness of the Socratic dialogue and performance on the test.

Kendall's Tau (τ)

Another nonparametric test designed to measure correlations among ordinal data is the Kendall's τ. Like Spearman's ρ, Kendall's τ relies on the ranking of data and features a range of -1.0 to 1.0. Unlike Spearmen's ρ, Kendall's τ does not rely on the difference in rankings. To present Kendall's τ, the data from Table IV can be sorted based on one of the variables. For this example, they are sorted by chemistry grade and are presented in Table V. Then, for each observation, a determination of the number of concordant and discordant pairs are made. For student 2, who earned a Chemistry grade of an A, one would examine each subsequent student. For each subsequent student, the Chemistry ranking is considered below or above that of student 2; similarly the Math SAT ranking is considered below or above that of student 2. If the relative placement for Chemistry (above or below the ranking) and Math SAT (above or below the ranking) is in agreement, it is considered a concordant pair; disagreement would be a discordant pair. Returning to student 2, evaluating student 1, the chemistry ranking for student 1 is below student 2 and the Math SAT ranking for student 1 is below student 2. This would be one concordant pair. Evaluating the remaining subsequent students it is found that, for student 2, all four subsequent students represent a concordant pair in this example and there are no discordant pairs.

Table V. Kendall's τ Example Data

Student	2	1	3	4	5
Chemistry Grade	A	B	C	D	F
Math SAT Score	720	560	610	490	580
Chemistry Ranking	5	4	3	2	1
Math SAT Ranking	5	2	4	1	3
Concordant pairs	4	1	2	0	
Discordant pairs	0	2	0	1	

Student 1 would be evaluated in the same way, focusing only on the subsequent students 3, 4 and 5 (the comparison of student 1 and student 2 has already been completed). In terms of student 1, evaluating student 3, the chemistry ranking is below, but the math ranking is above, which represents a discordant pair. Student 4 is a concordant pair with student 1 as student 4 is below on both metrics. Student 5 is a discordant pair with student 1 as this student is below on Chemistry and above on Math SAT. The process is continued with each student compared to the subsequent students. The last student, Student 5, has no subsequent students to compare with so no pairs can be computed for this student. Kendall's τ can then be calculated using the formula (11):

$$\tau = \frac{Sum\ of\ Concordant\ Pairs - Sum\ of\ Discordant\ Pairs}{n(n-1)/2}$$

For this example, $\tau = 0.4$. As with any test relying on ranks, a tie in rankings can become problematic. Midrank averages can be assigned. Problems, however, arise in the significance testing and the ability to model the distribution. A modified version of Kendall's τ, known as Kendall's τ-b, has been developed for instances where ties are common .

In selecting between Kendall's τ and Spearman's ρ, it has been pointed out that Kendall's τ approximates a normal distribution better, which would make for more reliable statistical significance testing. Kendall's τ also features the modification designed for working with extensive ties where Spearman's ρ does not. The advantage of Spearman's ρ lies in the methodological emphasis on the extent of disagreement with large disagreements in ranking receiving a stronger weight. The Kendall's τ methodology of concordant and discordant pairs does not provide a measure of disagreement strength. If it is important to emphasize large disagreements between the two variables, Spearman's ρ may be preferred (6).

Logistic Regression

Multiple regression is among the most useful statistical tools in CER. It allows for a determination of the role of multiple independent variables in impacting a dependent variable and for comparing the relative impact of each independent variable. Multiple regression typically fits data into the form

$$y = b_0 + b_1 x_1 + b_2 x_2 \ldots$$

where y is the dependent variable, each x represents a separate independent variable, b_0 represents the expected y value when all of the values for x equal zero, and the subsequent b values represent the impact of each respective independent variable. Logistic regression is an adaptation of multiple regression meant to predict a dichotomous dependent variable, which has only two possible values (e.g., pass or fail). Using multiple regression in such a scenario would violate the assumptions of multiple regression and underestimate the impact of the independent variables (15). Other types of logistic regression, such as multinomial logistic regression and ordered logistic regression, are available to model more complex outcome variables but are beyond the scope of this chapter.

In logistic regression the overall model is adapted toward making a prediction of probability P using the equation:

$$\ln\left(\frac{P}{1-P}\right) = b_0 + b_1 x_1 + b_2 x_2 \dots$$

The expression P/(1-P) is known as the odds ratio, and the expression ln(P/(1-P)) is the logit. This equation can be rearranged to:

$$P = \frac{1}{1+e^{-b_0-b_1 x_1 - b_2 x_2 \dots}}$$

if the researcher is interested in estimating the probabilities given a combination of values for the independent variables.

As an example, hypothetical data were collected on General Chemistry students with Math SAT scores and whether reform teaching was present (reform = 1 when present and 0 when not). The dependent variable was whether they completed the course (1 for yes, 0 for no). The output of a logistic regression on the data is presented in Table VI below.

Table VI. Output from Logistic Regression

	B	S.E.	Wald	df	Sig.	Exp(B)
Reform	1.011	0.240	17.803	1	<0.001	2.750
SAT Math	0.004734	0.001699	7.759	1	0.005	1.005
Constant	−1.741	0.927	3.523	1	0.060	0.175

Researchers can rely on the statistically significant results to evaluate whether the relationship between each independent variable and the dependent variable can be attributed to chance. The coefficients, under column "B", can be used in the equation above to estimate probabilities. The S.E. column, which indicates the standard error of each coefficient, is calculated based on the differences between the observed and the predicted values on the outcome variable and can provide context to the coefficients. The significance testing on the coefficient is based on the coefficient divided by the standard error and placed on the normal distribution. The Wald statistic is calculated as the coefficient divided by the standard error with the resulting quotient squared and is meant to follow the chi-square distribution. The Wald statistic is meant to evaluate the coefficient but the statistical power of this measure has been problematic (15, 16).

The values given in the "Exp(B)" column provide an indication of the impact of each independent variable, by describing how a change of one unit of a particular independent variable affects the odds ratio. For example, the Exp(B) for reform indicates that students with the reform, while controlling for SAT Math, have an odds ratio 2.750 times greater than the odds ratio for students without the reform. A common error is to report that students with the reform were 2.750 times more likely to complete the course. This statement is incorrect as the phrase "more likely" refers to relationships in probability (15) and changes in probability are not linear throughout a logistic regression. A chart of the predicted probabilities versus Math SAT is shown in Figure 1 and this visual presentation shows it is not possible to provide a single value to relate the probabilities of the two groups across the range of Math SAT scores.

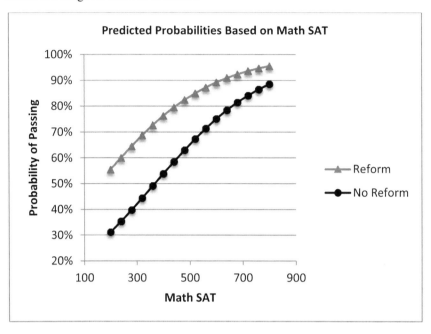

Figure 1. Predicted probabilities based on logistic regression equation.

In addition to the output in Table VI, it is recommended to include information evaluating the overall model (*17*). One technique is to assess the accuracy of the model's predictions. For this, a researcher would need to specify a cut-off value to turn the model's probabilities into predictions. A cut-off point for this example may be 0.5, so that cases where the probability is greater than 0.5 can be predicted to complete the course and those below are predicted to not complete it. The predictions can then be placed against the actual outcome in a classification table.

The Hosmer-Lemeshow (H-L) test also offers an indication of goodness of fit of the model, which tests the null hypothesis that the model is a good fit. If the results of the test have a p-value greater than the accepted error rate, typically 0.05, it is an indication that the model is tenable (*17*). As failure to reject the null hypothesis is the goal of the test, sufficient statistical power is needed for the results to be valid. Sample sizes of at least 400 are recommended for this test to have sufficient statistical power (*16*). Other measures are developed to approximate the common R^2 metric from multiple regression. The Cox and Snell R^2 is an example, but is problematic because it does not have a maximum of one. The Nagelkerke R^2 adjusts the Cox and Snell R^2 by dividing it by the maximum possible, thereby making for a maximum of one (*16*). Neither value is a measure of proportion of variance as the conventional R^2 is or corresponds to predictive efficiency and thus should only be presented as a supplemental to the previous measures (*16, 17*).

Emenike et al. (*18*) used logisitic regression in the analysis of a chemistry faculty survey on assessment practices. First, it was used to evaluate the association of demographic factors with the year surveys were administered prior to combining two cohorts. Second, logistic regression was used to investigate the impact of demographic variables on, separately, awareness of departmental assessment efforts and self-ratings of contributions to departmental assessment efforts. Mills and Sweeney (*19*) used logistic regression to relate performance on Exam 1 in General Chemistry to successful completion of the course. The relationship was then used as evidence that Exam 1 can serve as a placement tool and to highlight the importance of the material in Exam 1 for students' successful completion of the course.

Repeated Measures

Wilcoxon Signed Rank Test

Repeated measures analysis, or a pre/post design are common in education research to demonstrate changes over time. In parametric statistics a paired t-test can be used to determine the statistical significance of pre/post changes. The nonparametric equivalent is the Wilcoxon Signed Rank test, also referred to as the Wilcoxon Matched Pairs test. The Wilcoxon Signed Rank test can be used on data that is ordinal scale or higher.

As a hypothetical example, consider an investigation into students' perceptions of groupwork over time. Researchers administer a Likert-style survey where one item pertaining to group work is rated by students on a seven point scale ranging from Strongly Disagree to Strongly Agree. The responses are coded 1 through 7, with 1 indicating Strongly Disagree. The instrument is administered twice over the course of the study and the fictional data is presented in Table VII.

Table VII. Wilcoxon Signed Rank Test

Student	Pre-score	Post-score	Post - Pre	|Post - Pre|	Sign of difference
1	2	6	4	4	+
2	4	5	1	1	+
3	3	1	-2	2	−
4	1	4	3	3	+
5	6	7	1	1	+
6	5	5	0	0	
7	4	5	1	1	+
8	3	2	−1	1	−

The post-score is subtracted from the pre-score and the information is then split into the absolute value of the difference and the sign of the difference, as shown in Table VII. The information is then re-sorted based on the absolute value of the difference, as shown in Table VIII, and the differences are ranked.

Table VIII. Sorted Data from the Wilcoxon Signed Rank Test

Student	|Post - Pre|	Sign of difference	Rank of |Post - Pre|	Sign * Rank
1	4	+	7	7
4	3	+	6	6
3	2	−	5	−5
2	1	+	2.5	2.5
5	1	+	2.5	2.5
7	1	+	2.5	2.5
8	1	−	2.5	−2.5
6	0			

In Table VIII, the absolute value of post - pre was a tie for students 2, 5, 7 and 8, who would have covered the ranks of 1, 2, 3 and 4. The assigned rank is then the average of the four possible ranks which is 2.5. Finally, the sign and the rank are reunited in the Sign * Rank column. The test statistic, W, is found by summing the Sign * Rank column in Table VIII and then taking the absolute value of the result. In this example, W = 13. Statistical significance is determined based on the value for W and the sample size. In this example, with 8 students, the sample size would be reported as 7 as the sample size does not count cases where the post - pre score equals 0 (6). The p-value for this example is 0.27 indicating there is insufficient evidence that the students score the group work differently between the two administrations.

Jennings et al. (20) administered a twelve item open-ended test related to Physical Chemistry topics. The test was administered in a pre / post fashion assessing students' knowledge before and after they viewed a DVD. Each item response was coded on a five point scale ranging from "No Understanding" to "Sound Understanding". The pre-scores for each of the first ten items, as well as the average of the ten items, were then compared to the post-scores using a Wilcoxon Signed Rank test. The results were statistically significant, at $p < 0.05$ for every item tested, indicative of learning gains. Using 0.05 as a cut-off for Type I error has been the convention. As this test was conducted eleven times within the same context, this leads to a 0.43 chance $(1 - 0.95^{11})$ that at least one Type I error was made within the set of tests. Lowering the Type I error threshold, to 0.01 for example, would limit the overall error probability, in this case to 0.10. Depending on the p-values observed, the change to a 0.01 threshold might have affected the interpretation of the results. The follow-up qualitative discussion of the shifts in the item responses is commendable and offers insight into the changes in student knowledge that the statistical test alone could not provide.

Kerby et al. (21) conducted an outreach event for children and administered two conceptual questions both before and after the event. The responses to the questions were coded as correct or incorrect, and the Wilcoxon Signed-Rank test was used to compare the pre-event responses to those post-event. The findings showed that the difference was statistically significant and therefore was not likely attributable to chance.

Comparison of Groups

Mann-Whitney Test

The Mann-Whitney test is designed for comparing two samples on ordinal data and has the independent samples t-test as the parametric counterpart. The Mann-Whitney test is differentiated from the Wilcoxon Signed Rank test as the former is designed for independent samples (e.g., separate groups of students), while the latter is designed for repeated measures within the same sample. The Mann-Whitney test is also known as the Wicoxon Rank Sum test.

An example of the Mann-Whitney test using educational data can be the hypothetical comparison of two pedagogies, one termed "Active learning" and the other "Cooperative learning". Class averages on a common exam were collected from each class and were not found to follow a normal distribution, suggesting the need for the nonparametric test. The fictional data are presented in Table IX, along with the overall rank of each class.

Table IX. Mann-Whitney Example

Class Number	Pedagogy	Class Average	Overall Rank
1	Active	81.0	9
2	Active	78.5	6
3	Active	68.9	3
4	Active	79.0	7
5	Active	72.0	5
6	Cooperative	65.2	1
7	Cooperative	71.1	4
8	Cooperative	90.3	10
9	Cooperative	68.3	2
10	Cooperative	80.3	8

The measure in the Mann-Whitney test is denoted by the letter S and can be solved using the following formula, where the sub-script 1 can denote either group and n is the total sample size (6).

$$S = 2 \sum Rank_1 - n_1(n + 1)$$

In this example, using the "Active" pedagogy group, the ranks summed to 30 and n_{active} = 5, making S equal to 5. The S score indicates the measure of the difference between two groups. The S score found can approximate a normal distribution to determine statistical significance with a mean of 0 and a variance of $n_1n_2(n+1)/3$. In this case, an observed S score of 5, and a variance of 91.67, leads to a p value of 0.60; consequently, there is insufficient evidence to reject the null hypothesis.

In CER projects, Cooper et al. (22) implemented the *Chemistry, Life, the Universe and Everything* curriculum with a General Chemistry class and a control group was created from a stratified sample of General Chemistry students to match the treatment group in terms of demographics and measures of college readiness and logical thinking. Students in both groups completed a set of common assessments on Lewis structures and the Mann-Whitney test was used to compare the performance of the two groups. The results showed that the treatment group out-performed the control group on the assessment, offering evidence to the effectiveness of the curriculum.

Hein (23), implemented Process-Oriented Guided Inquiry Learning (POGIL) in an organic chemistry course and compared student performance against an antecedent student cohort. The outcome measure, percentiled rank on an ACS exam, was measured with both cohorts. The comparison was conducted using the Mann-Whitney test and a significant difference indicated the POGIL students outperformed the antecedent cohort. Similarily, Mahalingam et al. (24) implemented group problem solving and performed a comparison with historical student performance using the Mann-Whitney test. The statistical test was used twice, first to compare two years that were both prior to implementation and found no statistical significance. Second, to compare the year with implementation to the prior year without the implementation and found the difference to be statistically significant. As an alternative, the Kruskal-Wallis test, discussed next, would allow for the comparison of the three years simultaneously.

Kruskal-Wallis Test

The Kruskal-Wallis test is designed to compare data from more than two independent samples. In this manner, the parametric equivalent for this test is the one-way analysis of variance (ANOVA) test. The Kruskal-Wallis test is designed to work for ordinal or higher data. To introduce the test, the hypothetical example from the Mann-Whitney test can be expanded to incorporate a third group. This data is presented in Table X with the third group having a "Traditional" (or lecture) pedagogy.

Table X. Kruskal-Wallis Example

Class Number	Pedagogy	Class Average	Overall Rank
1	Active	81.0	13
2	Active	78.5	9
3	Active	68.9	5
4	Active	79.0	11
5	Active	72.0	8
6	Cooperative	65.2	2
7	Cooperative	71.1	7
8	Cooperative	90.3	15
9	Cooperative	68.3	4
10	Cooperative	80.3	12
11	Traditional	69.1	6
12	Traditional	62.0	1
13	Traditional	83.3	14
14	Traditional	67.5	3
15	Traditional	78.9	10

Like many statistical tests, the Kruskal-Wallis test can be thought of as a ratio of signal to noise. First, determine the average rank using the formula $(N+1)/2$, where N is the overall sample size, in this case 15, making an overall average rank of 8. Next, beginning with the "Active" group, the sum of ranks is 46. If the ranks were evenly distributed, then the five members of the active group should have a hypothetical sum of 40 (given the overall average rank of 8). The signal from the group is determined by squaring the difference between the sum of rank and the hypothetical sum of rank, and dividing it by the sample size of the group. In this case, the "Active" learning group has a signal of: $(46 - 40)^2 / 5 = 7.2$. Using the same procedure, the "Cooperative" group has a sum of ranks of 40, and thus a signal of 0. The "Traditional" or lecture group has a sum of ranks of 34 and thus a signal of $(34 - 40)^2 / 5 = 7.2$. The total signal is the sum of each group, which in this case is 14.4. The noise provides a description of what would occur randomly, and can be found by taking the sum of squares of the difference between each rank and the average overall rank: $noise = \sum(rank - average\ rank)^2$. With this data the noise is found to be 280. The test statistic K equals the signal times $(N - 1)$ divided by the noise, and with this data

$$K = \frac{signal * (N - 1)}{noise} = \frac{14.4 * (15 - 1)}{280} = 0.72$$

The K value follows a χ^2 distribution in determining significance with degrees of freedom equal to the number of groups minus one.

In CER projects, Bell and Volckmann (25) used the Kruskal-Wallis test to show that students' prior chemistry experience, as determined by self-report, influenced the students' confidence ratings of successfully solving chemistry questions, both at the beginning and end of the course. Jeon et al. (26) compared three instructional strategies: a control group, modeling a problem solving strategy for use individually, and modeling the strategy with students working in heterogenous pairs. The Kruskal-Wallis test was used to compare the three groups on their performance on different sections of a problem-solving test. The Kruskal-Wallis test was chosen in this case because of both the small sample size and the scoring rubric on the problem-solving test employing an ordinal scale. The results showed differences between groups on three outcomes: conceptual knowledge, recalling the related law and mathematical execution.

Conclusions and Further Readings

There are nonparametric tests for many of the common analyses performed in CER, and it is hoped that the hypothetical examples presented in this chapter illustrate the utility for nonparametric statistics in the field. Nonparametric tests serve a unique purpose in analyzing nominal and ordinal level data, which are commonplace in CER measures. Moreover, they have utility in interval and ratio data when the normality assumption is not tenable.

It is hoped that this introduction provides a suitable background to the rationale, methodology, and interpretation of each of the statistical techniques presented and ultimately serves to encourage readers to incorporate these techniques within their own research. For readers who are interested in more information on the underlying mathematics and exceptions, Leach (6) and Gibbons (5) each wrote useful and accessible texts on the majority of tests presented here. For logistic regression, Pedhauzer (15) and Cohen et al. (16) each have written an authoritative text on regression that include a chapter on logistic regression. It is also worth noting that each branch of statistics represents a dynamic field and readers are encouraged to continually examine new journals and textbooks in the field and consider collaborations with statisticians.

References

1. Bunce, D. M., Cole, R. S., Eds.; *Nuts and Bolts of Chemical Education Research*; ACS Symposium Series 976; American Chemical Society: Washington, DC, 2008.
2. Stevens, S. S. *Science* **1946**, *103*, 677–680.
3. Velleman, P. F.; Wilkinson, L. *Am. Stat.* **1993**, *47*, 65–72.
4. Stevens, J. *Intermediate Statistics: A Modern Approach*; Lawrence Erlbaum Associates, Inc.: Mahwah, NJ, 1999.
5. Gibbons, J. D. *Nonparametric Statistics: An Introduction*; Sage Publications, Inc.: Newbury Park, CA, 1992.
6. Leach, C. *Introduction to Statistics: A Nonparametric Approach for the Social Sciences*; John Wiley & Sons, Ltd.: Chichester, U.K., 1979.

7. Gron, L. U.; Bradley, S. B.; McKenzie, J. R.; Shinn, S. E.; Teague, M. W. *J. Chem. Educ.* **2013**, *90*, 694–699.

8. Chase, A.; Pakhira, D.; Stains, M. *J. Chem. Educ.* **2013**, *90*, 409–416.

9. Bergin, A.; Sharp, K.; Gatlin, T. A.; Villalta-Cerdas, A.; Gower, A.; Sandi-Urena, S. *J. Chem. Educ.* **2012**, *90*, 289–295.

10. Cohen, J. *Statistical Power Analysis for the Behavioral Sciences*, 2nd ed.; Lawrence Erlbaum Associates, Inc.: Hillsdale, NJ, 1988.

11. Conover, W. J. *Practical Nonparametric Statistics*, 2nd ed.; John Wiley & Sons: New York, 1980.

12. Christain, B. N.; Yeziersky, E. J. *Chem. Educ. Res. Pract.* **2012**, *13*, 384–393.

13. Hibling, C.; Barke, H. *Chem. Educ.: Res. Pract. Eur.* **2000**, *1*, 365–374.

14. Obenland, C. A.; Munson, A. H.; Hutchinson, J. S. *Chem. Educ. Res. Pract.* **2013**, *14*, 73–80.

15. Pedhauzer, E. J. *Multiple Regression in Behavioral Research*, 3rd ed.; Harcourt College Publishers: Orlando, FL, 1997.

16. Cohen, J.; Cohen, P.; West, S. G.; Aiken, L. S. *Applied Multiple Regression /Correlation Analysis for the Behavioral Sciences*, 3rd ed.; Lawrence Erlbaum Associates, Inc.: Mahwah, NJ, 2003.

17. Peng, C. J.; Lee, K. L.; Ingersoll, G. M. *J. Educ. Res.* **2002**, *96*, 3–14.

18. Emenike, M. E.; Schroeder, J.; Murphy, K.; Holme, T. *J. Chem. Educ.* **2013**, *90*, 561–567.

19. Mills, P.; Sweeney, W. *J. Chem. Educ.* **2009**, *86*, 738–743.

20. Jennings, K. T.; Epp, E. M.; Weaver, G. C. *Chem. Educ. Res. Pract.* **2007**, *8*, 308–326.

21. Kerby, H. W.; Cantor, J.; Weiland, M.; Babiarz, C.; Kerby, A. *J. Chem. Educ.* **2010**, *87*, 1024–1030.

22. Cooper, M. M.; Underwood, S. M.; Hilley, C. Z.; Klymknowsky, M. W. *J. Chem. Educ.* **2012**, *89*, 1351–1357.

23. Hein, S. M. *J. Chem. Educ.* **2012**, *89*, 860–864.

24. Mahalingam, M.; Schaefer, F.; Morlino, E. *J. Chem. Educ.* **2008**, *85*, 1577–1581.

25. Bell, P.; Volckmann, D. *J. Chem. Educ.* **2011**, *88*, 1469–1476.

26. Jeon, K.; Huffman, D.; Noh, T. *J. Chem. Educ.* **2005**, *82*, 1558–1564.

Chapter 8

Using the Statistical Program R Instead of SPSS To Analyze Data

Hui Tang[*,1] and Pengsheng Ji[2]

[1]Department of Chemistry, University of Georgia,
Athens, Georgia 30602, United States
[2]Department of Statistics, University of Georgia,
Athens, Georgia 30602, United States
*E-mail: huitang@uga.edu

R is a programming language and computing environment for statistics and graphics. It has been widely used in many academic fields including psychology, ecology and bioinformatics. Because it is free and powerful, R can be an attractive alternative to commercial statistical packages such as SPSS for chemistry education researchers. This chapter introduces both advantages and some basic features of R as well as applications of R in chemistry education research.

Introduction

Traditional statistical software packages such as SPSS are popular in the area of chemistry education research (CER). However, with the development of CER, some advanced statistical methods have been reported and various new methods are expected to be applied in this field. Many researchers have found that not all of the statistical analyses they plan to carry out can be performed easily in software like SPSS. A few years ago, one of the authors of this chapter was looking for tools to analyze scanpath data in an eye-tracking study in which students viewed NMR signals and molecular structures. He learned that comparing groups of scanpaths statistically was very difficult in SPSS; on the other hand, the software R made the task simple when using the existing functions provided by R's add-on packages. Details of this analysis will be described in the last section of this chapter.

This chapter provides an insight into R and discusses how R can be applied in CER. There are four sections in the chapter: (1) advantages and disadvantages of R, (2) introduction to using R, (3) basic statistical analysis in R and SPSS, and (4) analyses that R can perform better and SPSS does not adequately handle. However, this chapter is not meant to serve as an R manual. Thus, many basic features of R are not introduced in this chapter. If interested in learning more about R, there are many books (*1*) and online tutorials. Those new to R may want to refer to the recommended readings at the end of this chapter. This chapter is geared to those who are familiar with basic statistics, whether experienced or new chemistry education researchers. Therefore, most common statistical methods used in CER will not be described in detail. Readers who need more statistical background can refer to statistics textbooks and the introduction of basic statistics in CER (*2–4*). There are many statistical methods that have been published in the CER literature. We will only discuss a few from articles that were published in the *Journal of Chemical Education* (JCE) Chemical Education Research feature. Performing these statistical analyses in both SPSS and R will be briefly described and compared.

Reasons To Learn and Use R

Examples of popular statistical packages include SPSS, SAS, R, Stata, Matlab, Systat, Minitab, and JMP. In the field of chemistry education research, SPSS is the most popular statistical software for analyzing quantitative data. For instance, in the last two years (May 2011 to August 2013), JCE published about 70 research articles categorized as "Chemical Education Research". Among these articles, 44 statistically analyzed quantitative data including 21 articles that explicitly reported the names of the statistical packages including SPSS (16 papers), SAS (1 paper), SYSTAT (1 paper), and spreadsheet programs (3 papers). One important reason that many chemistry education researchers prefer SPSS is its intuitive interface. Basic statistical analyses for CER can be performed easily by several mouse clicks on the menus and dialog boxes in the software. However, there are some problems that SPSS does not adequately handle. An example is the Levenshtein distance and permutation test, which will be discussed in more detail later. In such cases, R is a much better choice, because R's add-on packages make it able to efficiently perform different kinds of analyses. In addition, there are many other advantages that help make R a very good statistical program to analyze chemistry education data.

The Advantages of R

R Is Free and Open-Source

First of all, R is free. Anyone can download and install R onto a personal computer free of charge. It runs on most operating systems, including Windows, Macintosh, and Unix/Linux. Researchers using R no longer need to worry about the availability and cost of the statistical package they use. In contrast, SPSS users who move to a new institution may find that there is no SPSS site license on campus. Accessibility and cost of software is an important factor, especially for graduate students. R is also useful for secondary school teachers who intend to carry out research but have no access to any commercial statistical software.

R is an open source programming language and is as accurate as SPSS or SAS software (5). That is, users can virtually read every detail of how each statistical function works. Like other successful open-source projects including Linux and MySQL, R has benefited from the "many-eyes" approach to code improvement because many experts in statistics and computation keep monitoring the source code. As a result, R has an extremely high standard of quality and numerical accuracy (5, 6). R also has open interfaces, meaning that it readily integrates with other applications and systems.

R Is Powerful and Flexible

The standard packages of R that are installed as default have most of the basic statistical functions that researchers need in chemistry education. Moreover, R has more than 4000 additional packages that cover a large range of analytical methods employed in diverse fields. As a programming language, the capability of R is not limited to statistics. It has been applied in computing, modeling, machine learning, and data mining. In addition, R has extensive capacities to generate high-quality graphics, which can be conveniently customized by specifying feature options including line styles, fonts, colors, axes, and titles. Graphs developed in R can be saved in a variety of formats such as pdf, png, jpeg, and postscript.

The flexibility of R makes it easy for users to interact with it. With different commands, users can obtain important intermediate outputs instead of a single final report. This helps the users understand both the principles and procedures of the statistical analyses they are carrying out. One can save every line of commands in R. Thus, it is convenient for users to keep a record of their previous work and re-use programs for similar analyses.

R Is Community-Backed

There are numerous mailing lists, forums and blogs that provide R resources, including tutorials, discussions, and troubleshooting. Examples are the R-help mailing list (https://stat.ethz.ch/mailman/listinfo/r-help), Local R User Group Directory (https://blog.revolutionanalytics.com/local-r-groups.html), Stack Overflow (https://stackoverflow.com), Cross Validated (https://stats.stackexchange.com) and Talk Stats (http://www.talkstats.com/). There have been a few blogs for R in chemistry, such as "The Chemical Statistician" (https://chemicalstatistician.wordpress.com) and R Code for Analytical Chemistry (https://sites.google.com/site/alisonappling/tools/r-scripts-for-analytical-chemistry). Before posting questions, one should enter a few keywords into Google or the dedicated R Seek (http://www.rseek.org), because there is a great chance that one will find helpful discussions already there. In short, R has a large and supportive community that consists of users, package developers, and book authors helping each other and also possibly shaping the future of statistical software.

Disadvantages

For a researcher who is familiar with the point-and-click interface of SPSS, an obvious disadvantage of R is the use of command lines. Although in most cases, either one or a few command lines are enough to perform a task in R, some people may still not be comfortable with entering commands instead of mouse clicking and dragging. Typing may also introduce typing errors and sometimes debugging can be a very frustrating process, especially for beginners (6, 7).

There are at least two partial solutions to this problem. On one hand, users can easily complete common tasks via the enhanced menus of a more user-friendly R interface like RStudio (http://www.rstudio.com), which has menus and dialog boxes for tasks such as changing the working directory, importing and viewing data, and installing packages. On the other hand, command typing can be minimized after users get used to the format of help files, mimic the examples, and reuse their own programs.

Introduction to Using R

In this chapter, all the names of R functions and packages are written in the Courier New font. R utilizes ">" as the command prompt after which one enters a command line. The prompt will be included sometimes for the purpose of demonstration when both commands and output are illustrated in this chapter; however, it does not need to be typed when one actually enters a command in R. Readers should be aware that the same statistical analysis can usually be conducted in multiple ways (i.e., different functions or packages) in R. We assume most readers using SPSS are familiar with its graphical user interface (GUI). All the R and SPSS menu names in this chapter are in Bold and arrows ("=>") are used to indicate the orders of menu clicking.

Installation and Interface

R can be downloaded from the Comprehensive R Archive Network (CRAN) at http://cran.r-project.org for Windows, Mac OS X, and Linux. Installation of R is straightforward. When opening R on Windows, a frame labeled "RGui" will appear on the screen (Figure 1).

However, this interface lacks menus for some basic tasks like importing data sets and often frustrates new users. Thus, we strongly recommend RStudio, an interface and R companion that can be downloaded from http://www.rstudio.com. After installing RStudio, whenever it is started, R will start automatically in the background. Figure 2 shows the RStudio interface, which consists of four panes. The bottom left is the interactive console where commands are entered after the command prompt (>) and executed when the Enter key is pressed. The top left is the script pane where multiple lines of commands can be written and saved, but are not executed until the **Run** button is clicked. The top right is the workspace pane where the names of data and values in the context can be shown. The bottom right pane has menus to open files, view plots, install packages, and find help information for functions. All panes can be rearranged and resized.

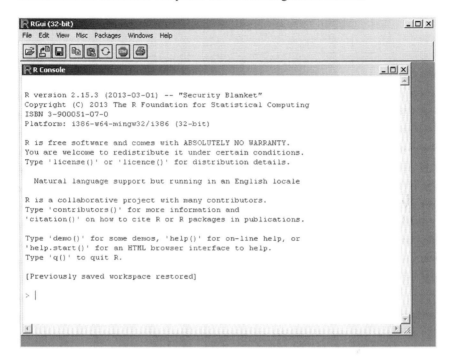

Figure 1. The R console window.

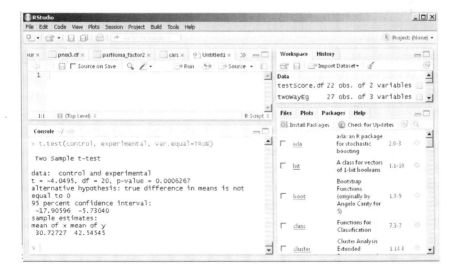

Figure 2. Screen shot of RStudio.

Packages

A package for R often consists of a set of functions for a specific task, such as linear models or graphics. A collection of basic packages will be automatically included when R is installed. The list of installed packages can be found in the bottom right pane of RStudio (Figure 2). The default packages are able to fulfill most common statistical requirements of chemistry education research. However, as the discipline develops, increasingly advanced methods are needed to analyze data. This may necessitate the installation of additional packages via **Tools** => **Install Packages** in RStudio.

There are some special add-on packages with an SPSS-like interface (*8*) that allow users to do basic data analysis by mouse-pointing and clicking. The examples include Rcmdr (*9*) and Deducer (*10*). It is understandable that SPSS users may feel more comfortable with these packages. However, it should be emphasized that the functions provided by these add-ons are very limited compared with SPSS or R commands. Therefore, these packages are not recommended.

Getting Help Globally and Locally

SPSS users may go through the menus to find the appropriate methods for their questions, while R users usually take a different approach. For example, to do a *t*-test, the users can first search in Google with the keywords "R function *t* test" and learn that t.test() is the function they are looking for. The users may have known how to use it from the linked pages, or they can look it up in the local help files in RStudio for more details. All the help pages have the same format, which includes Description, Usage, Arguments and Examples. Most of the time, one does not have to read these documents thoroughly to learn how to use these commands. For example, by looking at Usage and Examples in the Help page of Student's *t*-Test, one can easily find out that t.test(x, y) is for two-sample comparisons while t.test(x, mu=10) is the one-sample version.

Importing Data Sets

R commands with various parameters can be very powerful and flexible in importing and preprocessing data, but can also be counter-intuitive or even frustrating for new users. We suggest preprocessing data, saving into text files (.txt or .csv) in Microsoft Excel and then importing data by clicking **Tools** => **Import Dataset** in RStudio. Then the equivalent commands will be automatically generated in the console and the data will be shown in a table. For example, users see commands similar to the following ones that read data from a file named "studentGrade.txt", import the data into a data frame called "grades.df", and show the data in a table in RStudio:

```
grades.df  =  read.table("studentGrades.txt",  header=TRUE)
View(grades.df)
```

Small amount of data can also be input from the keyboard. Suppose two groups of students (11 in each group) took an exam in a research study. The researchers can create two vectors in R (named *control* and *experimental* in the following example) and input the exam scores from the keyboard into the corresponding vectors:

```
control = c(31,19,28,41,30,19,26,35,34,40,35)
experimental = c(37,44,42,35,50,38,46,50,45,49,32)
```

Output

Output in R is plain text by default. An example of an independent *t*-test output from the above dataset is shown below with a command for this *t*-test in the first line, which compares the means of the exam scores in the two groups. Before conducting this *t*-test, the readers should copy the above code that defines the dataset (exam scores) and paste it to RStudio.

```
> t.test(control, experimental)

Welch Two Sample t-test

data: t.test(control, and experimental)
t = -4.0495, df = 19.491, p-value = 0.0006549
alternative hypothesis: true difference in means is not
equal to 0
95 percent confidence interval:
 -17.916178  -5.720185
sample estimates:
mean of x    mean of y
 30.72727    42.54545
```

The output reports the *t* value, degree of freedom, and the *p* value. The two groups of grades are significantly different since p < 0.05, which is supported by the fact that zero (or no difference) is not contained within the 95% confidence interval. Note that the "Welch" *t*-test in this example, which assumes the variances are not equal, is the default *t*-test in R. A *t*-test with two equal variances can be performed by adding a parameter var.equal=TRUE, as shown in Figure 2.

R Files

When exiting R, a prompt window will appear asking to save the current workspace image that contains all the objects, functions, data, and loaded packages. The workspace image file, which has the extension *.RData, can be launched next time by simply clicking on it. Furthermore, all the commands typed in the console window or script window can be saved into an R history file with the extension *.Rhistory. This file can be opened by a common text editor so that one may check the history of work without opening R. One can also copy the commands in the history file and paste them into R to minimize repeated typing.

Fundamental Statistics in SPSS and R

The statistical methods selected in this section are frequently used and reported in the CER literature. We have chosen "Chemical Education Research" articles published in the *Journal of Chemical Education* as examples to illustrate the applications of these methods and how these analyses can be performed easily in both SPSS and R.

Descriptive Statistics and Graphics

The first step of data analysis is usually exploratory data analysis (EDA). EDA involves descriptive statistics and graphics that summarize and visualize the data. This serves several important purposes, such as checking statistical assumptions, identifying outliers and anomalies, discovering relationships among variables, and formulating valid hypotheses (*11*, *12*).

The mean and median are the two common measures of central tendencies. Measures of spread include the minimum, maximum, quantiles, range, variance, and standard deviation. R provides a simple function for each of these measures. Example commands for some of these functions are listed in Table I. In SPSS, these statistics can be generated by selecting **Analyze** => **Descriptive Statistics** => **Descriptives** or **Frequencies**.

Table I. R Command for Some Descriptive Statistics

Function	Description
mean (x)	Arithmetic average
median (x)	Median
sd (x)	Standard deviation
var (x)	Variance
length (x)	Number of values
quantile (x)	Quantiles
summary (x)	Mean, median and quantiles in one

x is a variable, just as in SPSS.

In addition to the numerical descriptive statistics, both R and SPSS provide straightforward ways to create tables and graphical representations of data. These tables and graphics allow researchers to visualize patterns of data, e.g., the distribution. Pie charts, bar charts, line charts, scatter plots, boxplots, and histograms are the most commonly used graphs. They all can be created under the **Graphs** menu in SPSS, while R can produce these graphs by using a single command for each. For example, the boxplot in Figure 3 can be obtained with the command:

```
boxplot(control, experimental, names=c("control",
"experimental"), ylab="Grade")
```

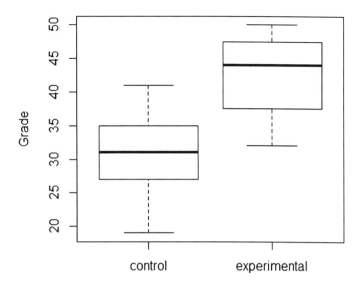

Figure 3. R boxplot.

Basic Inferential Statistics

Analysis of Difference and Association

Finding differences and associations between variables is the essential problem for quantitative data analysis in CER. In the first volume of this book series, Sanger (*2*) summarized some basic inferential statistics that researchers applied in visualization studies in chemistry. These statistical tests include correlation coefficients (Pearson's *r*), *t*-tests (one-sample, independent and dependent or paired), analysis of variance (ANOVA), post-hoc tests (*2*, *3*), tests of proportions (*z* values), and chi-square tests (tests of goodness of fit and tests of independence or association) (*2*, *4*). All these analyses can easily be conducted in both SPSS and R. Table II lists the commands for some basic inferential statistics.

Table II. R Commands for Some Inferential Statistics

Function	*Description*
cor.test (x,y)	Correlation of variables x and y
t.test (posttest~group)	Independent t-test with unequal variances: compare the difference in posttest between two groups
aov (grade~group)	One-way ANOVA: compare the difference in grade between two or more groups
chisq.test (chi.data)	Chi-squared test (chi.data is the name of the data)

Regressions

In addition to the above inferential statistics, regression is also an important tool to explore relationships among variables. It is applied to make predictions of the dependent variable from one or more independent variables. For instance, Rath *et al.* (*13*) used linear regression to explore how variables such as high school GPA and SAT scores contributed to university students' final grades in chemistry courses. In a study of examining gender differences in mental rotation tasks of chemistry representations, Stieff (*14*) used linear regression models to investigate the relationship between response time and angular disparity in two rotation tests. In SPSS, the **Linear Regression** window appears when one clicks **Analyze** => **Regression** => **Linear**. In R, linear regression analysis can be performed with the lm function. The following example is shown in the R help file when users enter ?lm in R. In the example, "weight" is the dependent variable and "group" is the independent variable:

```
lm.D9 = lm(weight ~ group)
summary(lm.D9)
```

The output includes coefficients, R-squared, residuals, and the residual standard error. Note that t and F statistics are also listed in the linear regression outputs of both SPSS and R. Here is part of the R output:

```
Coefficients:
            Estimate Std. Error t value Pr(>|t|)
(Intercept)   5.0320     0.2202  22.850 9.55e-15 ***
groupTrt     -0.3710     0.3114  -1.191    0.249

Residual standard error: 0.6964 on 18 degrees of freedom
Multiple R-squared: 0.07308, Adjusted R-squared:  0.02158
F-statistic: 1.419 on 1 and 18 DF,  p-value: 0.249
```

Sometimes the dependent variable is dichotomous instead of continuous. For instance, the outcome is either "correct" or "incorrect". In this case, logistic regression is appropriate for predicting a binary outcome from predictor variables. See the chapter by Lewis (*4*) in this volume for more on logistic regression. Schuttlefield *et al.* (*15*) used logistic regression to determine the overall difficulty of the complexity factors in gas law problems, as well as the relative difficulty of each variable within the same complexity factor, and how these factors influenced students' performance. Emenike *et al.* (*16*) used logistic regression analyses to identify differences among groups of participants for responses to an assessment survey. In SPSS, the logistic regression analysis is performed by selecting **Binary Logistic** or **Multinomial Logistic** in the **Regression** menu; while in R, the function glm can be used for logistic regressions.

Reliability Analysis

Another useful inferential statistic that has been frequently reported in CER is Cronbach's alpha (α) (*17–23*). It evaluates the internal consistency reliability of items in an instrument. A larger Cronbach's alpha value means a higher level of internal reliability. Usually $\alpha \geq 0.70$ is required as a minimum value for a satisfactory internal reliability of a designed instrument (24, 25). In SPSS, Cronbach's alpha is computed under the **Analyze** => **Scale** => **Reliability Analysis** menu. Several R packages provide functions to compute Cronbach's alpha, e.g., the function cronbach.alpha in the package ltm.

Nonparametric Statistics

When data are ordinal (ranks) or when assumptions such as normality are violated, nonparametric statistics should be applied (4). In the CER papers mentioned at the beginning of this section, two nonparametric statistics are commonly used. One is Spearman's rho (ρ) as an alternative to Pearson's r for correlation coefficients (18–26). The other is the Wilcoxon test, which is the nonparametric equivalent of the paired or the two-sample t-test (26–28); the latter is also known as the Mann-Whitney U test. In SPSS, Spearman's rho (ρ) can be obtained by selecting the **Spearman** box instead of **Pearson** in the **Correlations** window. The Mann-Whitney test and the Wilcoxon test can be performed under the **Analyze** => **Nonparametric Tests** menu. In R, the cor.test function can be used to attain Spearman's rho and the significance of the correlation. All variants of the Wilcoxon-Mann-Whitney test can be conducted by the function wilcox.test. For example, the one-sample version can be wilcox.test(control) and the two-sample version can be wilcox.test(control, experimental).

Questions Better Answered by R

If one is using SPSS, there is a good chance of encountering its constraints. For example, data preprocessing (cleaning, integration, etc.), Bayesian and complex survey analysis are difficult or impossible in SPSS. But they are both handled well in R because of R's extension packages. We discuss a few specific issues below.

Effect Size

Most researchers report statistical significance when inferential statistics are involved for data analysis. Statistical significance indicates whether an effect (difference or association) exists, but it does not show the magnitude of the effect (29, 30). It is the effect size that quantitatively interprets the extent of the effect. There are two types of effect size. The d family of effect size represents the magnitude of difference; while the r family tells the strength of association (29, 30). In the CER literature of interest investigated for this chapter mentioned earlier, few researchers have reported effect size (20, 21, 31–34). Of these six articles, half of them reported Cohen's d (21, 31, 34), which is the most commonly reported effect size measure for the t-test. The other articles reported phi (φ, for chi-squared tests) (32), omega squared (ω^2, for ANCOVA) (33), and f^2 (for correlation or for multiple regression) (*20, 21*). The American Psychological Association (APA) (*29, 30*) has emphasized the importance of reporting effect size and its confidence interval. We would make the same suggestion to the authors in CER.

SPSS can generate outputs of many measures of effect size in the *r* family, such as *r* for correlation coefficients, φ, eta^2 (η^2) for ANOVA, and R^2 for multiple regressions. However, SPSS does not generate outputs of Cohen's *d*. It must be calculated manually (*12, 35*). On the other hand, extension packages are available in R to calculate various effect size statistics. For example, the compute.es package consists of about twenty functions for common effect size estimates. Users can choose different functions to compute effect sizes for different statistics, including Chi-square tests, *t*-tests, and ANOVA (*36*). For instance, the function tes converts a *t* value to Cohen's *d*, calculates the variance of *d*, and more. Here, we use the *t* value in Figure 2 (11 is the sample size in each of the two groups):

```
> tes(-4.0495, 11, 11)

    $MeanDifference
            d        var.d                 g        var.g

    -1.7267126   0.2495804    -1.6611412   0.2309848
```

Rasch Model

The Rasch model is one of the most broadly recognized models under the Item Response Theory approach (*37, 38*). It models the probability of a person's correct answer on an item as a function of the person's ability and the item difficulty. The Rasch model has been applied in education for decades to construct and evaluate tests or surveys. Recently in CER, Wei, *et. al.* (*39*) employed the Rasch model to develop an instrument based on computer modeling and examining high-school students' conceptual understanding of matter. Barbera (*18*) applied the Rasch model analysis in investigating the characters of the Chemical Concept Inventory and the relationship between item difficulties and students' abilities in answering the questions in the instrument.

SPSS does not provide Rasch analysis directly. Thus, the analysis has to be conducted either in specialized software such Winsteps (*18, 39, 40*), or by writing dozens of lines of complex command in SPSS syntax (*41*). In R, several packages are available for Rasch analysis. Examples of these packages are eRm (*42*), psychomix (*43*), and ltm (*44*). Recently, the latter has been adopted by SPSS (version 17.0 and above) in its extension commands, which requires installation of R and/or another software such as Python.

Permutation Tests

Eye-tracking technology has begun to be applied in CER (*45–49*). One of the important variables in eye-tracking studies is scanpath. A scanpath is a series of eye gazes and is, in general, written in the form of a string consisting of characters. Sequence alignment algorithms such as the Levenshtein distance (LD) are used to quantitatively analyze the difference between two individual scanpaths. The LD

146

is obtained by calculating the minimum number of operations needed to transform one string into the other by inserting, deleting, or replacing characters. When there are two groups of scanpaths, each scanpath must be compared with all the other scanpaths, within the group or between groups, to obtain LD values. Therefore, a typical *t*-test or Wilcoxon test is not proper to statistically analyze the differences between scanpaths by comparing the LD values between two groups. To solve this problem, Feusner and Lukoff (*50*) proposed a permutation test to compare two groups of scanpaths. Tang et al. (*48*) later extended this analytical method to scanpaths with different lengths. It was applied in exploring students' eye movement patterns when students were asked to relate NMR signals and organic molecular structures.

The idea behind permutation tests is that new samples are repeatedly generated from the original sample. Each new sample has the same values as those in the original data set, but with different combinations of values in the experimental groups. Statistics can be calculated for each new sample and then the *p*-value can be obtained by comparing the new statistics with the statistics of the original sample.

In R, the sample function can be used to generate new samples. Meanwhile, functions in R were utilized by Tang *et al.* (*48*) to compute Levenshtein distances. As a simple example, suppose there are 3 scanpaths in one group and 3 scanpaths in the other group, which are all kept in a vector or variable called "sp". To compute the Levenshtein distance between each pair of scanpaths from the two groups, the adist function can be employed. The LD values are calculated using a for loop and then stored in a 3×3 matrix called "ld":

```
sp=c("abcdbc","bcabda","cdabcd","dcbaba","cbacba","dbcbaa")
n = 3
m = 3
ld = matrix(0,n, m)
for (i in 1:n){
    ld[i, ] = adist(sp[i], sp[(n+1):(n+m)])
}
```

Entering ld will display the nine LD values:

```
> ld
      [,1] [,2] [,3]
[1,]    5    4    4
[2,]    3    4    3
[3,]    4    4    5
```

With Levenshtein distance values being calculated, permutation tests to compare two groups of scanpaths can be conducted in R. On the contrary, SPSS does not provide LD calculation in its menus. To accomplish the same LD-permutation task, hundreds of lines of code have to be written in SPSS syntax.

Summary

We have discussed some advantages of R over SPSS, and introduced some fundamental aspects of R. We hope that readers will be more confident in overcoming the unease of typing a few lines of commands. It will not take long for one to realize that going through some well-structured R Help files or tutorials may not be harder than going through the SPSS menus and manual. As R becomes more popular in academia and industry, users are sincerely invited to join the most interactive community of statistical software to share their own experiences.

Recommended Readings

There are enormous free materials for R on the Internet. Based our experience in learning and teaching R, we recommend the beginners first take some online tutorials so that they can easily copy, run, and modify the programs. Soon they may find a quick reference card with only a few pages is very handy, and a short but comprehensive book is very helpful.

Online Materials

Gardener, M. Using R for Statistical Analyses – Introduction. http://www.gardenersown.co.uk/Education/Lectures/R/ (accessed May 2014).

Kabacoff, R. Quick-R for SAS/SPSS/Stata Users. http://www.statmethods.net/ (accessed May 2014).

Maindonald, J. H. Using R for Data Analysis and Graphics. http://cran.r-project.org/doc/contrib/usingR.pdf (accessed May 2014).

Baggott, M. R Reference Card V2. http://cran.r-project.org/doc/contrib/Baggott-refcard-v2.pdf (accessed May 2014)

Books

Pace, L. *Beginning R: An Introduction to Statistical Programming*; Apress: New York, 2012.

Kabacoff, R. I. *R in Action: Data Analysis and Graphics*; R. Manning Publications Co.: Shelter Island, NY, 2011.

Acknowledgments

The authors thank Elizabeth Day and Lisa Kendhammer for their comments and corrections on the manuscripts. We thank the editors and reviewers for their constructive suggestions to improve this chapter.

Note: The example code to calculate Levenshtein distances on page 147 has been modified from the original code published on July 31, 2014. The revised code was published on October 15, 2014.

References

1. Books Related to R. http://www.r-project.org/doc/bib/R-books.html (accessed May 2014).

2. Sanger, M. J. Using Inferential Statistics To Answer Quantitative Chemical Education Research Questions. In *Nuts and Bolts of Chemical Education Research*; Bunce, D. M., Cole, R. S., Eds.; ACS Symposium Series 976; American Chemical Society: Washington, DC, 2008; Chapter 8.

3. Pentecost, T. C. Introduction to the Use of Analysis of Variance in Chemistry Education Research. In *Tools of Chemistry Education Research*; Bunce, D. M., Cole, R. S., Eds.; ACS Symposium Series 1166; American Chemical Society: Washington, DC, 2014; Chapter 6.

4. Lewis, S. E. An Introduction to Nonparametric Statistics in Chemistry Education Research. In *Tools of Chemistry Education Research*; Bunce, D. M., Cole, R. S., Eds.; ACS Symposium Series 1166; American Chemical Society: Washington, DC, 2014; Chapter 7.

5. Muenchen, R. A. *R for SAS and SPSS Users*, 2nd ed.; Springer: New York, 2011.

6. Fox, J.; Andersen, R. Using the R Statistical Computing Environment To Teach Social Statistics Courses, 2005. http://www.unt.edu/rss/Teaching-with-R.pdf (accessed May 2014)

7. Zuur, A. F.; Ieno, E. N.; Meesters, E. H. *A Beginner's Guide to R*; Springer: New York, 2009.

8. R GUI Projects. http://www.sciviews.org/_rgui/index.html (accessed May 2014).

9. Fox, J. *J. Stat. Software* **2005**, *14* (9), 1–42.

10. Fellows, I. *J. Stat. Software* **2012**, *49* (8), 1–15.

11. Behrens, J. T. *Psychol. Meth.* **1997**, *2* (2), 131–160.

12. Morgan, G. A.; Leech, N. L.; Gloeckner, G. W.; Barrett, K. C. *SPSS for Introductory Statistics: Use and Interpretation*, 2nd ed.; Lawrence Erlbaum Associates: Mahwah, NJ, 2004.

13. Rath, K. A.; Peterfreund, A.; Bayliss, F.; Runquist, E.; Simonis, U. *J. Chem. Educ.* **2012**, *89* (4), 449–455.

14. Stieff, M. *J. Chem. Educ.* **2013**, *90* (2), 165–170.

15. Schuttlefield, J.; Kirk, J.; Pienta, N.; Tang, H. *J. Chem. Educ.* **2012**, *89* (5), 586–591.

16. Emenike, M. E; Schroeder, J.; Murphy, K; Holme, T. *J. Chem. Educ.* **2013**, *90* (5), 561–567.

17. Bruck, A. D.; Towns, M. *J. Chem. Educ.* **2013**, *90* (6), 685–693.

18. Barbera, J. *J. Chem. Educ.* **2013**, *90* (5), 546–553.

19. Heredia, K.; Lewis, J. E. *J. Chem. Educ.* **2012**, *89* (4), 436–441.

20. Knaus, K.; Murphy, K.; Blecking, A.; Holme, T. *J. Chem. Educ.* **2011**, *88* (5), 554–560.

21. Xu, X.; Lewis, J. *J. Chem. Educ.* **2011**, *88* (5), 561–568.

22. Cheung, D. *J. Chem. Educ.* **2011**, *88* (11), 1462–1468.

23. Bell, P.; Volckmann, D. *J. Chem. Educ.* **2011**, *88* (11), 1469–1476.

24. Cortina, J. M. *J. Appl. Psychol.* **1993**, *78* (1), 98–104.

25. Nunnally, J. C.; Bernstein, I. H. *Psychometric Theory*, 3rd ed.; McGraw-Hill: New York, 1994.

26. Herrington, D. G.; Luxford, K.; Yezierski, E. J. *J. Chem. Educ.* **2012**, *89* (4), 442–448.

27. Cooper, M. M.; Underwood, S. M.; Hilley, C. Z.; Klymkowsky, M. W. *J. Chem. Educ.* **2012**, *89* (11), 1351–1357.

28. Bauer, C. F.; Cole, R. S. *J. Chem. Educ.* **2012**, *89* (9), 1104–1108.

29. Cumming, G.; Fidler, F.; Kalinowski, P.; Lai, J. *Aust. J. Psychol.* **2012**, *64*, 138–146.

30. Fritz, C. O.; Morris, P. E.; Richler, J. J. *J. Exp. Psychol. Gen.* **2012**, *141*, 2–18.

31. Bridle, C. A; Yezierski, E. J. *J. Chem. Educ.* **2012**, *89* (2), 192–198.

32. Grove, N. P.; Cooper, M. M.; Cox, E. L. *J. Chem. Educ.* **2012**, *89* (7), 850–853.

33. Hall, M. V.; Wilson, L. A.; Sanger, M. J. *J. Chem. Educ.* **2012**, *89* (9), 1109–1113.

34. Lewis, S. E. *J. Chem. Educ.* **2011**, *88* (6), 703–707.

35. Leech, N. L.; Barrett, K. C.; Morgan, G. A. *SPSS for Intermediate Statistics Use and Interpretation*, 2nd ed.; Lawrence Erlbaum Associates: Mahwah, NJ, 2004.

36. Package 'compute.es' http://cran.r-project.org/web/packages/compute.es/ compute.es.pdf (accessed July 29, 2013).

37. Baker, F. B.; Kim, S.-H. *Item Response Theory: Parameter Estimation Techniques*, 2nd ed.; Marcel-Dekker: New York, 2004.

38. Scantlebury, K.; Boone, W. J. Designing Tests and Surveys for Chemical Education Research. In *Nuts and Bolts of Chemical Education Research*; Bunce, D. M., Cole, R. S., Eds.; ACS Symposium Series 976; American Chemical Society: Washington, DC, 2008; Chapter 10.

39. Wei, S.; Liu, X.; Wang, Z.; Wang, X. *J. Chem. Educ.* **2012**, *89* (3), 335–345.

40. Pentecost, T. C.; Barbera, J. *J. Chem. Educ.* **2013**, *90* (7), 839–845.

41. TenVergert, E.; Gillespie, M.; Kingma, J. *Behav. Res. Meth. Instr.* **1993**, *25* (3), 350–359.

42. Mair, P.; Hatzinger, R. *J. Stat. Software* **2007**, *20* (9), 1–20.

43. Frick, H.; Strobl, C.; Leisch, F.; Zeileis, A. *J. Stat. Software* **2012**, *48* (7), 1–25.

44. Rizopoulos, D. *J. Stat. Software* **2006**, *17* (5), 1–25.

45. Havanki, K. L.; VandenPlas, J. R. Eye Tracking Methodology for Chemistry Education Research. In *Tools of Chemistry Education Research*; Bunce, D. M., Cole, R. S., Eds.; ACS Symposium Series 1166; American Chemical Society: Washington, DC, 2014; Chapter 11.

46. Stieff, M.; Hegarty, M.; Deslongchamps, G. *Cognition Instruct.* **2011**, *29* (1), 123–145.

47. Tang, H.; Pienta, N. *J. Chem. Educ.* **2012**, *89* (8), 988–994.

48. Tang, H.; Topczewski, J. J.; Topczewski, A. M.; Pienta, N. J. Permutation test for groups of scanpaths using normalized Levenshtein distances and application in NMR questions. In *Proceedings of the Symposium on Eye*

Tracking Research and Applications, Santa Barbara, CA, March 28–30, 2012; ACM Press: New York, pp 169–172.

49. Williamson, V. M.; Hegarty, M.; Deslongchamps, G.; Williamson, K. C., III; Shultz, M. J. *J. Chem. Educ.* **2013**, *90* (2), 159–164.

50. Feusner, M.; Lukoff, B. 2008. Testing for statistically significant differences between groups of scan patterns. In *Proceedings of the Symposium on Eye Tracking Research and Applications*, Savannah, GA, March 26–28, 2008; ACM Press: New York, pp 43–46.

Cognitive-Based Tools for
Chemistry Education Research

Chapter 9

Designing Assessment Tools To Measure Students' Conceptual Knowledge of Chemistry

Stacey Lowery Bretz*

Department of Chemistry and Biochemistry, Miami University, Oxford, Ohio 45056, United States
*E-mail: bretzsl@miamioh.edu

The misconceptions that students (and teachers) hold about chemistry and the structure and properties of matter are documented extensively in the literature. Most of these reports were generated through clinical interviews with a small number of students and the subsequent meticulous analysis of their words, thoughts, and drawings. Concept inventories and diagnostic assessments enable teachers and researchers to assess large numbers of students regarding their chemistry misconceptions. This chapter discusses methodological choices to be made when designing such assessment tools and includes an appendix of chemistry concept inventories listed by topic.

Introduction

Assessment has a long, rich history in chemistry education (*1*). From an article in the very first issue of the *Journal of Chemical Education* (*2*) to the ACS Exams Institute (*3*) which is now nearly 80 years old, chemistry teachers have long been interested in measuring what their students do and do not know. Bauer, Cole, and Walter draw a distinction between measuring what happens in a course vs. the outcomes of a course (*4*). This chapter discusses the design of assessments that focus upon *unintended* student learning outcomes, namely the misconceptions that students have about the concepts and principles of chemistry. These assessment tools are commonly known as concept inventories or diagnostic assessments. In 2008, Libarkin (*5*) summarized the development of concept inventories across a variety of science disciplines in a commissioned paper for the National Research Council's *Promising Practices* report (*6*).

While some call these inaccurate ideas 'misconceptions,' others argue for the term 'alternative conceptions' (7). Wandersee, Mintzes, and Novak (7) distinguish *nomothetic* terms such as naïve conceptions, prescientific conceptions, and misconceptions from *ideographic* terms such as children's science, intuitive beliefs, and alternative conceptions. The key distinction between nomothetic and ideographic knowledge is that the former are compared to correct scientific information while the latter explores the explanations constructed by students to make sense of their experiences. It is not the purpose of this chapter to argue the epistemological and philosophical differences between these two stances. Rather, what is important is to realize that both views have exerted methodological influences upon discipline-based education researchers, including those in chemistry education research (CER). When researchers take the stance that students' views ought to be compared against those of experts, they tend to adopt experimental methods. Likewise, when researchers wish to investigate students' ideas, rather than impose the scientific community's knowledge as a framework for comparisons, then interviews, observations, and student self-reports become prominent methods for collecting data. Both stances are valuable to deepening our understanding of students' thinking. When it comes to developing assessment tools, the distinctions between these views and methods have been blurred, with most studies using a combination of both.

Design Considerations: Exemplar Assessment Tools

After that first article in the *Journal of Chemical Education*, chemists over the next 50 years were almost exclusively concerned with what facts and theories students ought to be taught in such curricula as ChemStudy (8) and the Chemical Bond Approach (9) created in the post-Sputnik era. Then, in the early 1970s, Derek Davenport authored a commentary (10), ostensibly about the importance of inorganic chemistry in the undergraduate curriculum. He shared an anecdote that entering graduate students in a chemistry Ph.D. program, despite having earned undergraduate degrees in chemistry, thought silver chloride was a pale green gas,. This one-page commentary is considered by many to be the first report of what now might be considered a misconception. Students were certainly never taught that silver chloride was a pale green gas. They knew enough chemistry to earn a chemistry degree and graduate college. Where could this unintended knowledge have come from? How did students learn information that was never taught? What additional ideas that would make experts cringe did students construct during their undergraduate chemistry experiences? These are some of the questions that today focus chemistry education research on documenting students' misconceptions and developing assessment tools to measure their prevalence.

Development Methods

There are several approaches to probing students' thinking and conceptual understanding, most of which can be traced back to Piaget's clinical interviews (11). This chapter focuses on the development of multiple-choice assessment tools geared toward measuring students' conceptual knowledge. As such, interviews are discussed only to the extent that they inform the development of a multiple-choice assessment tool.

In order to explore the range of possibilities for constructing such a tool, four exemplars are discussed below to highlight the variety of procedures used in the design of such tools. In chronological order of their development, the four exemplars are

- Covalent Bonding and Structure Diagnostic (*12*)
- Chemistry Concept Inventory (*13*)
- Foundational Concepts before Biochemistry Coursework Instrument (*14*)
- Enzyme-Substrate Interactions Concept Inventory (*15*)

This chapter discusses the numerous elements involved in developing an assessment tool focused on misconceptions for a chemistry classroom — from content selection to classroom implementation. In each phase of development, multiple examples are provided, drawing heavily on the four assessments listed above. Additional references in the literature are noted for detailed methodological discussions beyond the scope of this chapter.

Content Selection

When it comes to delineating what content will be assessed and what content is beyond the scope of interest, there are multiple approaches to identifying the boundaries. Some researchers focus on a narrow concept and what content ought to be learned regarding one particular topic, while others focus on what content might be prerequisite to learning new content. Still others assess conceptual knowledge across multiple concepts within a single course. Examples of each of these are described below.

Treagust outlines a 10-step procedure for developing diagnostic instruments (*16*), the first four of which involve specifying the content. The central tenet for identifying the content necessary to develop a diagnostic tool according to Treagust requires one or more experts (in this particular case, a chemistry education researcher) to identify the essential propositional knowledge statements and connect them to one another by creating a concept map. It bears noting that not every proposition directly correlates to one item on the diagnostic. For example, although there were 33 propositional statements identified when creating the Covalent Bonding and Structure Diagnostic, the tool itself consists of just thirteen items.

Villafañe (*14*) and colleagues collaborated with an expert community of biochemistry instructors when designing their assessment tool. Together, they identified five core concepts in general chemistry (bond energy, free energy, London dispersion forces, hydrogen bonding, and pH/pK_a) and three in biology (alpha helix, amino acids, and protein function) that were considered to be among the prerequisites to learning biochemistry.

When Mulford and Robinson (*13*) created the Chemistry Concept Inventory (CCI), they triangulated several sources of information to identify the focus of their assessment. First, they were interested in measuring what prior knowledge students brought with them when they enrolled in a university general chemistry course. Second, they generated a list of possible concepts by surveying general chemistry textbooks, reports calling for change in the general chemistry curriculum (*17*, *18*), the general chemistry exam from the ACS Examinations Institute (*3*), and the voluminous literature on chemistry misconceptions. Unlike Treagust's focus on one particular concept, or Villafane's emphasis on prerequisite knowledge for an upper division course, the CCI measured misconceptions on several concepts that students were expected to learn in a general chemistry course.

While the CCI was criticized by some chemistry faculty for including particulate images in the items and answer choices, it was in many ways "ahead of its time". However, particulate images would soon be ubiquitous. Within the next decade, Alex Johnstone (*19*) would be honored with the ACS Award for Achievement in Research on the Teaching and Learning of Chemistry, in part for his significant contributions toward demonstrating the importance that students understand particulate representations of matter. When developing the Enzyme-Substrate Interactions Concept Inventory (ESICI), Bretz and Linenberger (*15*) chose to identify the particulate content of their assessment by focusing upon students' confusion when trying to interpret multiple representations of enzyme-substrate interactions, often resulting in cognitive dissonance (*20*) on the part of the student.

Eliciting Students' Ideas

Treagust (*12*) identifies three keys steps for gathering information about students' misconceptions. First, a thorough review of the literature is warranted, as is the case with any research project. What has previously been reported regarding students' thoughts and misconceptions about the content of interest? Second, an individual clinical interview is conducted with each student in the sample. Students are asked open-ended questions, and their responses are probed for clarity, consistency, and comparison to expert-like responses. These interviews are digitally recorded, to facilitate the production of verbatim transcripts. Third, the collection of transcripts is analyzed to identify patterns and themes using constant comparative analysis (*21*).

Item Design

Given the different methods for identifying content and eliciting student thinking, it is not surprising to learn that there are variations when it comes to writing items. In Treagust's model, given the importance of propositional statements, each item must be directly correlated to one or more propositional statements. The multiple choice item with distractors from the interviews is presented to students, along with a request for the students to share, in a free-response format, their reasons for choosing their answer. These reasons are then used to create "two-tier" items in a subsequent version of the assessment tool. The first item in a two-tier asks students to share *what* they think; the second item asks students to share *why* they think as they do. If interviews and free responses indicate that students harbor multiple misconceptions about a key propositional knowledge statement, then the distractors will also reflect those multiple misconceptions.

Mulford and Robinson (*13*) drew inspiration from the literature regarding misconceptions, creating 7 items directly from tasks used in interview protocols designed for eliciting student ideas. The remaining CCI items were created from interviews and the research literature as with Treagust's model.

Villafañe and colleagues took a different approach to writing items. Rather than crafting one item for each misconception, they sought to build in redundancy in their instrument from the beginning. Each misconception was measured by a set of three items, created to measure student understanding regarding one of the eight prerequisite ideas for biochemistry. The distractors for each set of three items were 'matched' to see if students would consistently select the same incorrect idea.

While Treagust, Villafañe, and Mulford all began with expert-identified content and drafted items in response, Bretz and Linenberger took a different tact when designing items for the ESICI. That is, rather than impose a "top-down" expert-driven content framework upon students, the content that ultimately was included in the ESICI emerged in an authentic, "bottom-up" process driven entirely by students' misconceptions about particulate representations of enzyme-substrate interactions. Distractors to generate "two-tier" items on the ESICI were not gleaned from open-ended written responses, but rather from the semi-structured interviews in which students were asked to not only discuss their understanding of multiple representations of enzyme-substrate interactions (*20*), but also to annotate the representations themselves using digital paper and pen technology (*22*).

Validity and Reliability of Data

Designing measurements requires that close attention be paid to ensuring the validity and the reliability of the data generated by the instrument. In some ways, validity and reliability are akin to the chemistry constructs of accuracy and precision, respectively (*23, 24*).

Validity Methods

Treagust developed his distractors by drawing upon the methods reported by Tamir (*25*) in which students provided answers to open–ended essay questions. In the case of distractors drawn in part or in whole from the students' written thoughts and ideas, the authenticity, and therefore *face validity*, is much higher than incorrect answers crafted by what Tamir called "professional test writers." Students can be helpful in improving validity not only before data are collected with an instrument, but afterwards as well. Mulford and Robinson (*13*) and

Bretz and Linenberger (*15*) conducted interviews with students after they had answered the items, in order to investigate if students understood and interpreted the language and syntax of the items as they were intended to be. This post-hoc analysis of face validity with students requires interviewing students who performed across a range of scores, being sure to include lower-performing students so as to avoid the error of validating only with students who have better content knowledge. Tamir recently published (*26*) a protocol for exploring the importance of students' justification of their choices when responding to multiple choice items.

To ensure the *content validity* of Treagust's propositional knowledge statements and concept maps (*12*), both were subjected to careful scrutiny by experts in the discipline, including both scientists with extensive content expertise and science educators. These experts were asked to scrutinize the content for omissions, errors, or any contradictions. Mulford and Robinson (*13*) also employed content experts to ensure content validity, examining the responses of chemistry graduate students and faculty with expertise in chemistry education research.

When establishing content validity for the data generated by assessments intended to measure students' misconceptions, one caution is in order. Subject matter experts can be susceptible to "expert blindspots". For example, in the development of the ESICI, Bretz and Linenberger subjected the items to expert review as described above. Multiple instructors raised concerns about the use of "lock-and-key" and "induced-fit" images, noting that they never used these words, but rather, focused on complementarity of sterics and charge when discussing enzyme-substrate interactions. An analysis of the data corpus from student interviews revealed that, while faculty might not use the words "lock-and-key" or "induced fit," their students certainly did. These phrases were already in the students' vocabulary from previous courses in chemistry and biology, and therefore, shaping their mental models. Mulford and Robinson faced similar criticism from faculty during their expert review about including particulate images in the items. Faculty are indeed experts in content, but are not always aware of the quality and quantity of prior knowledge that students bring with them.

Villafañe (*14*) discusses the use of both exploratory factor analysis and confirmatory factor analysis to examine the internal structure, i.e., *construct validity*, of an instrument. Given their emphasis on writing three items for each of the eight concepts, factor analysis was a tool well suited to providing evidence that Villafane.had indeed succeeded in building in what they called "replicate trials" within one measure.

Lastly, asking students with different backgrounds (e.g., general chemistry students vs. organic chemistry students) provides the opportunity to establish *concurrent validity*, i.e., a measure of whether students with more instruction (and therefore, hopefully more knowledge) perform better than those with weaker backgrounds or less instruction. For example, Bretz and Linenberger (*15*) analyzed responses on the ESICI according to the students' self-reported majors including nutrition/exercise science, prehealth professions, biology, chemistry, and biochemistry because each of these majors has had different levels of science instruction.

Reliability Methods

After all the items have been created, Treagust (*12*) recommends creating what he calls a "specification grid" as one last check to ensure each item still tightly corresponds to both the propositional knowledge statements and the concept map. This grid "closes the loop," so to speak, to ensure internal consistency, i.e., *internal reliability*, between the assessment tool and the development process.

Measuring *external reliability* is important to demonstrating that students are consistent in choosing their responses, i.e., students are not randomly guessing. Villafañe and colleagues (*14*) measured consistency of responses through the design of their instrument—three items per concept and matched wrong answer choices for each of the three items.

A test-retest design also affords the opportunity to examine how consistent students' responses are over time. Bretz and Linenberger (*15*) administered the ESICI twice to the same group of students, with the administrations of the instrument separated in time by one month. Both the descriptive statistics (mean, median, standard deviation, skew) and a Wilcoxon signed ranks test indicated no significant difference between the students' responses. That is to say, the incorrect ideas the students held when answering the ESICI the first time were stable and remained constant when students responded for a second time.

A third method for exploring the consistency of student responses involves asking students to indicate their confidence about each response. Caleon and Subramaniam (*27, 28*) first introduced the confidence measure as a Likert scale by creating a "four-tier" instrument that consisted of four components: what, confidence level, why, confidence level. Students indicated their confidence level on a nominal scale: just guessing, very unconfident, unconfident, confident, very confident, or absolutely confident. Collecting data about students' confidence permits an analysis of confidence when correct vs. confidence when incorrect. McClary and Bretz (*29*) developed a diagnostic tool about acid strength of organic acids and subsequently modified Caleon and Subramaniam's confidence scale to an ordinal scale of 0% confident (just guessing) to 100% confident (absolutely certain) in order to permit a more quantitative treatment of the data. A plot of item difficulty vs. student confidence lead McClary and Bretz to the conclusion that confidence varied little, despite differences in item difficulty. That is, students do not know what they do not know.

Limitations

While establishing validity is an important prerequisite to establishing reliability, it is important to note that data collected to investigate validity and reliability in the initial creation of the instrument do not subsequently establish validity and reliability forever after. That is to say, the instrument is not reliable, nor is it valid for all circumstances and populations. Validity and reliability are characteristics of data, not the instrument used to collect the data. Each time data are collected, the validity and reliability of that data must be re-established (*30, 31*).

Furthermore, when scrutinizing the literature and examining the results of administering an assessment tool focused on misconceptions, it is important to ascertain how similar the sample of students from which data are to be collected is to the samples previously reported in the literature. Were the results reported for students in secondary school or university settings? For university chemistry majors or nonchemistry majors? High school chemistry teachers? No instrument is ever "perfect" in the sense that it requires no further modifications for use in a different circumstance or population. Each successive administration with different students in different settings can be expected to reveal nuanced differences in understanding.

While the statistical calculation of the Cronbach alpha (*32*) affords the opportunity to quantify internal consistency, i.e., reliability, this number has been the subject of recent skepticism with regard to interpreting its significance for measuring misconceptions (*29, 33*). A threshold alpha value of 0.7 is typically used as a cut-off to suggest that the items are internally consistent. This is reasonable when high inter-item correlations are expected. However, when measuring misconceptions where knowledge is fragmented in students' minds, expecting highly correlated responses is optimistic at best. Furthermore, given the development processes described above whereby multiple distractors for one item can represent multiple misconceptions, it is implausible to suggest that how a student responds to one item ought to be highly correlated to how that same student responds to another item—particularly if the assessment covers multiple concepts. Lasry and colleagues (*34*) have collected data to challenge the reliability of individual items, despite the overall reliability of the assessment tool as a collection of items.

These same considerations limit the value of factor analysis to indicate validity in that it is most useful when a researcher has the expectation that questions and students' responses to those questions will correspond to one another. However, cluster analysis (*35*) has recently emerged as a technique of some interest given its focus on grouping students who reason with similar models, as opposed to grouping questions as is typically done in factor analysis (*36*). Publications by Adams and Weiman (*33*), Ding and Beichner (*37*), and Arjoon and colleagues (*24*) explore the benefits and shortcomings of multiple methods for establishing the validity and reliability of data in the development of assessment measures.

Measuring What and How Much Chemistry Is Learned

The first level of analysis is simply reporting the percentages of students who have a scientifically correct understanding, and the percentages of students who choose each of the major misconceptions. Means with standard deviations, medians with ranges, and histograms are all useful methods for reporting data—for all of the items on the inventory as a whole, as well as for each individual item.

When two-tier items are used, percentages can be stratified by the response to both the *what* question and the *why* question. For example, if the *what* question has 4 possible answers, and the *why* question has 4 possible answers, there are 16 possible response patterns (see Table 1). With an instrument consisting of a dozen or more questions, the number of unique response options chosen by students quickly multiplies. Cluster analysis is a useful tool for distinguishing among the common reasoning patterns or models used by students amongst such a large number of possible responses.

Table 1. Sixteen unique possible response patterns for students to two-tier questions with four responses each

		What? (Tier 1)			
		A	B	C	D
Why ? (Tier 2)	A	1	2	3	4
	B	5	6	7	8
	C	9	10	11	12
	D	13	14	15	16

If the assessment is given to multiple demographics (e.g., students in general chemistry and students in organic chemistry, majors vs. non-majors, etc.), then each of these analyses may warrant comparisons across demographics, assuming the sample size is large enough to justify such comparisons. In some studies, such as the development of the CCI by Mulford and Robinson (*13*), researchers purposefully chose to exclude students who were repeating the course from their data analysis.

Pre-post designs to measure "value added" or knowledge gains can be done with anonymous data by reporting the means for the entire data set and creating histograms of pre-scores and post-scores. However, if students provide identifying information of some kind (email address, a 4 digit code, etc.), then student data can be paired from pre- to post-test, allowing the calculation of gain at the level of each individual student. Scatter plots of pre-scores vs. post-scores can identify students who improved and students who declined. Normalized gains (*38*) can be calculated to determine what fraction of the possible gain was achieved. Rasch analysis, which simultaneously examines item difficulty and student ability, has also been used to examine learning gains in chemistry students (*39, 40*). Determining if differences or gains are statistically significant requires paying careful attention to sample size, establishing equivalence of samples, power, and the reporting of effect sizes. Readers are directed to Lewis and Lewis (*41*) for a detailed, thorough discussion of the common errors when using a t-test when trying to establish statistically significant differences in chemistry education research.

One recent report in the literature (*42*) cautions that the act of asking students to respond to assessment tools that contain not just incorrect answers, but also misconceptions, as responses can in fact be generative of misconceptions in those students. More research is needed to explore the generalizability of this finding.

Recommendations for Classroom Teachers

While interviewing students can provide a rich data set and numerous insights into their thinking, it is not a practical choice for most teachers. An individual interview can easily last 30–60 minutes, even when focused on limited content or specific representations. Transcribing and analyzing multiple interviews to find themes or patterns in the students' thinking takes a great deal of time. Therefore, using assessment tools that are grounded in the analysis of students' open-ended responses provides chemistry teachers with a practical, more efficient alternative to access both the range and prevalence of their students' thoughts. These tests can be administered as paper-and-pencil tests and are easy to score. Data can be collected in a lecture or classroom setting, or administered to students in the laboratory setting to answer before beginning their experiment for the week.

These assessment tools could be administered as a diagnostic to measure what incorrect prior knowledge students bring with them into a course from life experience and/or previous instruction. As Villafañe and colleagues note, "students' incorrect ideas from previous courses…could hinder their learning … since they would be unable to correctly apply their knowledge to new contexts." ((*14*), p. 210) However, simply knowing what incorrect ideas a classroom of students holds about the behavior of atoms and molecules is not enough. A teacher cannot simply tell students their ideas are ill-informed and proceed to teach as though such ideas can be replaced with expert knowledge. Once a teacher is aware of what her students already know, she must design instruction accordingly (*43*). Students need to encounter discrepant events (*44*) to realize the inadequacy of their thinking and to construct more powerful models.

These assessments can also be used to measure what students learn as a result of instruction by measuring students' understanding with a post-test, or perhaps even the gain measured by the differences between pre- and post-administrations of the inventory. Collecting data on students' misconceptions after instruction provides one measure of the quality of instruction, especially if an intervention was designed to help students confront and re-structure their thinking based on the results of measuring their prior knowledge. Comparisons can be made within the same semester for one group of students, or from one year to the next using a historical control to measure gains due to changes in pedagogy or curriculum.

Recommendations for Chemistry Education Researchers

As researchers consider what new assessment tools are warranted (e.g., content areas where student understanding is not yet reported in the literature), collaboration with professional societies to identify the most foundational and cross-cutting ideas will be important. The eight concepts that frame Villafañe and colleagues' work were cited by the American Society for Biochemistry and Molecular Biology (*45*) as essential for students to learn in a biochemistry course.

The ACS Examinations Institute has recently published their methodology for creating the Anchoring Concepts Content Map (ACCM) (*46*). Each ACCM consists of 10 big ideas that cut across the entirety of chemistry, followed by enduring understandings, sub-disciplinary articulations, and specific content details. The ACCMs for general chemistry (*47*) and organic chemistry (*48*) have been published. ACCMs for physical chemistry, analytical chemistry, and inorganic chemistry are under development and could be used to explore students' misconceptions and development assessment items using any of the methods described above.

Regardless of the content focus of new instruments and research studies, chemistry education researchers need to pay careful attention to methodological choices and analytical decisions. This chapter outlines several choices for researchers with regard to eliciting students' ideas and item design. While there is no "one right way" to design a concept inventory, design choices do shape the validity and reliability of the data and the claims that researchers are able to make on the basis of their data.

Appendix
Chemistry Diagnostic Assessments and Concept Inventories

Atomic Emission and Flame Tests

- Mayo, A. V. Atomic Emission Misconceptions as Investigated through Student Interviews and Measured by the Flame Test Concept Inventory. Ph.D. Dissertation, Miami University, Oxford, OH, 2013.

Biochemistry (multiple concepts)

- Villafañe, S.; Bailey, C.; Loertscher, J.; Minderhout, V.; Lewis, J. E. *Biochem. Molec. Biol. Educ.*, **2011**, *89*, 102-109.

Bonding

- Luxford, C. J.; Bretz, S. L. *J. Chem. Educ.*, **2014**, 91(3), 312-320.
- Peterson, R. F.; Treagust, D. F.; Garnett, P. *Res. Sci. Educ.*, **1986**, *16*, 40-48.

Chemical Reactions/Light/Heat

- Artdej. R.; Ratanaroutai, T.; Coll, R.; Thongpanchang, T. *Res. Sci. Teach. Educ.*, **2010**, *28*(2), 167-183.
- Chandrasegaran, A. L.; Treagust, D. F.; Mocerino, M. *Chem. Educ. Res. Pract.*, **2007**, *8*, 293-307.
- Jensen, J. D. Students' Understandings of Acid-Base Reactions Investigated through their Classification Schemes and the Acid-Base

Reactions Concept Inventory. Ph.D. Dissertation, Miami University, Oxford, OH, 2013.
- Linke, R. D.; Venz, M. I. 1978. *Res. Sci. Educ.*, **1979**, *9*, 103-109.
- Wren, D.; Barbera, J. *J. Chem. Educ.*, **2013**, *90*(12), 1590-1601.

Enzyme-Substrate Interactions

- Bretz, S. L.; Linenberger, K .J. *Biochem. Molec. Biol. Educ.*, **2012**, *40*(4), 229-233.

Equilibrium

- Banerjee, A. C. *Int. J. Sci. Educ.*, **1991**, *13*(4), 487-494.
- Voska, K. W.; Heikkinen, H. W. *J. Res. Sci. Teach.*, **2000**, *37*(2), 160-176.

General Chemistry (multiple concepts)

- Krause, S.; Birk, J.; Bauer, R.; Jenkins, B.; Pavelich, M. 34th ASEE/IEEE Frontiers in Education Conference, October 20-23, 2004, Savannah, GA.
- Mulford, D. R.; Robinson, W. R. *J. Chem. Educ.*, **2002**, *79*(6), 739-744.

Inorganic Qualitative Analysis

- Tan, K. C. D.; Khang, N. G.; Chia, L. S.; Treagust, D. F. *J. Res. Sci. Teach.*, **2002**, *39*(4), 283-301.

Ionization Energy

- Chan, K.-C. D.; Taber, K. S.; Goh, N.-K.; Chia, L.-S. *Chem. Educ. Res. Pract.*, **2005**, *6*, 180-197.

Organic Acid Strength

- McClary, L. M.; Bretz, S. L. *Int. J. Sci. Educ.*, **2012**, *34*(5), 2317-2341.

Particulate Nature of Matter

- Nyachwaya, J. M.; Mohamed, A.-R.; Roehrig, G. H.; Wood, N. B.; Kern, A. L.; Schneider, J. L. *Chem. Educ. Res. Pract.*, **2011**, *12*, 121-132.

Redox Reactions

- Brandriet, A. R.; Bretz, S. L. *J. Chem. Educ.*, **2014**, in press

Structure of Matter/Changes of State/Solubility/Solutions

- Adadan, E.; Savasci, F. *Int. J. Sci. Educ.*, **2012**, *34*(4), 513-544.

•   Linke, R. D.; Venz, M. I. *Res. Sci. Educ.*, **1978**, *8*, 183-193.

References

1.   Bretz, S. L. A Chronology of Assessment in Chemistry Education. In *Trajectories of Chemistry Education Innovation and Reform*; Holme, T., Cooper, M. M., Varma-Nelson, P., Eds.; ACS Symposium Series 1145, American Chemical Society; Washington, DC, 2013; Chapter 10.

2.   Cornog, J.; Colbert, J. C. *J. Chem. Educ.* **1924**, *1*, 5–12.

3.   American Chemical Society Examinations Institute (ACS Exams). http://chemexams.chem.iastate.edu/about/short_history.cfm (accessed April 12, 2014).

4.   Bauer, C. F.; Cole, R. S.; Walter, M. F. Assessment of Student Learning: Guidance for Instructors. In *Nuts and Bolts of Chemical Education Research*; Bunce, D. M., Cole, R. S., Eds. ACS Symposium Series 976; American Chemical Society, Washington DC, 2008; Chapter 12.

5.   Libarkin, J. Concept Inventories in Higher Education Science. In *Promising Practices in Undergraduate Science, Technology, Engineering, and Mathematics Education: Summary of Two Workshops*; Proceedings of the National Research Council's Workshop Linking Evidence to Promising Practies in STEM Undergraduate Education, Washington, DC, October 13−14, 2008.

6.   National Research Council. *Promising Practices in Undergraduate Science, Technology, Engineering, and Mathematics Education: Summary of Two Workshops*; National Academies Press: Washington, D.C., 2011.

7.   Wandersee, J. H.; Mintzes, J. J.; Novak, J. D. Research on Alternative Conceptions in Science. In *Handbook of Research on Science Teaching and Learning*; Gabel, D., Ed.; Macmillan Publishing Co.: New York, 1994; pp 177−210.

8.   Merrill, R. J.; Ridgway, D. W. *The CHEM Study Story*; W.H. Freeman: San Francisco, 1969.

9.   Strong, L. E.; Wilson, M. K. *J. Chem. Educ.* **1958**, *35* (2), 56–58.

10. Davenport, D. A. *J. Chem. Educ.* **1970**, *47* (4), 271.

11. Piaget, J. *The Child's Conception of the World*; Routledge: London, 1929.

12. Peterson, R. F.; Treagust, D. F.; Garnett, P. *Res. Sci. Educ.* **1986**, *16*, 40–48.

13. Mulford, D. R.; Robinson, W. R. *J. Chem. Educ.* **2002**, *79* (6), 739–744.

14. Villafañe, S.; Bailey, C.; Loertscher, J.; Minderhout, V.; Lewis, J. E. *Biochem. Molec. Biol. Educ.* **2011**, *89*, 102–109.

15. Bretz, S. L.; Linenberger, K. J. *Biochem. Molec. Biol. Educ.* **2012**, *40* (4), 229–233.

16. Treagust, D. F. *Intl. J. Sci. Educ.* **1988**, *10* (2), 159–169.

17. Taft, H. *J. Chem. Educ.* **1992**, *67*, 241–247.

18. Spencer, J. N. *J. Chem. Educ.* **1992**, *69*, 182–186.

19. Johnstone, A. H. *J. Chem. Educ.* **2010**, *87* (1), 22–29.

20. Linenberger, K. J.; Bretz, S. L. *Chem. Educ. Res. Pract.* **2012**, *13*, 172–178.

21. Strauss, A.; Corbin, J. *Basics of Qualitative Research: Techniques and Procedures for Developing Grounded Theory*, 3rd ed.; Sage Publications, Inc.: Thousand Oaks, CA: 2007.

22. Linenberger, K. J.; Bretz, S. L. *J. Coll. Sci. Teach.* **2012**, *42* (1), 45–49.

23. Emenike, M.; Raker, J. R.; Holme, T. *J. Chem. Educ.* **2013**, *90* (9), 1130–1136.

24. Arjoon, J. A.; Xu, X.; Lewis, J. E. *Chem. Educ.* **2013**, *90* (5), 536–545.

25. Tamir, P. *J. Biol. Educ.* **1971**, *5*, 305–307.

26. Tamir, P. *Int. J. Sci. Educ.* **1990**, *12* (5), 563–573.

27. Caleon, I.; Subramaniam, R. *Res. Sci. Educ.* **2010**, *40*, 313–337.

28. Caleon, I.; Subramaniam, R. *Int. J. Sci. Educ.* **2010**, *32* (7), 939–961.

29. McClary, L. M.; Bretz, S. L. *Intl. J. Sci. Educ.* **2012**, *34* (5), 2317–2341.

30. Guba, E. G.; Lincoln, Y. S. *Fourth Generation Evaluation*; Sage Publications, Inc.: Thousand Oaks, CA, 1989.

31. Lincoln, Y. S.; Guba, E. G. *Naturalistic Inquiry*; Sage Publications, Inc.: Thousand Oaks, CA, 1985.

32. Cronbach, L. J. *Psychometrika* **1951**, *16*, 197–334.

33. Adams, W. K.; Weiman, C. E. *Int. J. Sci. Educ.* **2011**, *33* (9), 1289–1312.

34. Lasry, N.; Rosenfield, S.; Dedic, H.; Dahan, A.; Reshef, O. *Am. J. Phys.* **2011**, *79* (9), 909–912.

35. Everitt, B. S.; Landau, S.; Leese, M.; Stahl, D. *Cluster Analysis*, 5th ed.; Wiley: West Sussex, 2011.

36. Jensen, J. D. Students' Understandings of Acid-Base Reactions Investigated through Their Classification Schemes and the Acid-Base Reactions Concept Inventory, Ph.D. Dissertation, Miami University, Oxford, OH, 2013.

37. Ding, L.; Beichner, R. *Phys. Rev. ST Phys. Educ. Res.* **2009**, *5*, 1–17.

38. Hake, R. R. *Am. J. Phys.* **1998**, *66* (1), 64–74.

39. Pentecost, T. C.; Barbera, J. *J. Chem. Educ.* **2013**, *90*, 839–845.

40. Herrmann-Abell, C. F.; DeBoer, G. E. *Chem. Educ. Res. Pract.* **2011**, *12*, 184–192.

41. Lewis, S. E.; Lewis, J. E. *J. Chem. Educ.* **2005**, *82* (9), 1408–1412.

42. Chang, C.-Y.; Yeh, T.-K.; Barufaldi, J. P. *Int. J. Sci. Educ.* **2010**, *32* (2), 265–282.

43. Ausubel, D. P.; Novak, J. D.; Hanesian, H. *Educational Psychology: A Cognitive View*, 2nd ed.; Werbel & Peck: New York, 1978.

44. von Glaserfeld, E. A Constructivist Approach to Teaching. In *Constructivism in Education*; Steffe, I. P., Gale, J., Eds.; Erlbaum: Hillsdale, NJ, 1995; pp 3–15.

45. Voet, J. G; Belle, E.; Boyer, R.; Boyel, J.; O'Leary, M; Zimmerman, J. *Biochem. Molec. Biol. Educ.* **2003**, *31*, 161–162.

46. Murphy, K.; Holme, T.; Zenisky, A.; Caruthers, H.; Knaus, K. *J. Chem. Educ.* **2012**, *89* (6), 715–720.

47. Holme, T.; Murphy, K. *J. Chem. Educ.* **2012**, *89* (6), 721–723.

48. Raker, J.; Holme, T.; Murphy, K. *J. Chem. Educ.* **2013** (90) (11), 1443–1445.

Chapter 10

Measuring Knowledge: Tools To Measure Students' Mental Organization of Chemistry Information

Kelly Y. Neiles*

Department of Chemistry, St. Mary's College of Maryland, St. Mary's City, Maryland 20686, United States
*E-mail: kyneiles@smcm.edu

The selection of tools to measure students' knowledge in chemistry is an incredibly important but difficult step in the designing of chemistry education research studies. While traditional content tests can provide information as to students' understanding of facts, they often miss the nuances in students' understanding of the complex relationships between the topics in chemistry. Tools that measure students' structural knowledge of chemistry concepts rather than their factual knowledge may provide richer data for researchers to utilize and interpret. This chapter will describe the use of measurement tools that create network representations of students' structural knowledge of chemistry concepts as a way to assess and better understand students' chemical knowledge.

Introduction

When designing a chemistry education research study one of the many important decisions a researcher will make is what student outcomes he or she will investigate. The outcomes desired drive the selection or creation of valid measurement tools used to measure these outcomes. In chemistry education research, this often involves the choice of an indicator of students' understanding or learning of chemistry concepts. Depending on the research question being evaluated, the researcher may need a measure of change in students' understanding or of gains in students' learning of chemistry concepts.

Students' understanding and knowledge of chemistry concepts are usually measured through the use of traditional content tests. These tests are often found in the form of multiple choice or open-ended questions. While these testing procedures provide important insights about the students' declarative knowledge (knowledge of the facts within a concept), they may not provide a complete picture of students' understanding in chemistry. Take for instance the following test question:

1. Select the best Lewis structure for NH_4^+.

$$
\text{a.} \quad \left[\begin{array}{c} H \\ | \\ H-N=H \\ | \\ H \end{array} \right]^{+}
$$

$$
\text{b.} \quad \left[\begin{array}{c} H \\ .| \\ H-\overset{..}{N}-H \\ | \\ H \end{array} \right]^{+}
$$

$$
\text{c.} \quad \left[\begin{array}{c} H \\ | \\ H-N-H \\ | \\ H \end{array} \right]^{+}
$$

Students who choose the correct answer (C) may do so because they have an accurate understanding of the chemistry concept (drawing Lewis dot structures, polyatomic ions, formal charges, etc.). They may, however, select that answer because they believe that molecules will form certain structures because they 'like to be balanced or symmetrical', a common misconception identified in a study of Lewis structures by Cooper, Grove, Underwood, and Klymkowsky (1). This misconception may lead students to select the correct answer on this particular question, but may result in difficulties answering future questions. From the scoring of this test a researcher may infer that the student's knowledge of this chemistry concept was complete and accurate, though this may not be the case. This disconnect between the measurement instrument and the student's chemical understanding may call for the use of an alternative measurement that provides further detail of the student's understanding, such as a measurement tool that evaluates the way the student stores his or her chemical knowledge. The use of an alternate measurement method that probes the details of a student's mental organization may result in more complete data being collected on the student's understanding. This could in turn provide more in-depth insights into the chemistry learning process.

Unfortunately, there are currently no instruments that can create an exact replication of a student's understanding or mental organization of chemistry concepts. Unlike many other areas of chemistry research, the sample (student) can't be placed in an instrument to provide a readable output for interpretation. The selection of research methods to investigate a student's understanding of chemistry is a very complex process that must be considered carefully so that the inferences made from the data are valid and reliable. This problem of how to measure a student's understanding and mental organization of chemical knowledge has long plagued the education researcher. To make informed decisions on the selection of measurement tools to investigate a student's understanding of chemistry concepts, chemistry education researchers must first recognize how the student represents or stores the information in his or her mind.

Structural Knowledge

A description of a student's mental organization of information, also known as structural knowledge, is described by Mayer's (2) theory of schemas, which includes four underlying points. First, the concept of schema (structural knowledge) is general. It can be applied to a wide variety of situations as a framework for understanding incoming information. In other words, we create structural knowledge in a variety of situations and with many different topics. Second, a schema is a description of knowledge. It exists in memory as something that a student knows. Third, a schema has structure that is organized around a theme or concept. Finally, a schema contains 'slots' that are filled by specific pieces of information. These four points describe the basis for how a student creates structural knowledge used to store information long term for later acquisition and use. These points also allow chemistry education researchers to use a student's structural knowledge as a measure of his or her understanding of chemistry information.

The theories of structural knowledge describe the processes people utilize to remember and use knowledge. In his theory of structural knowledge, Bartlett (3) describes the act of acquiring new information as requiring the student to assimilate new material into his or her existing concepts. The outcome does not result in a duplication of the new information, but instead a new product in which the student's current structural knowledge and incoming information are combined into something that is meaningful to the student. In this process, the new information, previous knowledge, context, personal experiences, and current goals of the student all come into play in the altering of the student's structural knowledge. The student changes the new information to fit his or her existing concepts, or changes his or her existing concepts to accommodate the new information. When these changes occur, details of the original information may be lost as the knowledge becomes more coherent to the individual.

In his paper on a model of learning as conceptual change, Hewson (4) investigated the conditions under which a person holding a set of conceptions on a topic would change these conceptions as a result of being confronted by new experiences. The conceptions would be altered either by incorporating these experiences or replacing them because of their inadequacy. The model proposed by Hewson emphasizes the importance of the person's existing understanding of the topics and the role these preconceptions play in the assimilation or accommodation of new information.

As expertise in a domain grows, through learning and experience, the elements of structural knowledge become increasingly interconnected (5, 6). Someone who is highly knowledgeable in a certain topic would thus have a highly integrated structural knowledge of the concepts within that topic. When a student learns new chemical information, his or her structural knowledge of the information changes. As a student gains more chemical knowledge, he or she alters the structural knowledge to accommodate the new information. We would thus expect a student's structural knowledge to become more like that of an expert as the student increases his or her chemical knowledge (7, 8). The changing of a student's structural knowledge may therefore be used as a measurement of the changes in understanding of chemistry concepts.

A measurement of student's structural knowledge may provide a more complete picture of the student's understanding of chemistry concepts than a purely fact-based content test. Measuring structural knowledge leads to a better understanding of the connections and hierarchical structure the student uses to store the chemical information for later retrieval and use. Unfortunately, unlike factual knowledge that can be measured using traditional content exams, a student's structural knowledge is much more difficult to access and measure. It requires the creation of networks that represent the student's understanding of various topics.

Network representations have been widely used in various areas of cognitive science as measures of memory retrieval and human performance (9–11). In these studies, a network representation of student knowledge is described as a graph or representation that includes two major components: points and lines. The points, often referred to as nodes or vertices, represent the main idea or key terms present in the concept. In Figure 1, the nodes included are Chemical Bonding, Ionic Bonding, Ions, etc. The lines represent connections between the nodes. In other published research reports, the lines may also be referred to as edges, links, bonds, ties, and/or relations.

Once created, network representations of a student's chemical knowledge can be used to assess the student's understanding of chemistry topics or used to assess any changes in his or her understanding due to some type of intervention. Before we discuss how to utilize these networks, we must determine how to create the network representations in the first place. The next section of this chapter will investigate two methods for creating network representations of a student's structural knowledge: concept mapping and proximity data techniques.

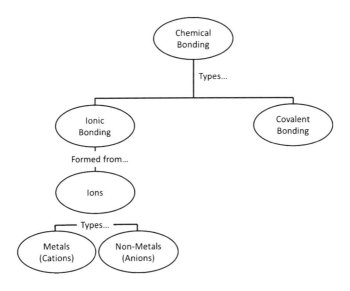

Figure 1. Example of a portion of a network representation of chemical bonding.

Concept Mapping

Concept maps have been shown to provide a measure of the structure of a student's knowledge in a certain topic or subject area (*12, 13*). The concept mapping principle works on the theories described above that describe knowledge as being structural in nature. Concept mapping can capture that structure in a graphical or network representation (*8, 13, 14*).

A concept map is a graphical representation consisting of nodes and lines, (*7*). In this representation, nodes are labels for important concepts (often keywords or terms) in a certain topic. The lines represent a relationship between a pair of nodes. The label on a line, if included, outlines how the two concepts are related.

As a student's understanding of a chemistry topic grows, his or her knowledge of the elements (nodes) in the topic become increasingly more interconnected (*5, 6*). We would thus expect a concept map created by a student with a full understanding (sometimes referred to as expertlike) of a chemistry concept to be more interconnected than that of a student with a novice understanding of the concept (*7*). This assumption allows concept maps to be used as an assessment of the completeness of a student's understanding of chemistry concepts. In a study by Francisco, Nahkleh, Nurrenbern, and Miller (*15*), the researchers investigated the connectedness of students' understanding in chemistry by investigating concept maps created by the students. Figure 2 shows examples from this study of a high quality concept map (highly connected nodes) and a low quality concept map (low number of connections.

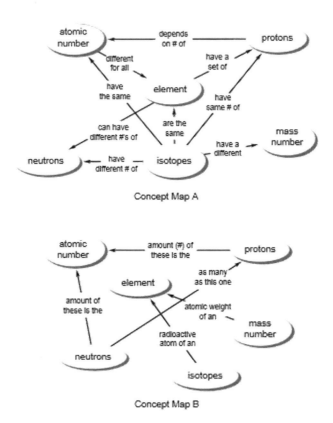

Figure 2. Examples of concept maps drawn by chemistry students. From Francisco, et. al. (15). Concept map A shows a connected knowledge structure; concept map B shows a lack of connections between closely related concepts. Reprinted with permission from Francisco, J.S., Nahkleh, M.B., Nurrenbern, S.C., Miller, M.L., J. Chem. Ed., 2002, 79, 248. Copyright 2002 American Chemical Society.

In this study, the researchers were able to assess the quality of students' understanding of the chemistry topics by evaluating the links present in the student-created concept maps. By studying the concept maps, the researchers were able to investigate the connectedness of students' understanding, something they may not have been able to determine using traditional measurements such as multiple choice tests.

Ruiz-Primo, Shavelson, and Schultz (7) describe three components that must be present for the use of concept maps as measures of student understanding, namely: 1) a task that invites the student to provide evidence of his or her knowledge structure of a chemistry concept; 2) a format for the student's response; and 3) a scoring system by which the student's concept map can be evaluated accurately and consistently. Without any one of these three components, the use of concept mapping cannot be considered assessment. There are, however, many variations in the use of concept mapping as assessment. These variations can be found in the tasks that the student is asked to complete, the format of the concept maps the student is asked to create, or the method of evaluation used to score the student's concept maps. See Ruiz-Primo, Shavelson, and Schultz (7) for a detailed explanation of the different types of concept mapping tasks used for assessments based on variations of the three components.

In the field of chemistry education, concept mapping has been used by many as a teaching pedagogy to help student understanding (16–18). The manner described here involves a shift in the perception of concept mapping tasks from learning tasks to tasks that can be used as an assessment. This shift was exemplified by the use of concept maps in the study by Francisco, et. al. (15). In this study the researchers used concept maps created by the students as an assessment of the students' understanding of chemistry concepts and how these understandings changed through the use of alternative study and assessment techniques. The concept maps were constructed by students as pre and post laboratory assignments. The researchers identified the links in the students' concept maps as correct, correct but noninformative, incorrect, or duplicate. These codes were then used to evaluate the concept maps through the following scoring algorithm:

$$\frac{\#\ correct\ (links) - \#\ wrong\ or\ noninformative\ (links)}{total\ \#\ of\ connections\ made}$$

Through the use of concept maps as an assessment, the researchers found that the conceptual understanding was a factor in the students' performance and that the alternative study methods described appeared to enhance students' ability to correctly solve complex problems. By using concept maps as assessments as seen in the study described here, researchers can create more detailed pictures of a student's understanding of chemistry topics than a content test can provide.

The use of concept mapping can, however, pose certain problems for the researcher. One issue is that the student must be able to evaluate his or her own understanding of a chemistry concept and try to reflect that understanding in the concept map task. This degree of reflection is often something that must be taught to the student through modeling by the instructor (this modeling should occur throughout the course multiple times and in different situations). The student also has to subjectively evaluate his or her own understanding and recreate that understanding in some type of concept map format (18). This introduces a degree of subjectivity into the data collection even if that subjectivity is coming from the students themselves.

Another point of subjectivity comes from the researchers' interpretations in scoring the concept map created by the students (18). Even when a strict rubric is utilized, as it should be, there is a degree of subjectivity in the evaluation of the concept maps. The researcher must infer the meaning behind the student's choices in the concept map and place value on the connections the student chooses to create. These difficulties in the use of concept maps have led to a search for a more objective measurement of students' structural knowledge. One result of this search was a method of creating network representations of students' understanding involving the use of something called proximity data, which is described in the next section.

Proximity Data Techniques

Proximity data can be used as an alternative to concept mapping when the researcher seeks a more objective data collection process because it takes both the student's and the researcher's interpretation of the student's understanding out of the creation of the network representation (*19*). Proximity data can be collected in a number of different ways. Essentially any time you can identify connections between people, places, ideas, concepts, etc., you can create proximity data. The amount of data collected in these methods often lends itself to electronic data collection methods that allow the data to be collected quickly, efficiently, and on a large scale.

One method of collecting proximity data that can be very useful in education research involves asking the student to make relationship and similarity judgments about a set of key terms that will later be used in the network as nodes. These judgments could be made in a number of different ways as outlined in Table 1.

Table 1. Proximity Data Collection Methods

Data Collection Method	*Procedure*
Pair-wise	Key terms are shown to the student in pairs. The student is asked to judge the relatedness of each pair of words on a Likert scale from 1 (completely unrelated) to 9 (completely related).
List-wise	On the right side of the screen the student is shown a list of key terms. On the left side of the screen a single key term appears. The student is then asked to select which of the key terms from the list (right side), the key term (left side) is most related to.
Clustering	The student is shown a computer screen with the key terms positioned in random order around the screen. The student is then asked to drag-and- drop the key terms so that their spacing from one another indicates the relationships the student believes the key terms to have with one another. The student is instructed to drag related terms closer together and unrelated terms farther apart.

Each of the methods described in Table 1 results in a set of proximity data. There are an infinite number of additional methods that could lead to proximity data appropriate for creating structural knowledge networks, therefore, an exhaustive list could not reasonably be included in this chapter. The three methods described in Table 1 were chosen to represent the most widely used data collection methods in education research. There are pros and cons to each of the methods described here, which are addressed more fully in a paper by Clariana and Wallace (*19*). The method chosen for data collection should optimize the creation of valid networks yet still work within the specifications and limitations of the study. The proximity data created through these methods will then be used by a computer program to create a network representation of the student's structural knowledge.

In a study by Acton, Johnson, and Smith (8), the researchers evaluated the ability of referent knowledge structures to discriminate subjects at different levels of domain expertise and to predict student performance on standard classroom measures. Specifically, they were interested in whether individual instructors, individual non-instructor experts, averaged experts, or averaged good students' structural knowledge networks were the best predictors of student performance. They created proximity data for each of these groups by having them complete a pair-wise relatedness task for the topics of interest to the study. The list of keyterms used in this task included 24 separate terms which resulted in 276 unique key term pairings. Participants were instructed to make their judgments relatively quickly (5-10 seconds) and extensive deliberation was discouraged. This resulted in the pair-wise relatedness task taking around one hour to complete. The proximity data was then transformed into a structural knowledge network by a computer program called Pathfinder (this study will be discussed further in the quantitative analysis of networks section of this chapter).

A computer program then algorithmically transforms the proximity data into network representations of structural knowledge. In this chapter, the Pathfinder program is used as an example to describe this process (20). The algorithm used by the Pathfinder program organizes data by eliminating those links that are not the minimum path between two concepts. Nodes in the Pathfinder network can be linked directly to one another or linked indirectly through a multi-node path. The pathfinder algorithm searches through the nodes to find the closest direct path between nodes. A link remains in the network only if it is the most direct path between two concepts. The most direct path could be a direct node-to-node link, or a multi-node path. As long as it is the shortest path, the link remains in the network. All other links between the two concepts are removed from the network by the computer program. The transformation from proximity data (in this example determined by the pair-wise task described above) to a Pathfinder network is illustrated below in Figure 3 with a data set taken from Neiles (21). In this figure, A-E represent various nodes or key terms within a chemistry concept and the network represents the relationships between those nodes within the students' mind.

In the pair-wise task, a student is asked to judge the relatedness of two words on a scale of 1 (completely unrelated) to 9 (completely related). These values are then subtracted from 10 so that a lower value (shorter connection) reflects a stronger relationship. These are the values shown in the proximity data table in Figure 3. The Pathfinder network of Figure 3 is the result of analyzing the proximity data using the algorithm in the Pathfinder program. In the network, direct links exist between nodes that are most closely related, for example between A and B or B and C. The length of these links also represents the strength of these connections. For example the connection between B and C is shorter then that between A and D representing a stronger connection between B and C. Those links that represent a multi-nodal path have the weakest proximity values, for example B and D or C and E. Through this process, Pathfinder is able to determine the most basic representation of the student's structural knowledge. The resulting network represents the student's knowledge of the relevant topic and how the key terms within that topic are linked together through relationships. One benefit of this process is that it does not force the student to create a hierarchical solution, however if a hierarchical representation exists in the student's knowledge structure, it will be included. The resulting network can then be used for an analysis of the student's structural knowledge of chemistry concepts.

Proximity Data

	A	B	C	D	E
A	0	1	5	3	4
B	1	0	2	6	7
C	5	2	0	8	9
D	3	6	8	0	7
E	4	7	9	7	0

Pathfinder Network

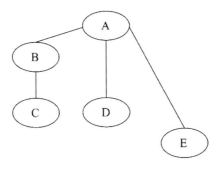

Figure 3. Creating a structural network from proximity data.

The data collection methods described here (concept mapping and proximity data techniques) are both valid techniques for creating network representations of a student's structural knowledge of chemistry concepts. The decision about which data collection technique to utilize depends on the research questions being evaluated and the constraints of the methodology of the study. For instance, if the research methodology does not allow for the students to have access to a computer, then concept mapping would be the better choice. Similarly, if objectivity of data collection is important to the research, then the proximity data method might be better. Once the networks are created, the analysis of these networks will be similar regardless of which data collection method was used (concept mapping or proximity data).

Network Analysis

Quantitative Network Analysis

Networks can be analyzed quantitatively by testing the network for internal consistency (coherency) and by comparing the network of interest (usually the student's network) to some other referent network. The referent network may be that of an expert, group of experts, or some other person with a degree of understanding of interest to the study (for instance a student who has performed successfully on the chemistry topic of interest in the past). A student's structural knowledge network can be compared to these referent networks to determine how similar or dissimilar the networks are to one another. The degree of similarity can be measured on a number of different variables (see Schvaneveldt (20) for a comprehensive list). In this chapter, three quantitative measurement variables will be discussed, namely, coherency, path length correlation, and neighborhood similarity, though others exist as well.

Coherency

Each structural knowledge network can first be analyzed for coherency, which is a reflection of the consistency of the data. The coherency of a set of proximity data is based on the assumption that the relatedness between a pair of items can be predicted by the relationship of the items to other items in the set. Coherency can range from a score of 0 (low coherency) to 1 (high coherency). Very low coherency (below 0.2) may indicate that the person whose data was used to create the network has a poor understanding of the chemistry concept (22). It can also be used as a validity measure when using expert networks as referent networks. If someone who has been identified as an expert creates a network with low coherency, then the researcher should seriously consider whether that person's data should be included in expert data. The incoherency of a network may indicate that the person does not have a full understanding of the concepts being tested.

Path Length Correlation

Path length correlation is a measure of the similarity of two networks (typically a student network and a referent network) based on the presence and strength of connections among nodes within the networks. The two networks being analyzed will receive a higher path length correlation if they have a greater number of similar links among nodes and if those links have similar strengths associated with them (strength represents the degree to which the person believes the two nodes are related with stronger relationships given a greater weight). Path length correlations are determined by the Pathfinder program and result in a score from 0 (completely dissimilar networks) to 1 (completely similar networks). Therefore, a high path length correlation score would indicate that the two networks being analyzed are very similar. If these networks are that of a student and an expert, then the high path length correlation indicates that the student's structural knowledge is similar to the expert's. For a full description of the mathematical calculations involved in determining path length correlations see Schvaneveldt (*20*).

Neighborhood Similarities

Neighborhood similarities are a measure of the similarity of two networks based on the degree to which the same node in both networks is surrounded by similar nodes. Essentially this is a measure of whether the groupings of nodes (neighborhoods) are similar between the two networks. Neighborhood similarities are also determined by the Pathfinder program and result in a score from 0 (completely dissimilar networks) to 1 (completely similar networks). For a full description of the mathematical calculations involved in determining path length correlations see Schvanevedlt (*20*).

Using these measures of similarity (coherency, path length correlations, and neighborhood similarities) researchers can measure the completeness and quality of a student's structural knowledge of a chemistry concept. In a study conducted by the author, students' structural knowledge of two chemistry topics (atoms, ions, and molecules, and stoichiometry) were assessed using a referent expert network. The following 16 key terms were used for the stoichiometry topic:

Reaction Stoichiometry	Balanced	Conversion
Chemical equation	Unbalanced	Limiting
Reactant	Mole	Excess
Product	Avogadro's number	Theoretical yield
Subscript	Molar mass	Actual yield
		Percent yield

Figure 4 shows the rating program used in the study. For the 16 key terms, the participant would indicate the relatedness of 119 pairs.

Seven experts' networks were averaged to create a referent network for each topic so that no one expert would unduly influence the structure of the networks. The referent expert network for the Stoichiometry topic is shown in Figure 5.

Each student's Pathfinder network for the two topics was evaluated for coherency and compared to the referent expert network based on path length correlation and neighborhood similarity. An example of a student network is shown in Figure 6.

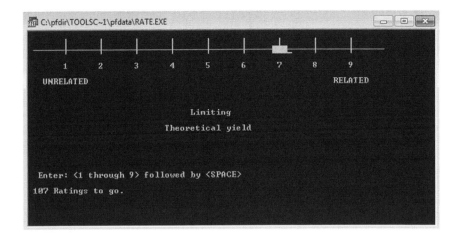

Figure 4. Rating program for collecting proximity data.

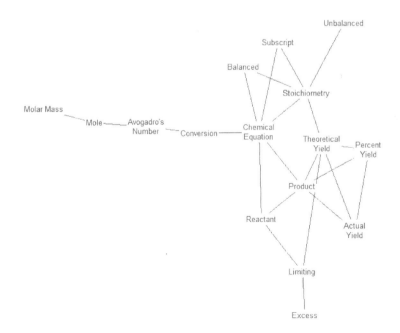

Figure 5. Averaged expert referent Pathfinder network.

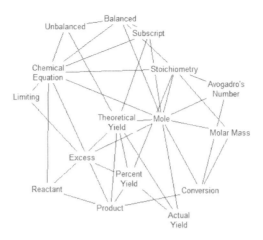

Figure 6. Student Pathfinder network for the stoichiometry topic.

After averaging these analyses across both chemistry topics, each student received three scores (coherency, path length correlation, and neighborhood similarity). These scores reflected the quality and 'expertlikeness' of the students' structural knowledge networks. Through this process the author was able to evaluate how the quality of a student's structural knowledge affected other cognitive processes, specifically his or her ability to read and understand chemistry texts. In using this method, the author did not have to evaluate the 'correctness' of the student's Pathfinder networks but instead compared them to the referent expert structures. The structural knowledge networks created using the Pathfinder program thus provided an objective measure of a cognitive process (chemical structural knowledge) that would otherwise have been difficult to evaluate.

In the study by Acton, Johnson, and Goldsmith (8) described previously, the researchers evaluated the ability of referent 'expert' networks to predict student performance (a description of the study is included in the 'proximity data techniques' section of this chapter). Once referent networks were created for each expert population (individual instructor, non-instructor individual expert, averaged experts, and averaged good students), the networks were compared for similarity. Each student's network was compared to and received a score for the degree of similarity between his or her network and each category of referent expert's networks. In this study, similarity was determined by evaluating neighborhood similarity (degree to which a concept has the same neighbors in two different networks). This resulted in each student receiving a separate similarity score for each expert category. These scores were then used as predictor variables for student performance on course exams. Three main conclusions were developed from the results of this investigation, namely, 1) instructor-based referents were equivalent to other experts in terms of predicting student exam scores, 2) there was substantial variability among experts, and 3) structures derived from both averaged experts and averaged best students provided valid referents, but the expert-based referent performed better in predicting student exam scores. This study, while not conducted in chemistry education, is a good example of how this type of assessment can be used to evaluate students' understanding or performance.

Both path length correlations and neighborhood similarities can be used to evaluate students' understanding of chemistry topics. They may also be used to group students into high, medium, and low performing groups by evaluating the distribution of scores within a study. When used as a grouping variable, these measures may be useful to evaluate differences in students' understanding or performance between groups.

The quantitative measures described here can provide important evaluation measures of the quality and completeness of a student's structural knowledge on a chemistry topic. These measures are based on the mathematical relationships underlying the nodes and links within the networks. In fact, the researcher need not even create a visualization of the student's structural knowledge such as those seen in Figure 5 and Figure 6 to perform this quantitative analysis. There are some research questions, however, that can only be answered by looking at the visualization of the student's networks. This evaluation involves using qualitative network analysis strategies in which the researcher interprets the visualizations of the student's structural knowledge networks instead of the mathematical relationships underlying the networks.

Qualitative Network Analysis

In qualitative network analysis, the researcher views visualizations of the student's structural knowledge like the networks shown in Figure 5 and Figure 6. The groupings and connections in these networks can be viewed and altered by the researcher to determine what relationships are present in the network and what they actually mean regarding the student's structural knowledge of the chemistry topic. These decisions about manipulating the representations are made by the researcher after extensive evaluation of the data through a qualitative data mining process (23). This process involves a spiraling approach where the researcher evaluates the nodes present multiple times and looks for re-occuring trends. Once a trend has been identified multiple times in the data, the researcher can state with some confidence that this is an overarching idea that may be used to group these nodes into a larger 'super node'. The researcher only makes decisions about interpreting the data based on this extensive data evaluation process or well developed theories from the literature. It is through this methodical process, as in all qualitative research, that potential bias imposed by the researcher is addressed and avoided.

One new open source program that can be used to evaluate networks qualitatively is GEPHI (24). GEPHI was created to help researchers evaluate networks with vast amounts of underlying data influencing the creation of the networks. For instance, a network in GEPHI may include hundreds or even thousands of nodes and tens of thousands of links. Though this program was created to accommodate these large networks, it is also a useful tool for manipulating the visualization of the smaller networks described in this chapter. GEPHI allows the researcher to undergo exploratory data analysis by investigating many different properties of the links and nodes within the network. For instance, this software allows the researcher to group a number of nodes together into a larger 'super node' so that the network becomes visually clearer. The ultimate goal of using GEPHI is to manipulate networks in such a way so that the researcher can create meaning from what he or she is seeing in the often dense networks. For example, consider the 77 node, 254 link GEPHI network in Figure 7.

This network could be a visualization of a student's structural knowledge of a chemistry concept. The visualization provides rich detail about the connections and relationships the student believes to be present between the nodes within the network. It may be, however, that this detailed network is too complex for easy interpretation. It may be easier to interpret this network if nodes grouped closely together are thought of as overarching 'super nodes'. A visualization of this interpretation can be seen in Figure 8.

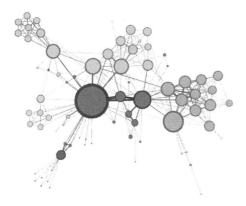

Figure 7. Visualization of network in GEPHI. (network is created using data from reference (24)).

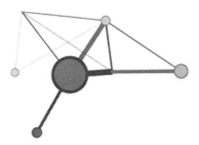

Figure 8. Visualization of GEPHI network after qualitative manipulation resulting in easier interpretation.

The visualization of the network shown in Figure 8 is based on the same data as that shown in Figure 7; however, the nodes in Figure 8 have been grouped together in the GEPHI program using different analysis tools. This results in a network with 7 'super nodes' and 12 links. The overarching 'super nodes' and links between them may be more interpretable than the network that contains 77 nodes. By manipulating the network using the qualitative analysis tool in GEPHI, the researcher is better able to interpret the student's structural knowledge representation.

There are currently no known studies in chemistry education research that explicitly utilize a program such as GEPHI in qualitative data analysis. There are studies in other fields, however, that illustrate the use of these methods and can viewed as exemplars. In a study by Kardes et. al. (25), the researchers used proximity data from social networking to investigate national funding in the United States to understand the collaboration patterns among researchers and institutions. The researchers used publically accessible grant funding information as proximity data. This resulted in a total of 279,862 entries for funded grants from 1976 to December 2011. The proximity data showed connections between institutions in three categories: PI collaborations, organization collaborations (between the organizations of the PIs), and state collaborations (between the states of the PIs). Networks were created for each of these three categories based on the proximity data collected. Due to time constraints, only the state network will be discussed in this chapter. The state networks created from the proximity data included 54 nodes and 1,289 edges (to illustrate how vast these networks can be the PI network included 104K nodes and 204K edges). Figure 9 illustrates the process the researchers went through to interpret the large amount of state data. Part a of Figure 9 shows that almost all of the nodes are well connected. Some states have many connections (indicated with a bold line). For instance there is frequent collaboration between New York (NY), California (CA), and Massachusetts (MA). The researchers then chose to analyze those states with frequent collaborations (part b). When they evaluated their data, they determined that if the number of collaborations was greater than 250 it would be considered a high number of collaborations. Part b of Figure 9 shows only those connections that represent 250 or more collaborations between those two states. The researchers created this new network using the GEPHI program.

Through this and other qualitative analysis of the data, the researchers found what they called the "six degrees of separation" in the state and organization collaboration networks. That is, they found large clusters of groups within the data, indicating researchers within the group tended to collaborate with other researchers within the group or in other large groups.

The type of analysis described in the Kardes, et. al. (25) study could be used in chemistry education research to evaluate large sets of student data. For instance, a large number of student-created networks on chemistry topics could be evaluated simultaneously to determine whether any clusters or largely weighted connections are present. These trends within the large data set could indicate widespread student understandings or misunderstandings of the chemistry concepts.

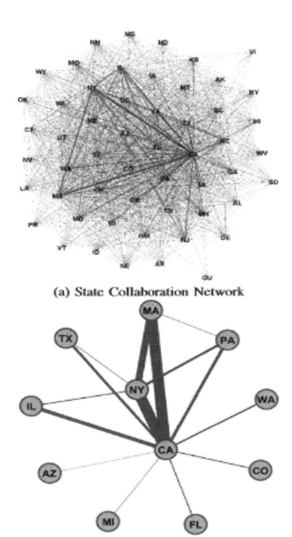

(a) State Collaboration Network

(b) State Frequent Collaboration Network

Figure 9. State collaboration networks from different perspectives. Reproduced with permission from reference (25).

The quantitative and qualitative network analysis methods described in this chapter can be used together or separately to investigate many aspects of students' structural knowledge. The choice of analysis methods should be driven by the research methodology and the research questions chosen for investigation in each study. Each analysis method can provide different but equally important insights into how students are storing chemical information for later retrieval and use.

Proximity data analysis can be a useful tool in that it provides a more objective measurement of students' understanding of chemistry concepts. The tasks necessary for collecting this data (for example relatedness judgments) can be cognitively taxing and time consuming for the students. However, the richness of the data collected using these methods may outweigh this downside, especially when a detailed understanding of students' structural knowledge is important to the goals of the chemistry education research.

Chemistry Education Research

Studying students' understanding of chemical concepts can often prove to be difficult. While a multiple choice test may seem like a simple test of students' content knowledge, it may not provide enough detail for the researcher to understand the nuances of the way students are storing this information. The measurement instruments discussed here provide more complex representations of the students' understanding that can better inform the researcher about how students are internalizing the chemical information. Tools that measure students' structural knowledge such as concept maps and proximity data techniques may provide important information about students' understanding that would not be found using traditional content tests.

As was shown here, concept maps and network analysis of proximity data can be used in chemistry education research to evaluate the depth and quality of students' structural knowledge. This type of analysis can provide both quantitative and qualitative data that could be used to evaluate students in many different types of chemistry education research studies. The analysis could involve comparing a student's network to a referent 'expert' network. It could also involve using students' path length correlation or neighborhood similarity scores to group students into high, medium, and low structural knowledge categories. These categories could then be compared on a number of variables such as final grades or use of online resources. Network analysis could also be used to determine if students' structural knowledge changes are due to an intervention. Students could be asked to complete a network creation task before and after an intervention. Quantitative or qualitative analysis could then be used to determine if any changes occurred due to the intervention.

Using network analysis to evaluate students' structural knowledge can provide important information to researchers interested in students' understanding in chemistry. These methods can be used by themselves or in conjunction with traditional content tests to provide rich data regarding the way students store chemical information. By using these methods, we can create research methodologies that investigate students' understanding in a deep and meaningful way.

Where to Find These Resources

Resource/ Method	References	Online Information	Cost
Concept Mapping	*7,18*	http://www.gradhacker.org/ 2011/11/07/tool-roundup-mind-mapping-software/ (This website has a good explanation of each mapping software (*27*)). There are also many free ipad/ipod/tablet apps that work well for concept mapping.	Many of the concept mapping resources are free or can be purchased for a fairly low price.
Proximity Data Collection	*19*	http://www.personal.psu.edu/rbc4/KUmapper.htm	This is free, open source software.
Pathfinder	*20*	http://interlinkinc.net/index.html	At the date of publication, Pathfinder costs approximately $200 ($100 for students).
GEPHI	*26*	https://gephi.org	This is free, open source software.

References

1. Cooper, M. M.; Grove, N.; Underwood, S. M.; Klymkowsky, M. W. *J. Chem. Ed.* **2010**, *87*, 869–874.
2. Mayer, R. E. *J. Res. Sci. Teach.* **1997**, *34*, 101.
3. Bartlett, F. *Remembering*; Cambridge University Press: Cambridge, U.K., 1932.
4. Hewson, P. W. *Eur. J. Sci. Educ.* **1981**, *3*, 383–396.
5. Glaser, R.; Bassok, M. *Annu. Rev. Psych.* **1989**, *40*, 631–666.
6. Shavelson, R. J. *J. Ed. Psych.* **1972**, *63*, 225–234.
7. Ruiz-Prima, M. A.; Shavelson, R. J. *J. Res. Sci. Teach.* **1996**, 569–600.
8. Acton, W. H.; Johnson, P. J.; Goldsmith, T. E. *J. Ed. Psych.* **1994**, *86*, 303–311.
9. Anderson *The Architecture of Cognition*; Harvard University Press: Cambridge, MA, 1983.
10. Collins, A. M.; Quillian, M. R. *J. Verb. Learn. Verb. Behav.* **1969**, *8*, 240–247.
11. Rumelhart, D. E., McClelland, J. L. *Parallel Distributed Processing: Explorations in the Microstructure of Cognition*; MIT Press: Cambridge, MA, 1986.

12. Lomask, M., Baron, J. B., Greig, J., Harrison, C. Paper presented at the *Annual Meeting of the National Association of Research in Science Teaching*, Cambridge, MA, 1992.
13. Novak, J. D. *J. Res. Sci. Teach.* **1990**, 937–949.
14. White, R. T., Gunstone, R. *Probing Understanding*; Falmer Press: New York, 1992.
15. Francisco, J. S.; Nahkleh, M. B.; Nurrenbern, S. C.; Miller, M. L. *J. Chem. Ed.* **2002**, *79*, 248.
16. Markow, P. G.; Lonning, R. A. *J. Res. Sci. Teach.* **1998**, *35*, 1015–1029.
17. Nakhleh, M. B. *J. Chem. Ed.* **1994**, *71* (3), 201–205.
18. McClure, J. R.; Sonak, B.; Suen, H. K. *J. Res. Sci. Teach.* **1999**, *36*, 475–492.
19. Clariana, R. B.; Wallace, PE. *Int. J. Instruc. Media* **2009**, *36*.
20. Schvaneveldt, R. W. *Pathfinder Associative Networks: Studies in Knowledge Organization*; Ablex Publishing Corporation: Norwood, NJ, 1990.
21. Neiles, K. Y. *An Investigation of the Effects of Reader Characteristics on Reading Comprehension of a General Chemistry Text*; A dissertation, 2012.
22. *Pathfinder 6.3 Software*, 2011.
23. Creswell, J. W. *Qualitative Inquiry and Research Design: Choosing Among Five Traditions*; Sage Publications, Inc.: Thousand Oaks, CA, 2012.
24. GEPHI, 2008 – 2011. Gephi Consortium. https://gephi.org.
25. Kardes, H., Sevinder, A., Gunes, M. H., Yuksel, M. *Proceeding of the 2012 International Conference on Advances in Social Networks Mining (ASONAM)*, pp 654–659.
26. Bastian, M., Heymann, S., Jacomy, M. *Gephi: An Open Source Software for Exploring and Manipulating Networks*; Association for the Advancement of Artificial Intelligence, 2009.
27. Zellner, A. *2011, Tool Roundup: Mind Mapping Software*, November 7, 2011. http://www.gradhacker.org/2011/11/07/tool-roundup-mind-mapping-software/.

Chapter 11

Eye Tracking Methodology for Chemistry Education Research

Katherine L. Havanki*,1 and Jessica R. VandenPlas2

1Department of Chemistry, The Catholic University of America, Washington, DC 20064, United States
2Department of Chemistry, Grand Valley State University, Allendale, Michigan 49401, United States
*E-mail: havanki@cua.edu

Eye tracking has long been a staple research tool in the field of psychology, and has recently begun to see applications in chemistry education research. This chapter will discuss the foundations of eye tracking research, including fundamentals of how eye tracking works, collection and analysis of eye tracking data, and research applications of eye tracking in chemistry education research and related disciplines.

Introduction

Advances in technology have made eye tracking more accessible to researchers from a variety of fields, including chemistry education. Recently, a number of eye tracking studies have begun to appear in the chemistry education research literature (1–4). With the ability to track both conscious and unconscious eye movements, this methodology provides researchers with valuable insights into the student experience that no other technique can capture. The goal of this chapter is to provide readers with a foundation to understand eye tracking studies, while also helping them decide if this technique is appropriate for their own research.

Basics of Vision

To begin any discussion of eye tracking, it is important to understand how visual information is processed. The human visual system takes in visual information from an object through the eyes (Figure 1), converts it to electrical impulses, and processes the impulses to generate an internal representation of the object in the brain.

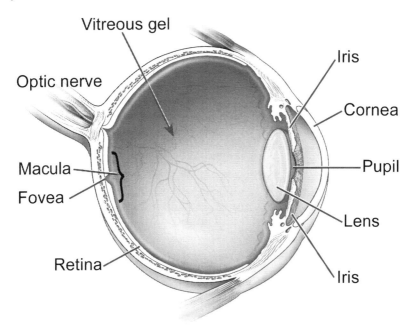

Figure 1. Structure of the human eye. Source: National Eye Institute/National Institutes of Health (NEI/NIH). http://www.nei.nih.gov/photo/eyean/images/ NEA09_72.jpg (accessed June 6, 2014).

Light rays reflected or emitted from an object enter the eye through the cornea. Light is refracted and passes through the pupil, the hole in the center of the colored part of the eye called the iris. The iris dilates and constricts the pupil, changing its shape to control the amount of light passing through to the lens. The lens further bends the light rays and focuses them on the back of the eye, or retina. The retina is made up of millions of light-sensitive nerve cells called cones and rods. Cones are responsible for processing details of an image and function best in bright light. These cells also perceive color and rapid changes in an image. Rods are more sensitive to light than cones and are used for low light vision. Although these cells have lower spatial resolution than the cones and do not discriminate colors, they are sensitive to movement, help in shape recognition, and are responsible for peripheral vision.

Cones are densely packed in a region of the eye called the fovea centralis, or fovea. The fovea is responsible for a viewer's sharp central vision. This region of the eye makes up approximately two percent of the visual field (5). In order to see a complex image clearly, the viewer's eyes must quickly move across the image and stop, if only briefly, to focus visual attention on different features of the image. In this way, light from different regions of the image falls on the fovea with each successive movement allowing the viewer to detect detail in the image. Information is also gathered from the remaining visual field outside the fovea. Since this region of the eye is composed primarily of rods with few cones, the peripheral vision is blurry and not as colorful. Visual information (both central and peripheral) is converted into electrical impulses that are carried to the brain by the optic nerve (6).

In the brain, working memory (WM) stores and manipulates visual information from successive eye movements. It also recalls declarative knowledge stored in long-term memory to develop an internal representation (*encoding*). It is this internal representation that is used for further cognitive processes, such as problem solving or image identification (for detailed information, see Baddeley (7)).

Relationship of Eye Movements and Cognitive Processes

The literature has shown that eye movement data can provide information on underlying cognitive processes, especially those used for encoding (for example: Just and Carpenter (8); Rayner, Raney, Pollatsek, Lorch and O'Brien (9); Rayner (10); Anderson, Bothell and Douglass (11)). The foundation of this connection relies on two assumptions. First, the immediacy assumption (8) states that the viewer immediately interprets the information that is viewed before the eye moves to the next fixation. Decisions are made as to how incoming visual information fits into the internal representation held in working memory before the next eye movement occurs. With each successive eye movement, different regions of the image are processed. Secondly, the eye-mind assumption proposes "that there is no appreciable lag between what is being fixated and what is being processed" (8). Therefore, the fixation time is a direct measure of processing time. It is these assumptions that allow researchers to connect eye movements to the input and processing of visual information (*visual attention*). Eye tracking collects information about eye movements and provides methods for analyzing a viewer's visual attention patterns.

Language of Eye Tracking

In order to discuss eye tracking, it is necessary to define a few terms. The vocabulary used to describe eye movements varies depending on the type of research carried out and, often, the field of study. For the purpose of this chapter, we have chosen to use the preferred terminology from the science education community. The following are some common terms used to describe eye movements:

Fixation – a pause in the eye movements, which focuses attention by repositioning the fovea on a specific area of a stationary visual object.

Dwell time (gaze duration or fixation duration) – total amount of time a viewer looked at a specific region of interest before moving to the next region of interest.

Total dwell time (total gaze duration or total fixation duration) – the total amount of time a viewer looked at a specific region of interest during the trial.

Saccade – rapid eye movements between fixations lasting between 10 ms and 100 ms.

Smooth pursuit – slow eye movements that keep an image stationary on the fovea, even if the visual object or viewer's head is moving.

Regress – to "re-fixate" on a previously viewed area of a stationary visual object. Most commonly used in reading research to describe when a reader's eyes move back to previously read text.

Scanpath – the series of fixations and saccades that occur when a viewer is exposed to a visual object

It is important to stress that this list is not complete and that the operational definitions of these terms may vary.

Types of Eye Trackers

Information about eye movements is gathered using an eye-tracker, a combination of hardware and software designed to collect and process data on eye position. In order to select an eye-tracker for research purposes, four main factors should be considered, namely: type of eye tracking, configuration of hardware, monocular versus binocular tracking, and pupil illumination.

Eye Tracking versus Gaze Tracking

Eye tracking is a general term for the methodology used to collect data on the movement of the eyes relative to head position or information about where a participant is looking and for how long, called "point of gaze". For some types of research, it is important to know how the eye moves relative to the head. For example, do an individual's eyes move when asked to recall a visual scene, absent of a visual stimulus? To answer this question, a system that is capable of tracking both eye and head position are necessary. Such systems include electro-oculography (EOG), scleral contact lenses, and video-oculography (VOG). For more in-depth information on how these trackers function, please see Holmqvist, Nystrom, Andersson, Dewhurst, Jarodzka and Van de Weijer (5). While these systems can identify eye movements, without additional hardware they cannot easily correlate these movements to an outside scene or stimuli. Therefore, they are not useful for identifying "point of gaze" (*where* a participant is looking, or *what* they may be fixated on). In order to associate an individual's eye movements with his or her point of gaze, video-based gaze tracking systems are commonly used. The most prevalent gaze trackers rely on pupil and corneal reflection tracking (5, 12). The video-based pupil-to-corneal reflection technique tracks two physiological targets–the pupil of the eye, and the reflection off the cornea of the eye generated by infrared light. Using these two points of reference allows the eye-tracker to use image-processing algorithms to separate true eye movements from head movements (*12*). In most discipline-based educational research, point of gaze tracking is far more important than tracking the position of the eyes alone; for this reason, this chapter will focus on video-based pupil-to-corneal reflection gaze trackers.

Configuration of Hardware

Video-based gaze trackers can take many different forms. The physical system and set-up of the hardware chosen will vary based on the goal of the research, but there are several common types of eye-trackers commercially available. These trackers all share basic components: a light source (generally infrared), a video detection camera, and some form of image processing hardware and software. The major difference in hardware set-up is how these components are placed relative to one another, and relative to the participant.

The largest basic distinction in video-based gaze trackers is whether the hardware components are head-mounted or table-mounted (static). Head-mounted trackers, as the name implies, place both the light source and camera on the head of the participant. These components are frequently built into glasses, helmets, or headbands, allowing them to move with the participant. Head-mounted trackers generally incorporate a second camera, or "scene camera", that records what the participant is seeing at a given point in time. A computer overlays the eye tracking data on the scene data to show in real-time what the participant viewed. This style of tracker allows for maximum flexibility in participant mobility and stimulus presentation. For researchers who want to study scene perception in authentic environments (i.e., interacting with laboratory equipment, in a lecture environment, or during small group work) head-mounted trackers are ideal.

Table-mounted, or static trackers, place all components in front of the participant rather than *on* the participant. The light source and camera are generally hidden from view within the eye tracker. There are two types of static trackers – contact trackers that are in close contact with the participant and remote trackers that set at a distance from the participant (*5*). Contact trackers restrict participant movements, using bite bars or forehead and chin rests to stabilize the head. This restriction allows for tracking with high precision and accuracy, but may be seen as cumbersome by the participants. Remote trackers are less precise, as head location is not controlled for without additional equipment. These trackers have the light source and cameras built directly into a computer monitor, allowing participants complete freedom of movement. In addition to participant comfort, remote trackers offer the additional advantage of increased accuracy in calibration against the stimulus, as both are fixed points. Such trackers work well with stimuli that can be displayed on a computer screen, namely, images, written information, or multi-media presentations. Static eye-trackers can also build the light source and camera into a piece of hardware separate from the monitor, allowing tracking against any computer monitor (desktop, laptop, or tablet) or other two-dimensional stimulus (book, magazine, or worksheet). This version of the eye-tracker is more portable and allows a wider array of stimulus options, but does require more attention to calibration and set-up.

A trade-off must be made when choosing a static tracker for research purposes. If high accuracy and precision are necessary, a contact tracker is a better choice; however, remote trackers offer participants greater comfort and improved correlation between eye position and stimulus. For most chemistry education research studies, remote trackers are sufficient. They allow participants to interact with the stimulus in a more authentic environment, without having to be as conscious of head movements or interactions with the tracker itself.

Monocular versus Binocular Tracking

Another factor to consider is the method of eye tracking. Both monocular and binocular gaze trackers are used in educational research. Monocular trackers collect data from only one eye, while binocular trackers collect data from both eyes simultaneously. For most individuals, it is assumed that both eyes will track together, following the same stimuli; therefore, monocular tracking is often used to save on resources and processing power. For individuals where this might not be the case, or for higher accuracy, binocular systems are preferred. Binocular systems are resource intensive, collecting a more robust data set.

Dark Pupil versus Bright Pupil

One final consideration for selecting an eye tracker is whether the system uses dark pupil or bright pupil tracking. Dark and bright pupil systems differ in where the light source lies with regards to the participant's optical path. If the light source is out of line with the optical path, the reflection of light off the pupil will also be out of line with the detection camera. This causes the pupil to appear dark. If the light source is in line with the optical path, the reflection off the pupil will be detected by the camera, making the pupil appear bright (similar to the "red eye" effect captured in normal photography). Bright pupil tracking works well to increase the contrast between pupil and iris. Therefore, it works for participants with a wider array of iris pigmentation or for participants with heavy lashes. However, bright pupil systems are sensitive to sunlight and require more stringent environmental controls than dark pupil systems. Over large populations, neither bright nor dark pupil systems have been shown to provide significantly better data (5), but the researcher should be aware of how their particular system operates so that they can compensate for the idiosyncrasies of pupil tracking within the system.

Types of Research

Due to advancements in computing and improvements in visual displays, the use of eye tracking to study how humans interact with visual information is expanding. This section will describe types of traditional research that employ eye tracking and provide a departure point for chemistry education research.

Usability

Usability is the most applied category of eye tracking research. Focusing on a given product, such as a computer interface or printed material, researchers determine how design elements affect a viewer's visual attention. Often referred to as "human factor studies" or "ergonomics", these studies use eye tracking along with other research techniques, such as interviewing or think-aloud protocols (13), to identify features that draw attention and possible reasons for viewing patterns. Products that have been reported on in the literature include websites (14–16), software (17, 18), printed media (19–22), and instrument panels (23). For the chemistry educator, usability studies could be conducted to determine the overall effectiveness of a textbook layout, website, or software application; guide the refinement of a computer interface or written materials; or study how students interact with the controls for analytical instruments in the laboratory setting.

Reading

Reading research focuses on the mechanics and cognitive processes involved in the comprehension of written language (for a review, see Rayner (24)). Eye movements during reading of English text are characterized by average fixation durations of 200-250 ms (25) and average saccade lengths of two degrees or seven to nine letter spaces (26). A number of factors can affect this visual attention pattern, including word length and frequency in the language. Rayner and McConkie (27) found that, while two to three letter words are skipped approximately 75% of the time, eight letter words are almost always fixated upon. Common words are processed quicker and have shorter fixation durations than words that appear less often in the English language (28, 29). Other factors that affect fixation durations include how the text is read (i.e., read aloud or self-reading (30)), typographical factors (i.e. letter spacing, print quality, and line length (31)), conceptual complexity (32), and syntactic complexity (33). It's even suggested that font choice can influence eye movement behavior. Fonts like Times New Roman are easier to encode because of their simple shapes and lead to "faster reading times, fewer fixations, and shorter durations" (34).

Reading research is also concerned with the cognitive processes that govern reading and comprehension. Several proposed models use eye movements to infer the underlying cognitive processes that support reading (for examples, see Just and Carpenter (8), Engbert, Longtin and Kliegl (35), Reichle, Rayner and Pollatsek (36), Gough (37), Rumelhart and Dornic (38), and Stanovich (39)). By analyzing eye fixation patterns, these models detail cognitive processes of encoding, lexical access, assignment of syntactic and semantic relationships, and creation of an internal representation. Eye tracking has also been used to develop domain specific models for comprehension, including arithmetic word problems (40) and organic chemistry equations (41).

In addition to providing information about comprehension, eye movement data can also reveal information about the reading habits of individuals. Differences in reading behaviors have been explored for a variety of populations, including successful and unsuccessful problem solvers (1, 40); readers with high working memory capacity and those with low working memory capacity (*42, 43*); dyslexic and normal readers (*44*); native language and non-native language readers (*45*); and domain experts verses novices (*46*).

With many in the chemistry education community interested in problem solving, reading research in chemistry is an important area of study. Before problem-solving strategies can be applied, a learner must encode the problem statement, process the visual information, and develop an internal representation. Identifying factors that affect comprehension are key to developing innovations that support problem solving in chemistry.

Scene Perception

Unlike reading English text where there is a well-defined task and an overall pattern for eye movements (left to right, top of the page to the bottom), looking at the environment does not exhibit the same well-defined patterns (*47*). Scene perception involves the identification of objects and their relative spatial positions in an environment. Stimuli used in this type of research are depictions of the real world (i.e., still images, paintings, movies) or the natural environment as encountered by the viewer. Compared to reading, average fixation durations and saccade lengths are longer, 300 ms and four to five degrees (*10*). However, viewers can gain an abstract meaning of the complete scene, or gist, within 40-100 ms (*48, 49*). The gist of a scene is determined in the first few fixations. All subsequent fixations create detail in the internal representation (*24*). It is important to note that fixation durations and saccade lengths vary with task and scene. It has been shown that eye movements are influenced by the task given to viewers (*50, 51*). For example, memorization tasks exhibit higher numbers of fixations than search tasks. The complexity of a scene also affects viewing patterns. As the number of elements increase and relationships between elements become more complicated, fixation frequency and duration also increase (*52*).

Henderson and Hollingworth (*53*) describe three levels at which scene perception can be studied. Low-level (early vision) perception deals with the saliency of a scene (i.e., color, intensity, brightness, texture, depth) and the detection of surfaces and edges. These features are used to create a *saliency map* that is used to guide attention for more detailed perception of the scene (*54, 55*). Intermediate level perception focuses on the shape and spatial arrangement of objects. Finally, high-level perception processes visual information in memory and assigns meaning.

Regardless of the level of perception to be studied, researchers use a variety of metrics to analyze eye movement. Researchers commonly report *first fixation duration* (duration of the initial fixation in a specific area), *first pass gaze duration* (sum of all the durations for fixations while viewing a specific area of the scene for the first time), and *second pass gaze duration* (sum of all the durations of fixations while viewing a specific area of the scene for the second time) in addition to global fixation data (i.e., total fixation duration). Henderson and Hollingworth (*53*) argue that focusing on these values, rather than global measurements of eye movement, isolates perception from other cognitive processes.

Scene perception has been used to study viewing patterns during activities, including driving (*56, 57*), viewing art (*58*) and natural scenes (*55*), recognizing faces (*59, 60*) and carrying out everyday activities (*61*). Similar techniques can be used to study how chemistry students use equipment, participate in lab, experience lecture, or work in small groups.

Visual Search

Visual search tasks require that a participant scan through a display of items to identify a target item from a larger collection of distractor items. For example, locating a specific person in a class picture is an example of visual search. Stimuli for this type of research include text, photographs, diagrams, and arrays of objects, shapes, or alphanumeric characters. Items in the array can differ in color, shape, motion, or orientation. Researchers vary the number of items in the display (*set size*) and record the time it takes for the participant to determine if a target is present or absent (*reaction time*) (for a review, see Wolfe (*62*)). As set size increases, reaction times typically increase. However, Treisman (*63*) suggests that, under certain circumstances, a target will "pop-out" and draw attention to itself in the collection of similar distractors. In this case the search is considered to occur in parallel (*64*), where all items in the display are processed in a single step, reducing reaction times. Other target items do not "pop-out" of the set and require a systematic search. These types of searches are considered serial, where attention is given to each item until the target is found. From this description, it would appear that visual search is a random act; however, participants have strategies that have developed through experience or use information from peripheral vision to guide the series of fixations. By varying characteristics of the display (i.e., number of distractors and targets, the size of the display, and the number of features for each item in the display), researchers are able to study how individuals guide attention and filter out visual information that is not related to the current task. The Feature Integration Theory (FIT), proposed by Treisman (*63*), describes how parallel and serial search behaviors work in concert. Initially, features such as color, shape, orientation, and movement are automatically analyzed using parallel processes (pre-attentive stage). In the second stage of FIT, these features are combined to create a master map that is used to guide attention to specific features that can be used for additional processing and identification.

Visual search has been explored in a variety of domains, including reading (*10*); diagrammatic representations, such as graphs, infographics, and schematics (*65, 66*); medical test results (*67, 68*); facial recognition (*69–71*); product labels (*72, 73*); sports (*74, 75*); and driving (*76, 77*). Visual search has also been incorporated into several models for the comprehension of diagrams (for examples, see Just and Carpenter (*78*), Koedinger and Anderson (*79*)).

Visual search research in chemistry education has been used to characterize student use of NMR spectra (*4*); use of representations of organic chemistry mechanisms in multi-representational displays (*3*); and the use of two organic chemistry visualizations (ball-and-stick and electrostatic potential maps) to answer questions about electron density, charge distribution, and mechanisms (*2*). Future applications of visual search techniques include active or binding site identification in molecular visualizations; interpretation of complex analytical spectra; and identification of key information in complex word problems.

Pupillometry

Pupillometry is the measurement of the diameter of the pupil. Tasks with a high cognitive demand have been correlated with a slight dilation of the pupils, known as the task-evoked pupillary response (TEPR) (*80–83*). The TEPR for cognitively demanding tasks is generally less than 0.5 mm of dilation, but can be accurately collected using a remote eye tracker (*80*). It has long been observed that pupil diameter can measure cognitive load during a memory task (*84, 85*), driving (*86*), mathematical computation (*87, 88*) or sentence processing (*89*). Unlike other types of research listed here, pupillometry is a measure of task difficulty and not processing time (*89*). Since the size of the pupil is also affected by brightness of the displayed image, physical effort, environmental conditions, and emotional arousal, careful design is required to ensure that these do not confound the measurement of cognitive task difficulty (for more information, see Chapter 11 in Holmqvist, Nystrom, Andersson, Dewhurst, Jarodzka and Van de Weijer (*5*)).

Eye Tracking: The Good, the Bad, and the Limitations

Like all research methods in a chemist's toolbox, eye tracking has advantages and disadvantages. This section details some pros and cons of using eye tracking as a research tool.

Advantages

Eye tracking has many advantages, in that it provides "an unobtrusive, sensitive, real-time behavioral index of ongoing visual and cognitive processing (*90*)" by providing the researcher with insight into the visual attention patterns of participants. Eye tracking does not typically require any special behavior on the part of the participants, and newer methods of data collection do not generally influence how a participant completes a given task. This allows the researcher to observe a participant's natural viewing behavior during a task, such as reading on a computer screen or driving a car. Depending on the sampling rate of the eye tracker, the temporal resolution of the data collected can be fairly good, offering a nearly continuous report of an individual's movements.

Another advantage of eye tracking is that this method collects data on behaviors that are difficult to articulate. During a complex task, some eye movements are automatic and unconsciously controlled, making it difficult for participants to remember and report on all the features of an object they viewed. By collecting information on all fixations, not just ones that are consciously remembered by the participant, it is possible to obtain more complete information on which features attract visual attention.

Finally, the eye tracker collects a large amount of data that can be used for analysis. In order to determine which regions of a stimulus are being fixated on, the eye tracker must collect data on the positions of the eye relative to the position of the stimulus, pupil diameter (to determine if the eye is open or closed), and a time stamp. With sampling frequencies ranging from 25 Hz (25 data points per second) to 2,000 Hz (2,000 data points per second), five-minute eye tracking session can produce 7,500 to 600,000 data sets. This is a considerable amount of numerical data that can be then used for quantitative analysis. Many consider this to be both an advantage and disadvantage to eye tracking. It is a considerable amount of data to manage; however, the robust numerical data can be further analyzed using a variety of statistical tools (i.e., t-Test, ANOVA, MANOVA, regression analysis, cluster analysis). A suite of statistical software often accompanies commercially available eye trackers to aid in the analysis.

Disadvantages

Like all research tools, eye tracking also has several disadvantages. As mentioned in the previous section, the eye tracker produces a considerable amount of data that needs to be stored and analyzed. While the analysis software included with many of the commercial eye trackers can handle simple statistical analyses (i.e. counts, frequencies, percentages), more complex analyses will require an expert with experience in data manipulation, aggregation, and statistical analysis.

Another disadvantage to this technique is the large time commitment required to carry out a study. Learning to collect data using an eye tracker can be easy. However, studies require careful planning, creation of stimuli, human subjects review (96), recruitment of an adequate number of suitable participants, time to complete individual tracking sessions, data reduction, and analysis.

Eye tracking can also be an expensive technique to deploy. While the cost of eye trackers has declined in recent years, a new eye tracking system can range from $3000 for a 50 Hz tracker (slow) to more than $80,000 for one with a sampling frequency of 250 Hz (moderate to fast). Not included in these figures are the additional costs for peripheral computers to run the tracker, adequate lab space, maintenance contracts for the tracker, and ongoing training for staff.

Lastly, eye tracking is subject to a wide range of technical problems. Along with the possibility of a computer hardware failure, several other factors can affect an eye tracking session. Eye trackers are sensitive to environmental factors, such as lighting conditions, vibrations, and electromagnetic fields (5). Loud noises and the temperature of the lab can be a source of distraction for participants during a session and have an effect on the data collected. Other technical problems relate to the suitability of participants for this type of research. Some characteristics of potential participants can make them poor candidates for eye tracking. Glasses, medical conditions related to the eye, shape of the eye slit, eye makeup such as mascara, and even the length of eyelashes can interfere with the eye tracker's ability to record data.

Considering the advantages and disadvantages of eye tracking, why would anyone choose to use this technique? If careful planning and considerations are made to the design and analysis of the data, eye tracking can provide a wealth of knowledge on visual attention and underlying processing. However, it is important to stress that there are the limitations to what information the eye tracker can provide about visual attention.

Limitations

This section identifies three major limitations to eye tracking that are important to this discussion. The first, and most important, is that eye tracking only gives the researcher evidence of what people viewed. Viewing patterns alone do not indicate conscious attention, understanding, or interest. For example, de Koning, Tabbers, Rikers and Paas (*91*) found that, although viewers spend a significant amount of time attending to visual cues meant to guide attention and increase understanding of an animation, there was limited understanding of the animation. For this reason, eye tracking is often paired with other qualitative or quantitative data. Common techniques used in conjunction with eye tracking include interviewing (i.e. think aloud, retroactive think aloud, parallel problems (*13*)); surveying or testing; conceptual mapping; and physiological measurements (i.e., fMRI, EEG, ERP). By triangulating the data, the researcher can then start to assign meaning to visual scan patterns.

Secondly, most eye tracking research only focuses on central (foveal) vision and does not collect data on peripheral vision. While central vision is important for processing detailed information about an object, it makes up only about two percent of the human visual field. Peripheral vision comprises the other 98%. Shape recognition occurs in the peripheral vision and contributes to the decision making process for subsequent fixations. Since peripheral vision is not accounted for in the eye tracking data, visualizations like heat maps (discussed later in this chapter) may be misleading.

The final limitation deals with experimental design. Research has shown that eye movements are task dependent and that fixation patterns depend on the directions given to the viewer. For example, Yarbus, Haigh and Rigss (*51*) illustrated this by asking participants to view paintings and perform specific tasks. The attention patterns of the participants changed as the task changed. Rayner, Rotello, Stewart, Keir and Duffy (*92*) also concluded that viewing time is dependent on the strategies associated with the instructions given to the participants. The task dependent nature of viewing patterns has major implications in experimental design and generalizability of a study. Participants must have a clear understanding of the task before eye tracking starts, and all participants in the same study must receive the same directions. Careful consideration must also be made when comparing data from different studies to ensure that the tasks are almost identical.

Experimental Design

As discussed, care must be taken during the design of the experiment in order to ensure useful data and applicable results. Good experimental design takes into consideration appropriate research questions, study type, sample population size, research protocols, and stimuli.

Research Questions

When one has access to an eye tracker, it is tempting to turn every research question into an eye tracking study. However, not all research questions lend themselves to an eye tracking investigation. As previously discussed, eye movements are recorded as an indirect measure of cognition. At its core, eye tracking is a specialized form of protocol analysis (93). Protocol analysis is a method of examining and classifying the series of actions (protocols) performed by an individual while he or she completes a task in order to study the cognition that underlies these actions. Eye tracking collects data on an overt behavior (eye movements). Researchers then make educated inferences about the thought processes or strategies behind these behaviors. The more closely research questions align with the overt, easily observed behavior (eye-movements), the fewer inferences the researcher has to make. This will result in a more successful study overall. For this reason, eye tracking research frequently focuses on overt visual attention, that is, *what* does an individual directly look at when presented with a variety of visual stimuli? This question can be further qualified to ask *when* the looking happens, for how long, or in what order. It is important to note that these questions all focus on the overt behavior of eye movements. Research questions that ask *why* a participant performs a certain action or makes a certain decision are much more difficult to answer, and generally cannot be answered with eye tracking alone. Similarly, questions dealing with affective measures (i.e., motivation, self-efficacy, or metacognition) are also not easy to directly address with eye tracking. It is important to note that the traditional types of eye tracking research previously discussed (usability research, scene perception, visual search, and reading research) focus on determining the features of an image that draw the attention of participants; the length of time participants spend viewing a particular feature or image; and the overall viewing patterns they exhibit for a particular image.

Study Type

As with any type of research, the research questions will dictate the design of the study. The type of study conducted will be largely determined by what, if anything, is already known about the research question. If the body of existing literature on the topic includes eye tracking research, this research can be used to guide the study design. Holmqvist and Nystrom (5) refer to this type of research as "highway research" because the path of travel is clear and relatively easy to follow. The traditional types of eye tracking research can provide existing paradigms, or lenses, that guide the design of a current study. This can save time and effort in deciding how many participants to use; how to design the stimuli; the eye tracking variables to collect; additional variables to collect; and the analysis scheme to use. Using this method of design also makes it easy to integrate new research into the existing paradigm.

However, eye tracking research is new to many fields, and directly applicable research may not exist in the literature. The best-case scenario in this situation is that there is a wide theoretical research base that supports predictions about participants' overt viewing behaviors. This theory-driven research offers many of the benefits of "highway research". The literature base should provide insights on variables to collect, research protocol, and the data analysis. Additional legwork to determine specific experimental parameters, such as sample population size or stimulus design, is required, however, the additional work will be reasonably targeted.

In the worst-case scenario, where little is known about the research question, it may be necessary to carry out explorative work. In this circumstance, the researcher may choose to conduct a pilot study. A pilot study can help a researcher visualize the data collected by the eye tracker; identify problems with the research protocol and stimuli; and develop an analysis plan. This type of study only requires two to three participants. Statistical analysis of this data is ill advised because of the small sample size, however, it can be used to guide the development of a larger study. The researcher may also opt for what Holmqvist and Nystrom (5) call a "fishing trip." This is essentially a large-scale pilot study, in which the researcher uses a large number of participants and various stimuli to collect data on as many variables as possible. This method has an economy of scale. Rather than running many small pilot studies that may prove inconclusive, the "fishing trip" collects a large amount of data at once. The researcher can then "fish" through the data for significant findings. The downsides to this type of study are numerous. Since there is little to draw from in the literature, developing effective stimuli is often difficult and time consuming. Procedures and protocols may not be adequate to address the research questions, leading to problems with data analysis. More importantly, the results of a fishing trip are generally post hoc explanations for the data. Given a large enough data set, it is usually possible to identify *some* significant effects. It is difficult to discern whether these effects truly exist or are replicable based on the ill-targeted experimental design. However, a fishing trip can be used like a large-scale pilot study, where significant results shape future work that is more targeted.

Sample Population Size

Regardless of the type of study planned, decisions relating to sample population size are an important part of the experimental design. As mentioned above, small sample populations (two to three participants) may be enough for a pilot study but will not provide a robust data set for statistical analysis. On the other hand, a large sample population can lead to logistical problems for the researcher. As discussed, eye tracking is incredibly data intensive. A single participant, tracked on a low-end instrument for 2-3 minutes, will provide thousands of data points. Larger sample sizes produce a significant amount of data that requires careful data management. In addition, extremely large data sets can make even minuscule differences in sample populations appear statistically significant, even when these differences have no practical significance in the real world. A balance must be struck between providing enough participants to give robust results and keeping the data set manageable.

The appropriate number of participants can be calculated using traditional statistical methods, taking into account acceptable α and β levels and expected effect size. However, this may still suggest a sample size that is not logistically feasible for eye tracking. In this case, a review of existing eye tracking studies in the chosen or related fields may provide guidance for acceptable sample population size. Recently published eye tracking studies in chemistry education research (Table I) show that studies between 9 and 27 participants are acceptable in this field. Even studies with just 4-5 participants are worthwhile and may provide statistically significant results. However, great care must be taken when discussing the generalizability of such small studies (5). Small eye tracking studies are more akin to qualitative studies and can be used as case studies to help steer future research.

Table I. Sample size for eye tracking studies in chemistry education

Reference	Topic	n
Tang and Pienta, 2012 (1)	Problem-solving (reading)	12
Tang, Topczewski, Topczewski and Pienta, 2012 (4)	NMR reading	12 expert 15 novice
Williamson, Hegarty, Deslongchamps, Williamson and Shultz, 2013 (2)	Problem-solving (image based)	9
Stieff, Hegarty and Deslongchamps, 2011 (3)	Interactive animations	10
Havanki, 2012 (41)	Reading organic chemistry equations	9 expert 19 novice
VandenPlas, 2008 (94)	Particulate level animations	5 expert 6 novice

Protocol and Stimuli

As discussed previously, eye movements are task dependent, so careful design of the stimuli and research protocol is key to success in eye tracking research. A good protocol must include all the activities that participants will be asked to complete (before, during, and after eye tracking), as well as the explicit instructions participants are given during the study. Most eye tracking protocols will include the collection of participant data in order to address important research variables or to triangulate the eye tracking data. This additional data may include pre-test and post-test scores, assessment of content knowledge, ability testing (spatial ability, logical thinking ability, etc.), affective measures (attitude, motivation, etc.), and demographic information (gender, age, etc.). The choice and timing of these tests should be determined by the research question and the existing literature base in the field.

In addition to the participant data, it may be necessary to collect data related to how the participants interacted with the stimuli to triangulate eye movement data and attribute meaning to specific viewing patterns. Because eye movement data is not a direct measure of cognition, additional data is necessary to determine the motives and underlying cognition that direct eye movements. This may include interviewing (*13*), observation (*95*), think-aloud protocol analysis (*13*), or retrospective reporting.

The protocol must also include the instructions given to the participants before or during the eye tracking task. This is crucial, as instructions have been shown to have a significant effect on how participants view an image (*51*). The instructions not only tell the participant how to interact with the stimuli but also specify the task or activity associated with it. While it is important to instruct participants on how to interact with the stimulus (buttons to press, how to operate drop-down menus, etc.), identifying the *goal* of the activity is crucial to the success of the study. Researchers need to strike a balance between giving explicit instructions and confounding results by biasing participants. Instructions should be given that lead participants to engage with the stimulus and perform the desired task, just as during a think-aloud interview or similar task. Overly vague instructions, such as "just view the images on the screen as they are presented", are not likely to engage participants, and their gaze patterns will probably be unfocused. However, instructions such as "You will be asked a series of questions after viewing this series of images" will cause the participant to engage more fully in the task, and provide a richer data set.

Just as important as the instructions, the stimulus must be purposely designed with the research questions in mind. Stimuli can include an image or series of images, videos, animations, questionnaires, tests, webpages, computer programs, and real-world objects such as textbooks or whiteboards. Regardless of its form, the stimulus must be engaging, drawing the participants to actively participate in the experiment. If participants are not interested in the stimulus, they may interact with it superficially and thus bias the data.

Some research questions may not allow the researcher full control over stimulus design, for example investigations into student interactions with textbooks, existing animations, or online homework systems. In these cases, it is important that the stimulus be as unbiased as possible towards the research design. It must be visually balanced, with all objects being presented equally. Several features of the stimulus can influence attention and should be part of a conscientious design of both the stimulus and the study itself. For example, participants are more likely to look in the center of the screen than the edges, so continually locating target item in the center of the screen would bias the eye tracking data. Other factors that will preferentially draw attention include differences in size, motion, color, and on-screen brightness (*5*). Hardware limitations are also important when designing stimuli. Parameters including screen size, monitor resolution, and supported image/video formats will dictate the way stimuli are presented. Low monitor resolution can make viewing difficult and pixelation can interfere with analysis. Ideally, objects in the stimulus are spread out over the entire screen, leaving space between them. This will facilitate mapping eye fixations on particular features of the stimulus, making it clear where the participant focused attention. Objects that overlap, have ill-defined edges, or are in close proximity to one another can complicate the analysis and should be avoided. When designing stimuli, it is important to balance all of these concerns.

Data Collection

After designing an eye tracking study, approval to conduct human subjects research must be obtained (96). Once approved, data collection can begin.

Like all analytical instruments, the eye tracker must be calibrated at the start of each eye tracking session. Assuming that the stimulus is displayed on a computer monitor, calibration allows the eye tracking software to correlate a participant's gaze with a specific point on the screen. During this process, the participants look at a series of points presented sequentially on the eye tracker's screen. The tracker's software constructs an algorithm allowing the system to correlate relative eye position to an on-screen location. Due to variations in facial structure and eye movements, each participant must be individually calibrated. Once calibration is complete, the data collection session can begin.

Depending on the eye tracker, raw data is collected for one eye, both eyes individually, or both eyes averaged together. Raw data typically includes the time each data point was recorded and the on-screen XY-coordinate for the eye(s) gaze. Some systems will give additional data, including distance of the eye(s) to the tracker, pupil-diameter, or a validity code indicating confidence in the system's estimate of the eye(s) measurements. Event data, such as mouse-clicks or key-presses made by the participant, and a running screen capture of the stimuli may also be recorded.

Once collected, raw data is processed through several steps. The first step involves evaluating the data quality and cleaning the data. Participants may blink, look off-screen, touch their face or eyes, or move their head in such a way that data collection is compromised. Compromised or "bad" data is generally discarded. Validity codes for each data point allow for relatively easy identification of these "bad" data points. Where possible, validity thresholds set by the researcher automatically filter this data. However, if validity information is not automatically calculated by the system, the researcher may have to manually inspect the data or construct an algorithm to identify anomalous data.

The next step in the processing of raw data is data reduction using a filter or series of filters to aggregate data. A common data reduction technique is to collapse the data into a series of fixations and saccades. When the eye maintains focus on the same XY-coordinates for a given period of time, it is identified as a fixation. In this way, several consecutive data points are collapsed into a single fixation event, reducing the data to a more manageable set of points. Most attentional studies ignore saccades, as vision is suppressed during eye movement. Data points identified as saccadic are typically removed from the analysis. To identify if a gaze point was part of a fixation or a saccade the researcher must identify fixation thresholds.

As part of the research protocol, the researcher must identify some minimum threshold for what is considered "fixation behavior". This threshold includes identifying the minimum length of time the eye must stay fixed on a given area, as well as the maximum size of that area. The minimum fixation time varies based on the type of research. Reading research tends to have different fixation times than image-based research. Salthouse and Ellis (97) report 50 to 100 milliseconds as the minimum fixation time necessary to process simple textual stimuli, while

Rayner (*28*) suggests times in the 200-300 millisecond range for fixations in reading research. Rayner further suggests mean fixation during visual search is in the range of 275 msec, however, if the task requires hand-eye coordination, a threshold of 400 msec or greater should be used. Existing literature should be consulted to determine the accepted fixation thresholds for the field of study.

To identify the area of fixation, Blignaut (*98*) suggests that a fixation radius of 1 degree captures the majority of fixation behavior. All gaze points that fall within 1 visual degree of one another during a defined time period are treated as a single fixation. Normal eye movements may drift or tremor during fixation. Allowing a slightly expanded fixation radius (rather than requiring all gaze points be made at identical XY-coordinates) accounts for these unconscious eye movements, as well as any measurement error. Many eye tracking systems report fixation radius in terms of pixels instead of visual degrees. The fixation radius in pixels is calculated using the distance of the participant's eye from the screen, the size of the screen, and the resolution of the monitor. For an individual seated 65 cm from a 17-inch computer monitor with a resolution of 1024 x 768 pixels, a 1 degree visual angle translates to about 33 pixels. It is important to note that, just like minimum fixation times, fixation radii can vary depending on the task, therefore these values should also be checked against relevant literature.

Instead of using a fixation threshold, fixations can also be identified by using saccades. Although saccades are not typically analyzed in attentional studies, some researchers define the parameters of a saccade. The time between individual saccades is defined as a fixation. To do this, the researcher defines a minimum acceleration and velocity for an eye movement to be identified as a saccade.

Both methods of filtering the data are acceptable, and may depend on the tracker's software. Cutoff values for fixations and saccades given in recently published chemistry education research studies are provided in Table II. Once fixations and saccades are identified, the final step in an eye tracking study is data analysis.

Table II. Fixation and Saccade Definitions in Chemistry Education Literature

Reference	Topic	Tracker Frequency	Fixation or Saccade
Tang and Pienta, 2012 (*1*)	Problem-solving (reading)	120 Hz	Fixation: Duration of 100 ms; Radius not reported
Tang, Topczewski, Topczewski and Pienta, 2012 (*4*)	NMR reading	120 Hz	Fixation: Duration of 100 ms; Radius not reported

Continued on next page.

Table II. (Continued). Fixation and Saccade Definitions in Chemistry Education Literature

Reference	Topic	Tracker Frequency	Fixation or Saccade
Williamson, Hegarty, Deslongchamps, Williamson and Shultz, 2013 (2)	Problem-solving (image based)	250 Hz	Saccade: 0.05° amplitude, 9500° s−2 acceleration threshold, 30° s−1 velocity threshold
Stieff, Hegarty and Deslongchamps, 2011 (3)	Interactive animations	250 Hz	Saccade: 0.05° amplitude, 9500° s−2 acceleration threshold, 30° s−1 velocity threshold
Havanki, 2012 (41)	Reading organic chemistry equations	120 Hz	Fixation: 50 pixel radius (cluster analysis)
VandenPlas, 2008 (94)	Particulate level animations	50 Hz	Fixation: Duration of 40 ms; 20 pixel radius

Data Analysis

There are a variety of data analysis methods reported in the literature. The selection of analysis methods depends on the research question. Below are three commonly used methods.

Area of Interest

For most research questions, simply identifying the XY-coordinates of where an on-screen gaze of fixation takes place is of little value. Researchers want to correlate fixations with features of the stimulus. One method is to identify areas of interest (AOIs), also called 'regions of interest' (ROIs) or 'zones', within the stimulus. This method is a data aggregation technique. The researcher defines regions of the stimulus he or she wishes to study. Consecutive fixations within the defined AOIs are aggregated into fixation groups, while fixations outside the AOIs are generally ignored. Fixation groups can then be further analyzed. Fixation durations and fixation frequencies are generally reported, giving the researcher an indication of the participants' attention on particular features of the stimulus. For example, a researcher might want to study how students read an online homework problem that contains both text and a diagram by comparing the time students spend reading the text to the time they spend viewing the diagram. In this case, the text and the diagram are identified as separate AOIs, as shown in Figure 2. Fixation times then can be compared for each AOI to determine if students spend more time reading text or viewing the diagram.

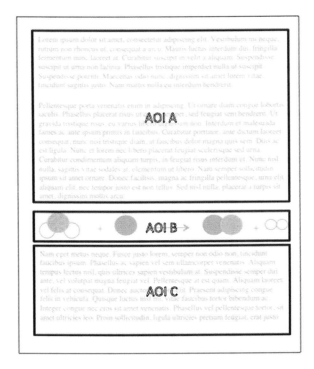

Figure 2. Example textbook selection showing three areas of interest.

AOIs can be any size. In the example above, the text could be broken down further, by paragraph, sentence, word, or even letter depending on the research questions being investigated. Similarly, a single diagram may be identified as a single AOI or broken down into individual pieces, identifying each piece as a separate AOI.

There are several methods for creating AOIs commonly found in the literature. The first is researcher-driven. Most eye tracking software includes a tool that allows the researcher to draw an AOI directly on the stimulus. While AOIs are commonly defined as rectangular shapes, many software packages allow for polygons to enclose irregularly shaped areas. It is important to note that, although AOIs can be defined before or after eye tracking data is collected, these regions are not visible to participants, and are only viewable by the researcher during analysis. The AOIs are not visible to the participants during tracking sessions.

Each stimulus presented will have its own set of AOIs. If the stimulus is not a static image, care must be taken to match the AOIs to what the participant views at any given point in time. For example, if the participant is viewing a computer animation, every frame of the animation will be slightly different, providing a different stimulus image with its own set of AOIs. These are referred to as dynamic AOIs because they "move" with the stimulus as the animation progresses.

AOIs can also be generated from the data using cluster analysis. This technique aggregates gaze data across multiple participants to create AOIs that represent areas with a high concentration of fixations. Unlike the area of interest tool, where the AOIs are selected by the researcher, this tool uses clustering algorithms to identify patterns of eye fixations. One such method uses the robust clustering algorithm proposed by Santella and DeCarlo (*99*). This type of iterative mean-shift analysis identifies dense regions of fixations or "clusters". To be considered part of the same cluster, the maximum distance between any two points must be below a pre-set threshold that supplies good resolution of the clusters and is outside the range of error for the eye-tracker. The resulting clusters are irregular shapes and may contain the stimulus, or simply white space that is not of interest to the researcher. The researcher must then use other information (i.e., interviews, theory, expert analysis) to ascribe meaning to the resulting clusters.

In general, the decision of how AOIs will be selected and the total number of AOIs identified for a given stimulus should be determined *a priori*, before data collection begins. Research questions may center on features of an image that draw participants' attention and the length of time they spend viewing these features. In this case, the researcher may already have identified particular features based on supporting literature or previous studies. Other research questions are better answered using data-driven (post hoc) AOIs. For example, if a research question asks "do students view the same regions of a diagram as their instructors during problem solving?" it would be difficult to define the AOIs correlating to the features the instructors view before data is collected. In this case, analyzing eye tracking data from a group of instructors will identify AOIs they viewed most often. Student data could then be analyzed against the instructor-defined AOIs to answer the research question.

Regardless of how AOIs are defined, care should be taken to position the AOIs in a way that facilitates the analysis. AOIs should be placed around objects of interest, ideally so that they do not overlap, and have some margin between them to avoid confusion in classifying gazes near the borders. The minimum size of an AOI is limited by the accuracy and precision of the eye tracking system (*5*). It is also advisable to provide a "buffer region" around an object of interest by making the AOI slightly larger than the object itself to account for small errors in measurement.

Once AOIs have been identified, recorded participant data can be analyzed with regard to these regions. The most basic questions addressed by this method of analysis are:

1. Did a participant view an AOI? That is, does the data contain one or more fixations within an identified AOI? To answer this, the total number of fixations made inside an AOI can be measured (*fixation frequency*).
2. How many times did a participant view an AOI? This can also be answered with fixation frequency.
3. How long did a participant view a feature with a given AOI? To answer this, the total amount of time the participant's gaze was inside the AOI (adding together all fixations within the AOI regardless of *when* they occurred to give a measure of *total fixation duration*), or the average length of fixation within the AOI (*average fixation duration*) may be reported.

The researcher may choose to compare all AOIs, compare one AOI to another AOI, or compare the AOI viewing patterns of different subpopulations. Depending on the research questions to be addressed, it may also be appropriate to report out descriptive statistics for the data, or to conduct some between- or within-groups statistical comparisons on the data (see Sanger (*100*) and Pentecost (*101*)).

Scanpath Analysis

The previous methods of analyzing the data study *where* participants focus their visual attention but do not take into account *when* participants are viewing a particular area or object. Research that focuses both on the spatial dimension and temporal dimension of visual attention will generally discuss eye movements in terms of a "scanpath". A scanpath is a chronological map of eye-movements, including both fixations and saccades, overlaid on the stimulus. Analyzing scanpath data requires using sequence-based analysis techniques for eye tracking data.

Many sequence-based methods begin by reducing the data further, from a long list of fixations to a simpler AOI string, which lists the sequence of a participant's AOI fixations in chronological order. Consider the following scanpath (Figure 3). This scanpath can be represented by the following string: AABBBBBCC. This string indicates that the participant first fixated on AOI A, followed by a second fixation on AOI A, then a series of five fixations on different regions of AOI B, and finally ending with two fixations on different regions of AOI C.

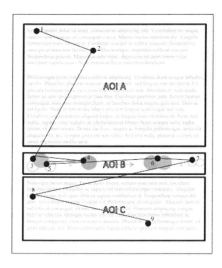

Figure 3. A scanpath overlaid on a stimulus. This scanpath is made up of 9 fixations.

Once the data for each participant is converted to a string, the strings can be compared. A common method used to directly compare two strings is the string-edit method (*102, 103*). This method counts the number of insertions or deletions required to make the strings match. The fewer insertions and deletions necessary, the "closer" the strings are to one another, and the more similar the scanpaths.

Another method of scanpath analysis focuses on the transitions that participants make between AOIs. Typically this method of research is limited to patterns that contain one or two transitions. For instance, consider patterns that contain one transition for the hypothetical AOI string given above, AABBBBBCC. The participant makes two fixations on AOI A (A-A) before transitioning from AOI A to AOI B (A-B). There are then five fixations on AOI B (B-B), before a transition from B to C (B-C). These transitions can tell the researcher how participants move their gaze within a stimulus. It can also tell a researcher about the relationships between AOIs and how viewing a particular AOI may trigger a reflexive view of another AOI. Transitions can be represented with a transition matrix, which lists all possible transitions between AOIs and tallies the total number of transitions made by a participant. Statistically, the significance of these transitions can be investigated, and Markov chain modeling can be used to predict the likelihood of a given transition based on previous behavior. For more technical details on these analysis techniques and an overview of other methods for data analysis, see Holmqvist, Nystrom, Andersson, Dewhurst, Jarodzka and Van de Weijer (*5*).

Heatmaps

No discussion of eye tracking analysis methods would be complete without mentioning heatmaps (Figure 4). Eye tracking data is frequently presented using this technique. Heatmaps are primarily a visualization tool and are similar to electron density plots. They aggregate fixation data and overlay the information on top of a stimulus image, using color to show "hot spots" (where many fixations were made) and "cool spots" (where relatively few fixations were made). Depending on the algorithm used to produce the heatmap, the definition of "hot spot" or "cool spot" can change. Heatmaps can also be used for quantifying or statistically analyzing the data. The discussion of this goes beyond the scope of this chapter, but a more in-depth presentation of these techniques is given in Holmqvist, Nystrom, Andersson, Dewhurst, Jarodzka and Van de Weijer (*5*).

Heatmaps can represent data for a single individual or a group of individuals who view the same stimulus. They are useful for identifying participant-defined AOIs, which can be particularly important within the context of a pilot study. They are also easy to read, making them the most common visualization used to convey results of an eye tracking study to a general audience. However, great care must be given when using heat maps. They only represent frequency data and do not convey the underlying meaning of the patterns shown. It is important to include additional data to put the heatmaps in context.

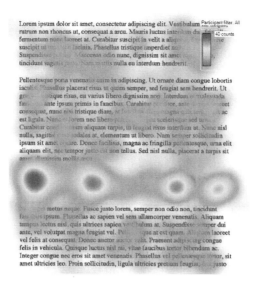

Figure 4. Heatmap overlaid on a stimulus. It shows the location and frequency of fixations. Red indicates regions with the highest number of fixations (40 fixations).

Conclusion

Eye tracking is a powerful research tool to add to the chemistry education researcher's arsenal. Although many considerations need to be addressed when designing and carrying out research using eye tracking, the insights that this methodology reveals about how individuals experience chemistry are invaluable. Drawing from existing research in usability, reading, scene perception, visual search and pupillometry, both inside and outside our domain, this versatile tool can be used to investigate a variety of topics. By combining tracking data with data from other techniques, like the ones discussed in this book (i.e., interviewing (*13*), observation (*95*), or surveying), we can gain a deeper understanding of the cognitive processes and strategies individuals use while learning, teaching, and practicing chemistry.

References

1. Tang, H.; Pienta, N. *J. Chem. Educ.* **2012**, *89*, 988–994.
2. Williamson, V. M.; Hegarty, M.; Deslongchamps, G.; Williamson, I.; Kenneth, C; Shultz, M. J. *J. Chem. Educ.* **2013**, *90*, 159–164.
3. Stieff, M.; Hegarty, M.; Deslongchamps, G. *Cognit. Instr.* **2011**, *29*, 123–145.
4. Tang, H.; Topczewski, J. J.; Topczewski, A. M.; Pienta, N. J. Permutation Test for Groups of Scanpaths Using normalized Levenshtein Distances and Application in NMR Questions. In *Proceedings of the Symposium on Eye Tracking Research and Applications*; ACM Digital Library: New York, 2012; pp 169−172.

5. Holmqvist, K.; Nystrom, M.; Andersson, R.; Dewhurst, R.; Jarodzka, H.; Van de Weijer, J. *Eye Tracking: A Comprehensive Guide to Methods and Measures*; Oxford University Press: Oxford, U.K., 2011.

6. Daw, N. *How Vision Works: The Physiological Mechanisms Behind What We See*; Oxford University Press: Oxford, U.K., 2012.

7. Baddeley, A. *Working Memory, Thought, and Action*; Oxford University Press: Oxford, U.K., 2007.

8. Just, M.; Carpenter, P. *Psychol. Rev.* **1980**, *87*, 329–354.

9. Rayner, K.; Raney, G. E.; Pollatsek, A.; Lorch, R. F.; O'Brien, R. J. In *Sources of Coherence in Reading*; Lawrence Erlbaum Associates, Inc.: Hillsdale, NJ, 1995.

10. Rayner, K. *Q. J. Exp. Psychol.* **2009**, *62*, 1457–1506.

11. Anderson, J. R.; Bothell, D.; Douglass, S. *Psychol. Sci.* **2004**, *15*, 225–231.

12. Duchowski, A. T. *Eye Tracking Methodology: Theory and Practice*; Springer: London, 2003.

13. Herrington, D. G.; Daubenmire, P. L. Using Interviews in CER Projects: Options, Considerations, and Limitations. In *Tools of Chemistry Education Research*; Bunce, D. M., Cole, R. S., Eds.; ACS Symposium Series 1166; American Chemical Society: Washington, DC, 2014; Chapter 3.

14. Nielsen, J.; Pernice, K. *Eyetracking Web Usability*; New Riders: Berkeley, CA, 2010.

15. McCarthy, J. D.; Sasse, M. A.; Riegelsberger, J. *People Comp.* **2004**, 401–414.

16. Cowen, L.; Ball, L. J.; Delin, J. In *People and Computers XVI − Memorable Yet Invisible*; Springer-Verlag: London, 2002; pp 317−335.

17. Law, B.; Atkins, M. S.; Kirkpatrick, A. E.; Lomax, A. J. C. In *Proceedings of the Symposium on Eye Tracking Research and Applications*; ACM Digital Library: New York, 2004.

18. Goldberg, J. H.; Kotval, X. P. *Int. J. Indust. Ergon.* **1999**, *24*, 631–645.

19. Holsanova, J.; Holmberg, N.; Holmqvist, K. *Appl. Cognit. Psychol.* **2009**, *23*, 1215–1226.

20. Wedel, M.; Pieters, R. *Eye Tracking for Visual Marketing*; now Publishers, Inc: Hanover, MA, 2008.

21. Holmqvist, K.; Wartenberg, C. *Lund University Cognit. Studies* **2005**, *127*, 1–21.

22. Radach, R.; Lemmer, S.; Vorstius, C.; Heller, D.; Radach, K. J. In *The Mind's Eye: Cognitive and Applied Aspects of Eye Movement Research*; Hyona, J., Radach, R., Deubel, H., Eds.; Elsevier: Amsterdam, 2003; pp 609−632.

23. Wickens, C. D.; Goh, J.; Helleberg, J.; Horrey, W. J.; Talleur, D. A. *Hum. Factors* **2003**, *45*, 360–380.

24. Rayner, K. *Psychol. Bull.* **1998**, *124*, 372–422.

25. Pollatsek, A.; Rayner, K.; Collins, W. E. *J. Exp. Psychol. Gen.* **1984**, *113*.

26. Morrison, R. E.; Rayner, K. *Atten. Percept. Psychophys.* **1981**, *30*, 395–396.

27. Rayner, K.; McConkie, G. W. *Vision Res.* **1976**, *16*, 829–837.

28. Rayner, K. *Psychol. Bull.* **1998**, *124*, 372–422.

29. Rayner, K.; Duffy, S. A. *Mem. Cognition* **1986**, *14*, 191–201.

30. Levy-Schoen, A. In *Eye Movements: Cognition and Visual Perception*; Fisher, D. S., Monty, R. A., Senders, J. W., Eds.; Lawrence Erlbaum Associates, Inc.: Hillsdale, NJ, 1981; pp 299−314.

31. Morrison, R. E.; Inhoff, A. W. *Visible Lang.* **1981**, *15*, 129–146.

32. Rayner, K.; Chace, K. H.; Slattery, T. J.; Ashby, J. *Sci. Stud. Reading* **2006**, *10*, 241–255.

33. Crain, S.; Shankweiler, D. In *Linguistic Complexity and Text Comprehension: Readability Issues Reconsidered*; Davison, A., Green, G. M., Eds.; Lawrence Erlbaum Associates, Inc.: Hillsdale, NJ, 1988; pp 167−192.

34. Slattery, T. J.; Rayner, K. *Appl. Cognit. Psychol.* **2009**, *S1*.

35. Engbert, R.; Longtin, A.; Kliegl, R. *Vision Res.* **2002**, *42*, 621–636.

36. Reichle, E. D.; Rayner, K.; Pollatsek, A. *Vision Res.* **1999**, *39*, 4403–4411.

37. Dornic, S. In *Handbook of Reading Research*; Pearson, P. D., Ed.; Routledge: New York, 1984; Vol. 1, pp 225−253.

38. Rumelhart, D. E.; Dornic, S. In *Attention and Performance*; Lawrence Erlbaum Associates, Inc.: Hillsdale, NJ, 1977.

39. Stanovich, K. E. *Reading Res. Q.* **1980**, *16*, 32–71.

40. Hegarty, M.; Mayer, R. E.; Monk, C. A. *J. Educ. Psychol.* **1995**, *87*, 18–32.

41. Havanki, K. Doctoral Thesis, The Catholic University of America, Washington, DC, 2012.

42. Osaka, N.; Osaka, M. *Am. J. Psychol.* **2002**, *115*, 501–513.

43. King, J.; Just, M. A. *J. Mem. Lang.* **1991**, *30*, 580–602.

44. Olson, R. K.; Kliegl, R.; Davidson, B. J. *J. Exp. Psychol. Hum. Percept.* **1983**, *9*.

45. Sun, F.; Morita, M.; Stark, L. W. *Percept. Psychophys.* **1985**, *37*, 502–506.

46. Rebert, G. *J. Educ. Psychol.* **1932**, *23*, 192–203.

47. Kennedy, A. In *Eye Movements and Visual Cognition*; Springer: New York, 1992; pp 379−396.

48. Castelhano, M. S.; Henderson, J. M. *J. Exp. Psychol. Hum. Percept.* **2008**, *34*.

49. Potter, M. C. *J. Exp. Psychol. Hum. Learn.* **1976**, *2*.

50. Castelhano, M. S.; Mack, M. L.; Henderson, J. M. *J. Vision* **2009**, *9*.

51. Yarbus, A. L.; Haigh, B.; Rigss, L. A. *Eye Movements and Vision*; Plenum Press: New York, 1967.

52. Vlaskamp, B. N. S.; Hooge, I. T. C. *Vision Res.* **2006**, *46*, 417–425.

53. Henderson, J. M.; Hollingworth, A. *Annu. Rev. Psychol.* **1999**, *50*, 243–271.

54. Koch, C.; Ullman, S. *Hum. Neurobiol.* **1985**, *4*, 219–27.

55. Underwood, G.; Foulsham, T.; van Loon, E.; Humphreys, L.; Bloyce, J. *Eur. J. Cognit. Psychol.* **2006**, *18*, 321–342.

56. Underwood, G.; Phelps, N.; Wright, C.; van Loon, E.; Galpin, A. *Opthalmic Physiol. Opt.* **2005**, *25*, 346–356.

57. Land, M. F.; Lee, D. N. *Nature* **1994**.

58. Henderson, J. M.; Hollingworth, A. In *Eye Guidance in Reading and Scene Perception*; Underwood, G., Ed.; Elsevier: New York, 1998; pp 269−203. .

59. Riby, D. M.; Hancock, P. J. B. *J. Autism Dev. Disord.* **2009**, *39*, 421–431.

60. Lewis, M. B.; Edmonds, A. J. *Perception* **2003**, *32*, 903–920.

61. Land, M.; Mennie, N.; Rusted, J. *Perception* **1999**, *28*, 1311–1328.
62. Wolfe, J. M. *Psychol. Sci.* **1998**, *9*, 33–39.
63. Treisman, A. *Sci. Am.* **1986**, *255*, 114–125.
64. Treisman, A. M.; Gelade, G. *Cognit. Psychol.* **1980**, *12*, 97–136.
65. Carpenter, P. A.; Shah, P. *J. Exp. Psychol. Appl.* **1998**, *4*.
66. Mayer, R. E. *Learn. Instr.* **2010**, *20*, 167–171.
67. Nodine, C. F.; Mello-Thoms, C.; Kundel, H. L.; Weinstein, S. P. *Am. J. Roentgenol.* **2002**, *179*, 917–923.
68. Augustyniak, P.; Tadeusiewicz, R. *Physiol. Meas.* **2006**, *27*, 597–608.
69. Schwarzer, G.; Huber, S.; Dümmler, T. *Mem. Cognit.* **2005**, *33*, 344–354.
70. Janik, S. W.; Wellens, A. R.; Goldberg, M. L.; Dell'Osso, L. F. *Percept. Motor Skill.* **1978**, *47*, 857–858.
71. Pelphrey, K. A.; Sasson, N. J.; Reznick, J. S.; Paul, G.; Goldman, B. D.; Piven, J. *J. Autism Dev. Disord.* **2002**, *32*, 249–261.
72. Goldberg, J. H.; Probart, C. K.; Zak, R. E. *Hum. Factors* **1999**, *41*, 425–437.
73. Krugman, D. M.; Fox, R. J.; Fletcher, J. E.; Fischer, P. M.; Rojas, T. H. *J. Advertising Res.* **1994**, *34*, 39–39.
74. Singer, R. N.; Cauraugh, J. H.; Chen, D.; Steinberg, G. M.; Frehlich, S. G. *J. Appl. Sport Psychol.* **1996**, *8*, 9–26.
75. Kato, T.; Fukuda, T. *Percept. Motor Skill.* **2002**, *94*, 380–386.
76. Chapman, P. R.; Underwood, G. *Perception* **1998**, *27*, 951–964.
77. Recarte, M. A.; Nunes, L. M. *J. Exp. Psychol. Appl.* **2000**, *6*.
78. Just, M. A.; Carpenter, P. A. *Cognit. Psychol.* **1976**, *8*, 441–480.
79. Koedinger, K.; Anderson, J. R. *Cognit. Sci.* **1990**, *14*, 511–550.
80. Klingner, J.; Kumar, R.; Hanrahan, P. Measuring the Task-Evoked Pupillary Response with a Remote Eye Tracker. In *ETRA; 08: Proceedings of the 2008 Symposium on Eye Tracking Research and Applications*; ACM Digital Library: New York, 2008.
81. Paas, F.; Tuovinen, J.; Tabbers, H.; Van Gerven, P. W. M. *Educ. Psychol.* **2003**, *38*, 63–71.
82. Schultheis, H.; Jameson, A. Assessing Cognitive Load in Adaptive Hypermedia Systems: Physiological and Behavioral Methods. In *Adaptive Hypermedia and Adaptive Web-Based Systems, Proceedings*; Springer: New York, 2004; pp 225–234.
83. Van Gerven, P. W. M.; Paas, F.; Van Merrienboer, J. J. G.; Schmidt, H. G. *Psychophysiology* **2004**, *41*, 167–174.
84. Kahneman, D.; Beatty, J. *Science* **1966**.
85. Beatty, J. *Psychol. Bull.* **1982**, *91*.
86. Palinko, O.; Kun, A. L.; Shyrokov, A.; Heeman, P. Estimating Cognitive Load Using Remote Eye Tracking in a Driving Simulator. In *Proceedings of the 2010 Symposium on Eye Tracking Research and Applications*; ACM Digital Library: New York, 2010; pp 141–144.
87. Hess, E. H.; Polt, J. M. *Science* **1964**, *143*, 1190–1192.
88. Landgraf, S.; Van der Meer, E.; Krueger, F. *ZDM* **2010**, *42*, 579–590.
89. Just, M. A.; Carpenter, P. A. *Can. J. Exp. Psychol.* **1993**, *47*.

90. Henderson, J. M.; Ferreira, F. In *The Interface of Language, Vision, and Action: Eye Movements and the Visual World*; Psychology Press: New York, 2004; pp 1–58.

91. de Koning, B. B.; Tabbers, H. K.; Rikers, R. M. J. P.; Paas, F. *Learn. Instr.* **2010**, *20*, 111–122.

92. Rayner, K.; Rotello, C. M.; Stewart, A. J.; Keir, J.; Duffy, S. A. *J. Exp. Psychol. Appl.* **2001**, *7*.

93. Salvucci, D.; Anderson, J. *Hum. Comp. Interaction* **2001**, *16*, 39–86.

94. VandenPlas, J. R. Doctoral Dissertation, The Catholic University of America, Washington, DC, 2008.

95. Yezierski, E. J. Observation as a Tool for Investigating Chemistry Teaching and Learning. In *Tools of Chemistry Education Research*; Bunce, D. M., Cole, R. S., Eds.; ACS Symposium Series 1166; American Chemical Society: Washington, DC, 2014; Chapter 2.

96. Bauer, C. F. Ethical Treatment of the Human Participants in Chemistry Education Research. In *Tools of Chemistry Education Research*; Bunce, D. M., Cole, R. S., Eds.; ACS Symposium Series 1166; American Chemical Society: Washington, DC, 2014; Chapter 15.

97. Salthouse, T. A.; Ellis, C. L. *Am. J. Psychol.* **1980**, *93*, 207–234.

98. Blignaut, P. *Atten. Percept. Psychophys.* **2009**, *71*, 881–895.

99. Santella, A.; DeCarlo, D. In *Proceedings of the 2004 Symposium on Eye Tracking Research and Applications*; ACM Digital Library: New York, 2004.

100. Sanger, M. J. Using Inferential Statistics to Answer Quantitative Chemical Education Research Questions. In *Nuts and Bolts of Chemical Education Research*; Bunce, D. M., Cole, R. S., Eds.; ACS Symposium Series 976; American Chemical Society: Washington, DC, 2008; Chapter 8.

101. Pentecost, T. C. Introduction to the Use of Analysis of Variance in Chemistry Education Research. In *Tools of Chemistry Education Research*; Bunce, D. M., Cole, R. S., Eds.; ACS Symposium Series 1166; American Chemical Society: Washington, DC, 2014; Chapter 6.

102. Brandt, S.; Stark, L. *J. Cognit. Neurosci.* **1997**, *9*, 27–38.

103. Levenshtein, V. I. *Soviet Phys.-Dokl.* **1966**, *10*, 707–710.

Chapter 12

A Short History of the Use of Technology To Model and Analyze Student Data for Teaching and Research

Melanie M. Cooper,[*,1] Sonia M. Underwood,[1] Sam P. Bryfczynski,[1] and Michael W. Klymkowsky[2]

[1]Department of Chemistry, Michigan State University, East Lansing, Michigan 48824, United States
[2]Department of Molecular, Cellular and Developmental Biology, University Colorado Boulder, Boulder, Colorado 80309, United States
[*]E-mail: mmc@msu.edu

The use of technology for teaching, learning and research has become almost ubiquitous in the chemistry classroom from student response systems, simulations and virtual environments, to online courses complete with assessments. The data generated by such activities can provide insight into how students learn and how we might provide environments that support learning. However, to take full advantage of the affordances of technology, the activities that students perfom must be meaningful and must generate useful data that can shed light on student learning and trajectories towards competence. In this paper we present examples from our work describing how we have used technology to investigate and assess student learning with large enrollment courses.

Introduction

Over the past few decades, technological approaches to teaching and learning have led to the development of a wide range of approaches to support learning including online homework systems (*1–3*), simulations (*4*), games (*5*), class response systems (*6, 7*) and even whole online courses where all of the readings, assignments, and tests are completed online (e.g. massive, open online courses – MOOCs) (*8, 9*). For the researcher, these technologies can provide a trove of data, that, if properly analyzed, could provide insights into a wide range of research questions.

Most of the work to date has been focused on whether the use of a particular technology improves learning outcomes for students. For example, studies have compared the use of online homework systems with traditional homework (*10–13*), how the use of clickers affects learning (*14, 15*), whether simulations improve conceptual understanding (*5*), and have compared online and blended courses with face-to-face courses (*16*). What is interesting about all these works is that the results are not conclusive; as was noted in the NRC DBER report (*17*) and in an extensive literature analysis (*16*), there is little strong evidence that technological aids to learning are effective in themselves. That is, *"The use of learning technology in itself does not improve learning outcomes. Rather, how technology is used matters more" (NRC DBER report)* (*17*). This is not to say that technological approaches to learning have no promise, but rather that the studies to date have not provided strong evidence. For example, there are almost no randomized control treatment studies on the effects of online vs face-to-face courses (*16*). As has been previously noted (*16*), many studies of online learning systems are conducted by parties who have something to gain from their findings, making any claims somewhat suspect (we would certainly not accept the claims of drug manufacturers without well designed clinical trials). It appears that merely incorporating technology into the classroom is not a guarantee of improved learning. For example, studies on the use of class response systems seem to indicate that it is implementation – to promote socially mediated learning – rather than the actual use of the technology that improves learning (*15*). Clearly, if we are to embrace the opportunities (and avoid the pitfalls) offered by online learning, it will become ever more important to learn how to assess student learning in technological settings. At the moment, the evidence required to make such decisions is lacking.

A second, much less investigated, area of research using technology involves its use to collect data that captures students' approaches as they answer questions and work through tutorials or simulations. That is, rather than investigating whether technology supported learning is more effective or efficient than traditional face-to-face instruction, it could be argued that an equally important use for such technologies is to investigate **how** students learn, how they solve problems, construct models and arguments, and how their trajectories to competence arise (*18*). This kind of analysis requires that we encourage students to answer questions that require them to construct responses, rather than being asked to choose from among a set of preconstructed responses. Constraining student actions by providing answers from which to choose, imposes conditions that may (almost certainly will) affect the questions we can ask about learning, and what we learn from those questions.

Although the promise of "big data" (*19*) may be highly seductive, it is important to understand that the nature of the online activity must be one in which students construct artifacts, rather than simply recognize them. No matter how many data points are collected, using commonly available, easily graded, types of assessments such as simple multiple choice questions cannot provide a particularly illuminating picture of student understanding. So, while a great deal of time and effort has been spent on developing intelligent systems that can provide each student with a customized set of assessments of increasing difficulty (*20, 21*), there is still a great need for technological systems that help students go beyond recognizing facts and using algorithms. These systems need to help students integrate ideas into conceptual understanding and construct models (diagrams, pictures and structures), arguments and explanations (*22*). The truism that assessments drive the enacted curriculum indicates that if assessments are unable to probe important skills and concepts, then what is assessed will become what is learned.

Systems That Allow Data Capture for Later Analysis

Technological supports to research data collection have been a mainstay of many studies for many years. For example, researchers routinely record both audio and video of student interviews or teaching activities for further analysis. See Herrington and Daubenmire's Chapter 3 as an example (*23*). More recently, it has become possible to record and replay students' writing with implements such as a Livescribe pen (*24*). In practice however, all of these qualitative data collection techniques are very time consuming to analyze and requires considerable expertise. In a typical study using recorded audio, video or drawings, only a few (usually less than 30) students' or teachers' data are analyzed. While these studies have provided important insights about teaching and learning, they are not as feasible for large numbers of study participants. Examples of the qualitative analysis process can also be found in Talanquer's Chapter 5 (*25*).

This chapter provides an overview of systems that use technology to investigate how data from large numbers of students can be used in a meaningful way to investigate how students develop scientific practices and skills. That is, the use of knowledge rather than the acquisition of knowledge.

Interactive Multi-Media Excercises (IMMEX)

There are a growing number of studies where technological approaches have been used to study teaching and learning for larger groups of students. One of the first such systems was *Interactive Multi-Media Exercises* (IMMEX) software, which allowed students to solve problems by choosing menu items that were tracked and stored in a database for further analysis (*26*). While the actions of the students were predicated on the menu items available, a typical problem required that students choose a sequence of as many as ten items to solve the problem, and the possible permutations were very large. These problems were "knowledge rich" and complex, and could not typically be solved by use of a heuristic or algorithm. There was not one way to solve the problem, meaning that there were multiple approaches to the solution. The sequences of students' actions were then clustered using data mining techniques (*27, 28*), which produced a number of problem solving strategies or models. This process is shown in Figure 1.

A typical IMMEX problem scenario for general chemistry (Hazmat) required students to identify an unknown compound by performing (virtual) tests (*29*). For example, students could perform a solubility or conductivity test to determine the properties present and identify the unknown compound. Ideally, students would use the results from each test to determine how to proceed, rather than randomly performing tests. Using the hidden Markov models (*30*) produced by the IMMEX analysis system, the students' strategies were determined and could be used to investigate how student problem solving abilities and strategies changed over time, and after interventions (*31*).

In our studies, we found that student's problem solving strategies stabilized after solving five problems (*31–33*). No amount of extra practice, even with targeted feedback, produced further improvements. However, if students were paired in problem solving dyads, where they were able to discuss their actions and had to jointly decide on how to proceed, most students significantly improved both their problem solving strategy and their ability (as measured by Item Response Theory – IRT) by about 10% after one group session. Moreover, these improvements were retained even after students returned to individual problem solving sessions (*31*). A more detailed discussion of the background and use of Hidden Markov modeling are provided in the original paper (*31*).

While almost all students improved equally in problem solving ability, there were two exceptions to this finding (*31*). Prior to the problem solving activity, students were given a test of logical thinking (specifically the Group Assessment of Logical Thinking – GALT (*34*)) and assigned to one of three groups: high (formal), medium (transitional) and low (concrete). If two students from the lowest category (i.e. those who had difficulty with such tasks as proportional reasoning or using data to make inferences) were paired, there was no improvement in problem solving – not a surprising finding. However, we also found that female students from the medium (transitional) group who were paired with a student in the lowest group improved significantly more than any other group, and ended up in the same problem solving category (i.e. with the same ability, and effective, efficient problem solving strategies) as students in the highest ability (see Figure 1, state 5). While much has been written on the advantages of collaborative learning, there are few examples like this study involving large numbers of students (around 800) showing how and where collaborative grouping can improve problem solving.

Unfortunately, the IMMEX software is no longer available for use and its discontinuation is a reminder of how many resources are needed to operate such software for research purposes. There are currently no other programs similar to IMMEX that are commonly available. The development of systems such as IMMEX require a great deal of expertise across a wide range of domains, from computer science to cognitive science. Disciplinary expertise and an understanding of psychometric techniques is also necessary. The inter- and cross-disciplinary teams required to construct and maintain a complex system such as IMMEX are few and far between, and the funding to support these systems is also difficult to maintain. Our intent here by including a discussion of IMMEX is to show what kinds of systems are possible.

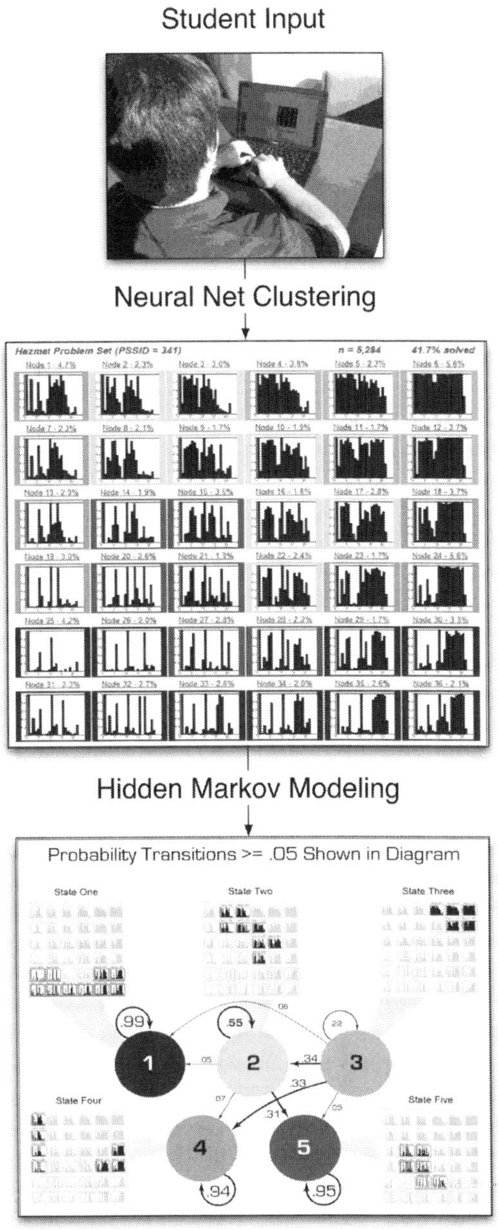

Student Input

Neural Net Clustering

Hidden Markov Modeling

Figure 1. Process for data collection and analysis using IMMEX. Adapted with permission from reference (31). Copyright 2008 American Chemical Society.

Moving from IMMEX to *OrganicPad*

While the IMMEX system provided a way to investigate how students solved quite complex problems and allowed the researcher to track and model student inputs, it was still somewhat limited in that all the possible actions had to be pre-programmed into the problem space. That is, using the Hazmat IMMEX example, each of the different unknowns must also contain pre-programmed results for the various simulated tests that students could perform (e.g. litmus paper testing, flame tests, solubility tests, and conductivity tests). While students had a great deal of choice, it was not possible to allow them to construct their answer "from scratch". The advent of tablet PCs, and even more so iPads and other tablets, has provided a more flexible interface on which students can write and draw directly. Our first foray into completely open-ended input was the development of a chemical structure drawing tool: *OrganicPad* (*35*).

OrganicPad is tablet PC software that can recognize and respond to free-form input of chemical structures; it has since morphed into a web-based cross-platform system as part of *beSocratic* (see discussion below). This system provides a natural environment in that students can construct *their* structural representations using a stylus, slate, or trackpad (*35*). *OrganicPad* can be used in a number of ways for teaching and research purposes: as a classroom response system to automatically grade and respond to students' structures (using its teacher-student interaction feature); as a formative assessment system where the system collects and grades students' structural drawings (using its quiz feature); or as a tutorial with multi-tiered assistance (using its pre-programmed contextual feedback feature) (*36, 37*). *OrganicPad* can recognize and respond to students' structural drawings in addition to providing feedback based on the students' input, as it was designed to guide students as they work to produce a reasonable structure (*36*). This feedback is multi-tiered in that initial feedback may be more general, but if a student does not respond with a correct action the feedback becomes increasingly more specific. For example, if a student's structure contains a carbon with six bonds, he/she might first be prompted to reflect on the number of valence electrons present for the structure. If the student continues to struggle, he/she would receive more specific feedback that carbon typically forms 4 bonds. A thorough description of *OrganicPad*'s various features are found in the user manual for the program (*37*). In all of these modes (in addition to recognizing and responding to student free-form structures), the students' input data are recorded and stored for later analysis. Using this replay feature, we were able to investigate the development of skills associated with drawing and using chemical structures.

For example, we used *OrganicPad* to record and analyze how students enrolled in organic chemistry developed Lewis structure drawing skills (*38*). We determined that students had great difficulty constructing any structure unless they were presented with structural cues (for example CH_3OH instead of CH_4O). Even with such cues, the success rate fell from around 80% to around 30% correct, when the number of carbon atoms increased from one to two (*38*). After supplementing this data with interviews and open-ended responses, we believe that the reason students have such difficulty with this task is that the "rules" for drawing structures are not based on any theoretical framework for learning.

Constructing Lewis structures is often seen (at least by students and sometimes by faculty) as an activity that has no meaning or purpose (for example, over half of all students at all levels did not report any use for Lewis structures beyond representing structural information) (*38*). That is, students were unable to see that the reason for learning to draw such structures is to use them as a predictive tool for how that substance may behave.

It should be noted that students can be quite successful in chemistry courses, even organic courses where structure-property relationships are central, by using heuristics instead of reasoning (*39*). That is, if students do not have a basic understanding of these ideas, they cannot use reasoning to predict answers, and must resort to heuristics and memorization.

In order to test this hypothesis, we designed a learning progression (*40–43*) for structure and properties, as part of a new curriculum – Chemistry, Life, the Universe and Everything (CLUE) (*44, 45*). Again we used *OrganicPad* to record and analyze how students constructed their representations of Lewis structures. Using the tagging feature for analyzing students' structures we were able not only to code the types of errors that were present in each student's structure, but also to develop a timeline for the actions of each student. For example, we could identify if the student was able to correctly connect the atoms to form a viable structure or if their structure contained too many bonds or electrons on carbon. We could analyze the order students constructed their structures, and whether this process was coherent or random. Using pre-programmed coding allowed us to analyze large numbers of students' structures since the program could recognize and compile identical structures. For example, 100 responses might be collapsed into 10 unique structures. This would give the researcher fewer representations to analyze and would help in the process of coding large data sets. In addition to coding students' common mistakes, we were also able to model the sequence of students' actions during this construction process. Markov modeling (*37, 46*) tracked each student's steps through the construction process. Based on similarities among structures, the user can determine which paths are most commonly taken for constructing structures. Again the reader is directed to the original literature for more details on how the modeling was performed.

These methods allowed us to compare students in the CLUE curriculum with a matched cohort. We found that the CLUE students were significantly more likely to construct reasonable Lewis structures than their traditional counterparts (*44*). It should be noted that without this technology we would have been unable to do this kind of comparison, not only because of the large numbers of students involved, but also because *OrganicPad* allowed us to track and model the students' input data.

In a similar manner, we used *OrganicPad* to investigate how students develop competence in constructing mechanisms in organic chemistry. Using the program's replay feature allowed us to create Markov models to identify the sequence in which students wrote mechanisms (for example, about 20% of students put the mechanistic arrows onto the reaction scheme after they had drawn it – that is, they did not use mechanisms for predictive purposes) (*47*). We were also able to determine that use of mechanisms did not affect the correctness of common reaction products, but when students were faced with a problem in which they did not know the answer, the students who used mechanistic reasoning were more likely to produce a reasonable product (*48*). Once again, it is important to emphasize that it was the data collection and analysis tools provided by *OrganicPad* that allowed us to do this research with large numbers of students, to model trajectories over time, and to compare students in different courses and stages in their development.

Moving beyond *OrganicPad*: The Development and Assessment of Science Practices

While *OrganicPad* has been useful in providing information that was previously unavailable, it is limited to concepts and skills associated with chemical structure drawing. We wanted to use the kinds of data collection and modeling techniques piloted with *OrganicPad* and extend them to other areas of chemistry (and science).

There is a growing recognition that a working understanding of science requires not only a grasp of the core disciplinary ideas, but also the ability to combine disciplinary knowledge with a range of science practices. The recent National Research Council Framework for STEM education (*22*) defines eight "science practices" including modeling, explanation and argumentation, while the Next Generation Science Standards (NGSS) (*49*) and the recently redesigned high school AP courses (*50*) link disciplinary knowledge and science practices into performance expectations that are assessed through tasks that involve the ability to accurately apply working knowledge to new systems (transfer). Both require students to construct drawings (by which we mean models, including graphs, and representations) and to use data, concepts, and reasoning to construct scientific explanations and arguments about specific phenomena. However, a major problem with this evolving approach to teaching science is that the assessments have not kept pace with the ideas and practices that we would like students to develop. If students are to learn to construct models, defend arguments and develop explanations, they must be provided with a learning environment that allows them to develop these skills. Ideally, formative assessment activities, where students receive meaningful feedback, would be part of this learning environment. That is, we need to develop systems that will be able to recognize and respond to student-constructed models, explanations and arguments. Furthermore, if we as researchers want to learn about how students develop these skills, and how to help students develop them, we need systems that allow us to collect appropriate data for analysis. The rest of this chapter will focus on how and why we developed *beSocratic* – a formative assessment system – and research activities that have resulted.

beSocratic: A Formative Assessment System for Free-Form Diagrams, Models, and Structures

beSocratic builds upon our work with *OrganicPad*; indeed, *OrganicPad* is now subsumed within the new system. *beSocratic* is designed to allow students to construct drawings, graphs and diagrams. However, systems that can recognize

– a priori – any drawing that students might construct have yet to be developed. Even recognizing relatively simple drawings is extremely difficult if there is no point of reference for comparison. It is relatively simple to develop a system that can recognize and respond to chemical structures, since there is an underlying architecture and set of rules that can be used by the programmer. Recognizing drawings requires a much more complex system, and while there are systems that can recognize simple drawings (for example *Cogsketch* (*51*)), they are based on highly sophisticated artificial intelligence systems and can be quite difficult to use, both for the instructor and the researcher. It was our aim to develop a system that would be easy to use, but would be powerful enough to recognize many types of diagrams, graphs and some drawings. Our goal was to hit the balance among flexibility, free-form input and ease of use. Currently over 100 activities have been authored in the system and have been administered as assessments, in-class activities, or homework. Access to *beSocratic,* along with pre-made activities, can be obtained by contacting the primary author of this chapter. A user guide for how to author *beSocratic* activities can be found on the website (*52*).

Description: *beSocratic is an online, cross-platform, intelligent tutoring system (21) designed for the recognition, evaluation and analysis of free-form student drawings (53). It consists of two main interfaces: 1. an instructor interface that allows for the development of activities and analysis of student data and 2. a student interface where activities are presented and completed. This system was specifically built as both a tool for instructors and researchers and has been used for both purposes.*

beSocratic activities are relatively easy to author (faculty have successfully developed activities after a two hour workshop). The authoring interface is purposefully designed to be reminiscent of *PowerPoint* or *Keynote*. That is, activities are developed as a series of slides, on which one or more *modules* are placed (Figure 2a). Most of the interactivity is gained through the *SocraticGraphs* module discussed below. Other modules allow positioning of text, images, text input boxes, drawing canvases, 3D molecular model viewers, *GraphPad* (*54*), and chemical structure drawing (*OrganicPad* module). All student input data, either drawing or text, is recorded for later analysis.

The *SocraticGraphs* module enables students to respond by drawing not only graphs, but also diagrams and pictures. Especially when used in conjunction with an underlying image, the range of responses that can be detected and responded to is quite broad (Figure 3). Student drawings are analyzed based on rules that have been pre-specified by the activity designer (i.e. researcher or instructor), such as the number of maxima/minima, area under the curve, slope, and intersections with coordinates or areas. The researcher/instructor can specify not only the shape of a curve, but also the number of curves and (if there is an underlying image) where student drawn responses should appear relative to that image. By using and combining these rules it is relatively simple to develop activities for drawing graphs and simple diagrams (Figure 2b).

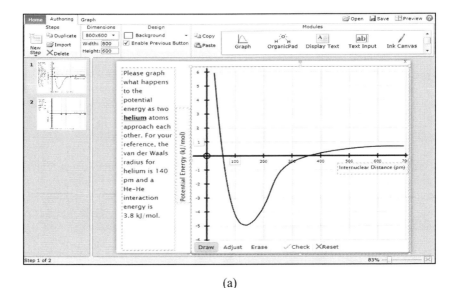

(a)

(b)

Figure 2. Screenshot of beSocratic: (a) authoring interface and (b) designing contextual feedback with SocraticGraphs module.

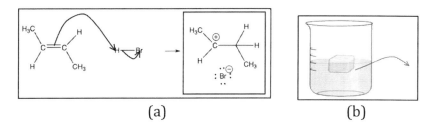

Figure 3. Examples of using SocraticGraphs with an underlying image: (a) arrow pushing in constructing mechanisms and (b) thermal energy transfer from the system to surroundings.

As with *OrganicPad*, students are presented with contextual feedback of increasing specificity when their submission does not meet one or more of the rules for a correct answer. Typically the feedback is designed to elicit reflection on the part of the student, rather than providing a correct/incorrect response. The feedback from the system usually requires a response from the student, either written or drawn, before the student can proceed. Even when a response is correct (or adequate), the student is also asked to explain their thinking in a pop-up text box before moving to the next page of the activity. This student response is also captured. An example of incorrect feedback is shown in Figure 4a, while Figure 4b displays a correct feedback prompt. This type of activity can provide rich data for both researcher and instructor. As we will discuss later, the data produced can be mined and modeled to investigate how students respond to such activities.

beSocratic also has a unique feature that allows students to edit a previously constructed drawing or text by making modifications according to their new understanding. More specifically, students can be presented with their initial model or explanation and asked to revise it by explaining or pointing out features that align with their new knowledge and those that do not. *beSocratic* can track not only the students' original submission but also their edits for later analysis. In fact this general procedure is applicable to a wide range of activities, including constructing models, explanations and arguments (see specific examples below).

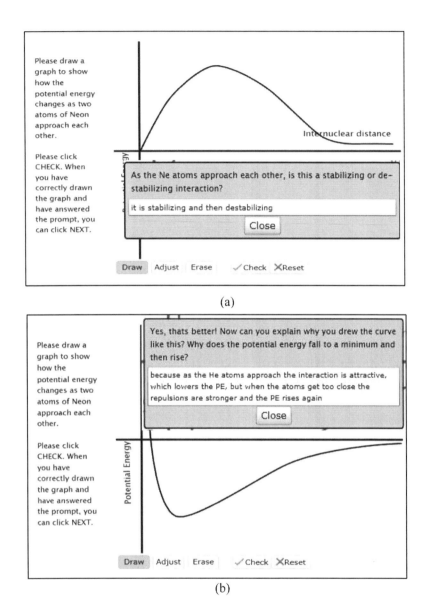

Please draw a graph to show how the potential energy changes as two atoms of Neon approach each other.

Please click CHECK. When you have correctly drawn the graph and have answered the prompt, you can click NEXT.

Internuclear distance

As the Ne atoms approach each other, is this a stabilizing or de-stabilizing interaction?

it is stabilizing and then destabilizing

Close

Draw   Adjust   Erase   ✓Check ✗Reset

(a)

Yes, thats better! Now can you explain why you drew the curve like this? Why does the potential energy fall to a minimum and then rise?

because as the He atoms approach the interaction is attractive, which lowers the PE, but when the atoms get too close the repulsions are stronger and the PE rises again

Close

Please draw a graph to show how the potential energy changes as two atoms of Neon approach each other.

Please click CHECK. When you have correctly drawn the graph and have answered the prompt, you can click NEXT.

Potential Energy

Draw   Adjust   Erase   ✓Check ✗Reset

(b)

Figure 4. Screenshots of feedback for (a) an incorrect student submission, and (b) a correct student submission.

beSocratic Data Analysis

There are a number of ways to visualize and analyze student submissions, both individually and for large data sets. For example, a specific student's response can be replayed so that sequences of their actions can be observed, coded or tagged for later analysis. This feature works with all modules in which students' submitted input, specifically including text, graphs and drawings. Alternatively, the grid view (Figure 5) presents a thumbnail of each student's final submission. Selecting a submission enlarges the thumbnail and provides the controls needed to replay the submission.

The grid view allows instructors and researchers to quickly scan the images for interesting submissions that may require further investigation or which might serve as useful (anonymous) exemplars for in-class discussion. *beSocratic* provides an excellent way to incorporate just-in-time teaching techniques (*55*). For example, in a large enrollment classroom setting we might have students vote (for example, by clicker) on the best answers and possible ways to improve the response.

beSocratic Post-Analysis Research Tools

beSocratic is capable of recording hundreds of student submissions for an activity. Since analyzing this volume of data can be very time consuming, we have integrated a set of post-analysis tools that helps facilitate the process in order to discover insights about student learning. Post-analysis in *beSocratic* is broken down into two stages: coding and clustering. During the coding phase, researchers are able to use *beSocratic* to attach codes to submissions either manually or, in some cases, automatically. For manual codes, researchers replay student submissions, pause the replays, and assign one or more custom codes at the paused position in the replay. In addition to these hand coded submissions, *beSocratic* can automatically assign codes for questions that used the *SocraticGraphs* module. In this case, the question's rules and feedback act as the codes themselves. It is important for us to code the replay instead of simply analysing the final submission. With the codes created in this way, *beSocratic* uses various clustering techniques (such as Hidden Markov Modeling (*30*)) to discover and visualize groups of students who have similar final answers and used similar actions to arrive at their final answer (Figure 6). By analyzing submissions in this way, *beSocratic* can quickly identify distinct strategies that students are using. Furthermore, *beSocratic* can use the results of the clustering (i.e. the strategies that students used) as input into future activities so that when a student is following a previously identified problematic strategy, the system can intervene with targeted feedback as an attempt to guide the student to a more correct strategy.

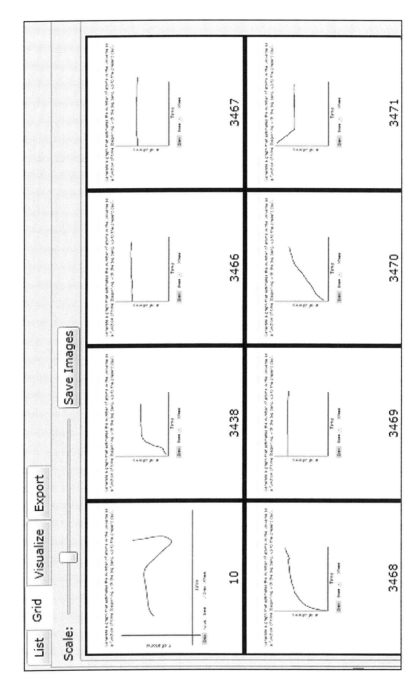

Figure 5. Grid view of a group of students' submissions for the question "Draw a graph of the number of atoms in the Universe vs time."

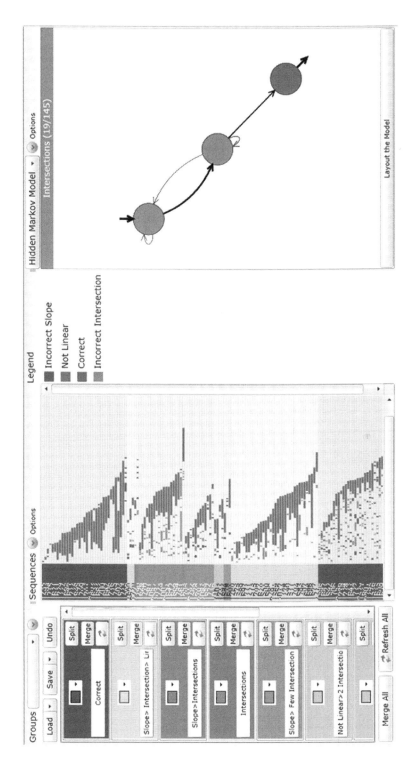

Figure 6. An example of hidden Markov modeling to determine students' strategies.

As previously noted, there is an increasing move towards assessments that require students to construct rather than recognize. There is also emerging evidence that student understanding cannot be captured by forced choice instruments that probe only one aspect of a complex concept. For example, Talanquer (*56*) has shown that many students use recognition heuristics to answer the kinds of ranking tasks that are often addressed in multiple choice questions. Furthermore, we have found that students who are able to correctly identify (for example) relative boiling points of compounds often use inappropriate reasoning (*39*). Instead of possessing a coherent model of a concept, many students utilize a rather loosely woven tapestry of facts, heuristics, and skills that may allow them to choose a correct answer without constructing a model or explanation. If we want to promote robust learning, we must provide students with appropriate learning materials that allow them to develop these skills, rather than accepting the results of shallow learning.

We present here two possible approaches to using *beSocratic* to support the development of conceptual understanding while, at the same time, emphasizing important science practices.

1. The Development of a Model To Describe the Energy Changes as Molecules or Atoms Interact

It is well known that many students believe bonds contain energy that is released when they are broken. This "misconception" is persistent, pervasive, and resistant to instruction (*57, 58*). While there are many reasons for this (*59*), it is clear that current approaches to teaching bond energy ideas are not effective. We are using *beSocratic* to develop approaches, using the construction of models and diagrams, to help students develop a more coherent concept of chemical energy involving the role of energy in atomic and molecular interactions. An early activity involves students constructing graphical representations of forces and energy changes (potential, kinetic and total) as two isolated atoms approach each other. As students construct these graphs, feedback is provided based on the underlying logic that has been supplied by the instructional designer. Figures 4a and 4b show a student's initial and final attempt to draw the potential energy (PE) curve as two neon atoms approach each other. After each attempt, the student is provided with a prompt (of increasing specificity if necessary), and finally asked to explain the shape of the curve they have drawn. Ultimately, they must construct not only the representation of the energy change, but also explain what that representation means.

Recall that models are important because they can be used to predict how a system will behave under different conditions. Typically in *beSocratic*, once the student has mastered the first activity they are asked to use that understanding in a new situation. So for example, in the interactions activity students are presented with their final PE curve of neon (*beSocratic* has a "copy previous" function that can copy a student's input from one screen to another) and asked to draw a new potential energy curve for the interaction of two argon atoms. That is, students are asked to show (and then explain) how the position of the minimum for argon is related to that of neon (Figure 7).

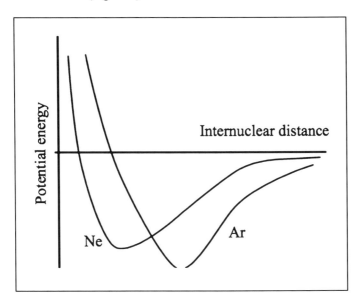

Figure 7. Development of a model for potential energy changes for interactions and bonding. A student's response indicating the potential energy change for the interaction of two argon atoms versus two neon atoms.

The students' responses for these activities can then be analyzed using a variety of techniques. For example, the students' graphing attempts can be clustered to determine the types of strategies present in that group (Figure 6). We can look at how student success on the initial activities follows through to the transfer activity by clustering the data from each activity. All of the student text responses have also been captured, providing a wealth of data that can be analyzed. While automated analysis of the text responses is not currently available, we are pursuing collaborations with other researchers to investigate this possibility (*60*).

2. Using beSocratic To Help Students Learn To Develop Arguments and Explanations

While *beSocratic* is not able to respond automatically to students' text submissions, there are a number of approaches we have used that involve student writing alone (that is, text without an associated drawing or model to describe). For example, we are using *beSocratic* to help students learn to construct arguments and explanations. Students might be asked to predict and explain their choice as to which of a given set of atoms is larger, has the highest ionization energy, or is the most acidic. Often when students are asked to do this kind of task, they will make a prediction (a claim), but are unable to explain why their answer is correct. Even if an explanation is given, it is often in the form of a heuristic (56), rather than one in which students use data or scientific knowledge to provide their reasoning and support their prediction. Therefore, when students are initially assigned such a task, we tend not to see a rich explanation of the phenomenon; in fact, little use of data or scientific understanding to support the argument is observed.

We have developed a series of activities in which students are asked the initial question in *beSocratic*, and then introduced to the idea of providing an explanation that has three components: a claim, the supporting data or evidence on which they are making this claim, and the reasoning that links these two. The student is then asked the initial question in three separate ways.

1. What is the claim you are making?
2. What data, evidence or scientific principle(s) are you using to support this claim?
3. How does this data/evidence/principle support this claim (i.e. what is the reasoning behind your answer)?

The student is then presented with their initial response, using the "copy previous" feature, and asked to edit their answer in light of the new reasoning. Students' initial and final responses can then be compared to evaluate the effectiveness of the activity. In this way, even though we cannot automatically analyze students' text input with *beSocratic*, we ask student to reflect on their own writing and improve it.

For example, in an activity where students were asked to explain the trend in atomic radius across a row in the periodic table, one student wrote: "*As you go across a row, the atomic radius decreases.*" After the activity she edited her response to read (changes from initial submission underlined for emphasis): "*As you go across a row, the atomic radius decreases. This is because the number of protons increases as you go across a row. The more protons that are present, the stronger pull the nucleus has on the electrons. Therefore, the radius is smaller because the electrons are pulled in closer to the nucleus.*" In this way we have encouraged the students to provide a much more full explanation of the phenomenon. This use of the affordances of the technology to prompt students to reflect and retrieve relevant information can make student thinking more accessible to the instructors and researchers.

Conclusion

In this chapter we have attempted to present a number of alternate approaches to the use of technology for research on teaching and learning. It is our firm belief that appropriate use of technology can provide information that will result in real improvements in our approaches to the development of research-based pedagogies. However, if we limit ourselves to approaches in which students are asked to complete forced choice assessments, the result will be that students will find it increasingly difficult to synthesize their ideas into coherent conceptual models.

The use of technology can help us learn how students develop not only knowledge, but also the science practices that allow them to use this knowledge in new situations.

References

1. Blackboard. http://www.blackboard.com/ (accessed September 1, 2013).
2. MasteringChemistry. http://masteringchemistry.com/site/index.html (accessed September 1, 2013).
3. The Purdue Online Writing Lab (OWL). http://owl.english.purdue.edu/ (accessed September 1, 2013).
4. PhET. http://phet.colorado.edu/ (accessed September 1, 2013).
5. *Learning Science through Computer Games and Simulations*; Honey, M. A., Hilton, M. L., Eds.; National Research Council, National Academies Press: Washington, DC, 2011.
6. i>clicker. http://www1.iclicker.com/ (accessed September 1, 2013).
7. Turning Technologies. http://www.turningtechnologies.com/ (accessed September 1, 2013).
8. MOOC Madness. The Chronicle of Higher Education. http://chronicle.com/ section/Online-Learning/623/ (accessed January 4, 2013).
9. Pappano, L. The Year of the MOOC. *The New York Times*, November 2, 2012.
10. Arasasingham, R. D.; Martorell, I.; McIntire, T. M. *J. Coll. Sci. Teach.* **2011**, *40*, 70–79.
11. Bonham, S. W.; Deardorff, D. L.; Beichner, R. J. *J. Res. Sci. Teach.* **2003**, *40*, 1050–1071.
12. Cole, R. S.; Todd, J. B. *J. Chem. Educ.* **2003**, *80*, 1338–1343.
13. Eichler, J. F.; Peeples, J. *J. Chem. Educ.* **2013**, *90*, 1137–1143.
14. Caldwell, J. E. *Cell Biol. Educ.* **2007**, *6*, 9–20.
15. MacArthur, J. R.; Jones, L. L. *Chem. Educ. Res. Pract.* **2008**, *9*, 187–195.
16. Lack, K. A. *Current Status of Research on Online Learning in Postsecondary Education*; Ithaka S+R: New York, 2013; p 72.
17. *Discipline-Based Education Research: Understanding and Improving Learning in Undergraduate Science and Engineering*; National Research Council, The National Academies Press: Washington, DC, 2012.
18. Lajoie, S. P. *Educ. Res.* **2003**, *32*, 21–25.
19. Manyika, J.; Chui, M.; Brown, B.; Bughin, J.; Dobbs, R.; Roxburgh, C.; Byers, A. H. *Big Data: The Next Frontier for Innovation, Competition, and Productivity*; McKinsey Global Institute: Atlanta, 2011.
20. Aleven, V.; Mclaren, B.; Roll, I.; Koedinger, K. *Int. J. Artif. Intell. Educ.* **2006**, *16*, 101–128.
21. Murray, T. In *Authoring Tools for Advanced Technology Learning Environments*; Murray, T., Blessing, S., Ainsworth, S., Eds.; Kluwer Academic Publishers: Dordrecht, 2003; pp 493–546.

22. *A Framework for K-12 Science Education: Practices, Crosscutting Concepts, and Core Ideas*; National Research Council, National Academies Press: Washington, DC, 2012.
23. Herrington, D. G.; Daubenmire, P. L. Using Interviews in CER Projects: Options, Considerations, and Limitations. In *Tools of Chemistry Education Research*; Bunce, D. M., Cole, R. S., Eds.; ACS Symposium Series 1166; American Chemical Society: Washington, DC, 2014; Chapter 3.
24. Linenberger, K. J.; Bretz, S. L. *J. Coll. Sci. Teach.* **2012**, *42*, 45–49.
25. Talanquer, V. Using Qualitative Analysis Software To Facilitate Qualitative Data Analysis. In *Tools of Chemistry Education Research*; Bunce, D. M., Cole, R. S., Eds.; ACS Symposium Series 1166; American Chemical Society: Washington, DC, 2014; Chapter 5.
26. Stevens, R.; Soller, A.; Cooper, M. M.; Sprang, M. In *Proceedings of the Intelligent Tutoring Systems 7th International Conference, Maceió, Alagoas, Brazil*; Lester, J. C., Vicari, R. M., Paraguaçu, F., Eds.; Springer: Berlin, 2004; pp 580–591.
27. Soller, A. *J. Comput. Assist. Learn.* **2004**, *20*, 212–223.
28. Soller, A.; Lesgold, A. In *Proceedings of the Conference on Artificial Intelligence in Education: Shaping the Future of Learning through Intelligent Technologies*; Hoppe, U., Verdejo, F., Kay, J., Eds.; IOS Press: Amsterdam, 2003; pp 253–260.
29. Case, E.; Stevens, R. H.; Cooper, M. M. *J. Coll. Sci. Teach.* **2007**, *36*, 42–47.
30. Rabiner, L. R. In *Proceedings of the IEEE*; 1989; pp 257–286.
31. Cooper, M. M.; Cox, C. T.; Nammouz, M.; Case, E.; Stevens, R. *J. Chem. Educ.* **2008**, *85*, 866–872.
32. Sandi-Urena, S.; Cooper, M. M.; Stevens, R. *J. Chem. Educ.* **2012**, *89*, 700–706.
33. Stevens, R.; Johnson, D. F.; Soller, A. *Cell Biol. Educ.* **2005**, *4*, 42–57.
34. Roadrangka, V. The Construction and Validation of the Group Assessment of Logical Thinking (GALT). Doctoral Dissertation, University of Georgia, 1985.
35. Cooper, M. M.; Grove, N. P.; Pargas, R.; Bryfczynski, S. P.; Gatlin, T. *Chem. Educ. Res. Pract.* **2009**, *10*, 296–301.
36. Cooper, M. M.; Underwood, S. M.; Grove, N.; Bryfczynski, S. P.; Pargas, R. In *Proccedings of ConfChem*; 2010.
37. Underwood, S. M. Bridging the Gap between Structures and Properites: An Investigation and Evaluation of Students' Representational Competence. Doctoral Dissertation, Clemson University, 2011.
38. Cooper, M. M.; Grove, N.; Underwood, S. M.; Klymkowsky, M. W. *J. Chem. Educ.* **2010**, *87*, 869–874.
39. Cooper, M. M.; Corley, L. M.; Underwood, S. M. *J. Res. Sci. Teach.* **2013**, *50*, 699–721.
40. Corcoran, T.; Mosher, F. A.; Rogat, A. *Learning Progressions in Science: An Evidence Based Approach to Reform*; RR-63; Consortium for Policy Research in Education: Teachers College, Columbia University, 2009.

41. Krajcik, J. S.; Sutherland, L. M.; Drago, K.; Merritt, J. In *Making It Tangible: Learning Outcomes in Science Education*; Bernholt, S., Neumann, K., Nentwig, P., Eds.; Waxmann: Münster, 2012; pp 261–284.

42. Schwarz, C. V.; Reiser, B. J.; Davis, E. A.; Kenyon, L.; Achér, A.; Fortus, D.; Shwartz, Y.; Hug, B.; Krajcik, J. S. *J. Res. Sci. Teach.* **2009**, *46*, 632–654.

43. Smith, C. L.; Wiser, M.; Anderson, C. W.; Krajcik, J. S. *Meas. Interdiscip. Res. Perspect.* **2006**, *4*, 1–98.

44. Cooper, M. M.; Underwood, S. M.; Hilley, C. Z.; Klymkowsky, M. W. *J. Chem. Educ.* **2012**, *89*, 1351–1357.

45. Cooper, M. M.; Klymkowsky, M. W. *J. Chem. Educ.* **2013**, *90*, 1116–1122.

46. Grinstead, C. M.; Snell, L. J. *Grinstead and Snell's Introduction to Probability*, July 4, 2006 ed.; The CHANCE Project; American Mathematical Society: Providence, RI, 2006.

47. Grove, N. P.; Cooper, M. M.; Rush, K. M. *J. Chem. Educ.* **2012**, *89*, 844–849.

48. Grove, N. P.; Cooper, M. M.; Cox, E. L. *J. Chem. Educ.* **2012**, *89*, 850–853.

49. *Next Generation Science Standards: For States, By States*; National Research Council, The National Academies Press: Washington, DC, 2013.

50. Science: College Board Standards for College Success. College Board. http://professionals.collegeboard.com/profdownload/cbscs-science-standards-2009.pdf (accessed June 29, 2012).

51. Forbus, K.; Usher, J.; Lovett, A.; Lockwood, K.; Wetzel, J. *Top. Cogn. Sci.* **2011**, *3*, 648–666.

52. *beSocratic*. http://besocratic.chemistry.msu.edu/ (accessed January 21, 2014).

53. Bryfczynski, S.; Pargas, R.; Cooper, M. M.; Klymkowsky, M. W. In *IADIS International Conference Mobile Learning*; Berlin, 2012.

54. Bryfczynski, S. P. BeSocratic: An Intelligent Tutoring System for the Recognition, Evaluation, and Analysis of Free-Form Student Input. Doctoral Dissertation, Clemson University, 2012.

55. Novak, G. M. *Am. J. Phys.* **1999**, *67*, 937–938.

56. Maeyer, J.; Talanquer, V. *Sci. Educ.* **2010**, *94*, 963–984.

57. Boo, H. K. *J. Res. Sci. Teach.* **1998**, *35*, 569–581.

58. Teichert, M. A.; Stacy, A. M. *J. Res. Sci. Teach.* **2002**, *39*, 464–496.

59. Cooper, M. M.; Klymkowsky, M. W. *CBE Life Sci. Educ.* **2013**, *12*, 306–312.

60. Automated Analysis of Constructed Response (AACR) Research Group. create4stem.msu.edu/project/aacr (accessed September 1, 2013).

Practical Issues for Planning, Conducting, and Publishing Chemistry Education Research

Chapter 13

A Two-Pronged Approach to Dealing with Nonsignificant Results

Diane M. Bunce*

Chemistry Department, The Catholic University of America, Washington, DC 20064, United States
*E-mail: Bunce@cua.edu

Chemistry education research requires a great deal of planning, coordination among students, faculty and institutions, as well as permission from the Institutional Review Board (IRB). The process is not quickly duplicated when an experiment yields statistically nonsignificant results. A check of the experimental design, analysis, and theoretical framework should be conducted before either accepting the nonsignificant results as valid or deciding that the experiment must be redesigned and re-implemented. This chapter discusses a two-pronged approach of Planning and Post-hoc analysis for dealing with nonsignificant research results before making that decision. Included in this approach is the importance of developing well constructed experimental designs that are not dependent on the results of a single low power statistical test and are grounded in appropriate theoretical frameworks. It also includes strategies for re-examination or re-analysis of nonsignificant results when a solid research design has been employed.

Introduction

As researchers, we understand that research is not complete until it is shared with others in the field through widely available publications. Publication in peer-reviewed journals, monographs, or books—either online or in print—is the time-honored way that professionals make their work available to others. But how often have you seen authors publish research that shows no significant results? Many researchers (and reviewers) seem to believe that if there are no statistically significant results, the research is not valid and therefore not worthy of publication. However, this is not necessarily true. If the research is carefully planned and analyzed, it is just as important to report nonsignificant results as it is to report significant results. This expectation is more prevalent in medicine than in studies on learning. For instance, according to Jaykaran et al. (1), of the 276 studies published in Pub Med journals in 2009, 18.4% described important primary outcomes that were not statistically significant. There are also articles published in other fields that report nonsignificant results for their main or primary effects (2).

Statisticians argue that nonsignificant p values don't necessarily mean that the null hypothesis has been proven. In other words, when you have a nonsignificant p value, the compared groups may not be statistically significantly different but that doesn't *prove* they are equal (1). Other intervening variables in the research methodology must be considered including adequate sample size, effect size (strength of association) (3), and power. It is important that researchers are aware of the experimental design parameters that can unduly affect one's ability to predict significance. Included among these parameters are the following three issues: (1) The ability to meet the assumptions of the statistical tests used, including normality of distribution or homogeneity of variance; (2) Effect size (partial eta squared), which is a measure of the degree to which the two variables are associated with one another; and (3) Power, which can be influenced by sample size, effect size, or alpha level (4).

The question remains—is research with nonsignificant differences valid and if so, is it publishable? The answer is a strong "it depends". It depends on issues of theoretical framework, experimental design, and analysis. Part of this discussion also relies on the researcher's ability to both explain and defend the research to editors, reviewers, and the reader. Understanding the issues of experimental design, theoretical frameworks, and analysis can empower a researcher to develop a manuscript that highlights the strengths and accurately describes the limitations of nonsignificant results. Tightly designed and knowledgeably explained research with nonsignificant results is worthy of publication because it can address important issues. Nonsignificant results do not necessarily mean no results.

Specific Issues To Be Addressed in Chemistry Education Research

It takes a great deal of time to plan and execute quality research in chemistry education. Both qualitative and quantitative studies involve time intensive protocols and analyses. Quantitative research designs typically require a good deal of time on the front end identifying variables, modifying or creating the tools to measure the variables, and determining the validity and reliability of the tools before using them. Database formats should be determined and data collected, entered into the database, and combined or modified before being analyzed. Statistical tests used in quantitative research must be selected and matched to the questions asked and the data collected. However, before these statistical tests can be legitimately employed, the assumptions of each test must be met or valid modifications made to the data, test, or interpretation.

Qualitative research is not a quick and necessarily less structured approach to research. There are "rules of engagement" for qualitative research designs that also must be addressed. In chemistry education research, these often include, but are not limited to, selecting the type of qualitative methodology used such as surveys, interviews, think-aloud sessions, observations, ethnographic studies, focus groups, or free response written answers. Qualitative research often generates a very large amount of data and the researcher must be prepared to handle and analyze such an abundance of information. Often inexperienced qualitative researchers become so bogged down with the data collected that they run out of time or perseverance to analyze it effectively.

These methodological issues are well within the realm of problems faced by researchers in any field of chemistry. However, there are other issues facing chemistry education researchers that are not typically faced by chemists but which are equally important in terms of being able to carry out the study of choice. The main issue here is that chemistry education researchers are dealing with human subjects with all the rights, protections, and availability issues that are associated with such research. Human subjects must be protected not only from bodily harm but also from misrepresentation of the purpose of the research they are participating in and violation of their privacy. The ethics of this type of research also include safeguarding subjects who are enrolled in school and are not to be put at undue risk of failure or decreased academic success by agreeing to participate in the research. For these and other reasons, all research that involves human subjects must be guided by the national rules for conducting such research. Most colleges, universities, school districts or other organizations have set up Institutional Review Boards (IRB's) whose role is to assure that all human subjects involved in research understand the risks and benefits of engaging in such research and are guaranteed the right to either join or withdraw from the research at any time without any detrimental consequences. The chapter by Bauer (5) in this book describes the IRB review process in more detail. However, it is important to realize that IRB guidelines may affect or alter the researcher's initial plan for collecting data. In some cases the question the researcher wants to investigate cannot be addressed by the data that is permissible to collect under IRB guidelines. In these cases, the question asked may need to be modified and in other cases the data collected may need to be aggregated or analyzed differently in order to accommodate both the researcher and the IRB. Putting all these pieces together requires creativity, patience, persistence, and the ability to compromise

When working with intact classes, it is also advisable to take into consideration the timeframe of academic quarters, semesters, or school years. If the experiment is not planned with this framework in mind, the student population, teacher, or content may change substantially due to preset academic timeframes. This poses a viable threat to data collection or interpretation. It therefore is necessary to plan the research within the academic timeframe of the population being investigated. This might mean that the actual planning of the research or preparation of the IRB proposal begins earlier than first anticipated.

In many institutions, IRB guidelines are interpreted to prohibit including the researcher's current classes in the research being conducted. As discussed by Bauer (5), this interpretation may vary by institution. Under such guidelines, the cooperation of other teachers may be necessary to provide access to a population that is not under the grade control of the researcher. Cooperation of other faculty can be a good opportunity to engage non-chemistry education researchers in a project to address common problems. Cooperation under these circumstances may be relatively easy to organize. However, in some situations other faculty may have little interest in the question being investigated and may not be particularly interested in being inconvenienced or having their workload increased to support the project at hand. The researcher must be sensitive to these concerns and also to the issue of the other faculty members' perception of the threat of judgment of their teaching abilities that such cooperation might entail. Cooperating teachers, as well as the subjects, must have their professional and personal rights protected in the research study. Open communication, safeguards to professional and personal reputations, minimization of additional workload and communication about the purpose and methodology of the research are the responsibility of the researcher in this situation. The researcher must also accept at face value the decision of colleagues not to participate in the research. Cooperation with other colleagues can add to the validity and feasibility of a chemistry education research project but such arrangements may take time, planning, and open communication.

In light of all the planning, permission, and cooperation that can be involved in setting up chemistry education research projects, it is especially discouraging when the results of a study are nonsignificant. Very often the IRB approval process, academic timeframe, and cooperation agreements with other teachers will not sustain a duplication or modification in a "do over" experiment. The researcher often will need to start the process over again from the beginning. This could delay a follow-up experiment by several months or even a year.

Taking these factors into consideration when planning a research project should help minimize organizational and timeframe problems that might add to the chance of nonsignificant differences in a quantitative or mixed design study. If these organization issues can be controlled, then there is a greater chance that nonsignificant results, if they occur, can be interpreted as valid.

Two-Pronged Approach to Dealing with Nonsignificant Results

There are two ways of handling the issue of nonsignificant research results. This two-pronged approach involves **Planning** and **Post-hoc analysis**. Planning involves developing an experimental design that identifies and controls confounding variables to help insure that nonsignificant results are valid. Post-hoc analysis includes re-examining both the data and analysis to check for accuracy, validity of statistical tests used, and possibilities for re-analyzing the data already collected to further address the original or related research questions.

Planning

Experimental Design and Theoretical Framework

Research is an iterative process. It is not actually carried out in the straight forward, efficient way it appears in the final publication. Through a formative evaluation process, decisions can be made throughout the research that were not anticipated at the start. However, the research process works best when the advance planning is thorough and well done. Thorough research planning includes asking an important question that has a strong theoretical framework and uses a multi-variable experimental design. The tools used to measure these variables should be identified, modified or created, and tested before they are used to collect actual data. The data collected, either qualitative, quantitative, or both, should approach the research question from different directions but all directions should add to the understanding of the research question. This approach sometimes takes the form of investigating smaller research questions whose results help explain the main question even if they don't directly or fully address that question. The conclusions from the investigation of any research question should be tied to the theoretical framework that drives the research. Without this connection to theory, each experiment is a stand alone "point of light" that doesn't advance our systematic understanding of the question asked. The following discussion includes several areas that should be taken into account in the planning process.

Asking Good Questions

It is important to ask important questions in research. Many times the important questions are not easy to investigate and researchers may opt to address the easier questions. An example of opting to address easier questions includes the situation when we really want to know if a new method of teaching helps students learn but we only collect data that investigates how students feel about the new method. It is true that how students respond to a teaching method has an effect on how well they learn, but investigating only how students feel is not the same as investigating if there are statistically significant learning gains with the new method. Learning gains are harder to measure than student opinion but student opinion alone doesn't fully address the question of learning.

Once the research question has been identified, the researcher should ask what "take-home" message he or she would like to report as a result of this research. Identifying the "take-home" message is a good way to insure that you are asking the questions you really want to investigate. The process of getting from your question to an experimental design and eventually to the "take-home" message has been discussed elsewhere (6).

To illustrate what is meant by asking important questions and comparing them to the take-home message, information is presented here for a specific study on the effect of online homework (HW) on student achievement in general chemistry. The question that initially drove the study was the following: Does the use of online HW affect student achievement in general chemistry? The take-home message was hypothesized to be either: (1) The use of online HW does increase student achievement and therefore, is worth the effort and expense or (2) The use of online homework does not increase student achievement and therefore, is not worth the effort and expense.

Theoretical Framework

Theoretical frameworks are commonly misunderstood by journal authors. The purpose of a theoretical framework is to situate the current research within a larger picture. In chemistry education, the larger picture is often how learning occurs within the student and what can be done to facilitate that learning. In order to investigate the larger picture, theories of how learning occurs may be needed to explain how the current research fits into the larger picture of learning in general. Theoretical frameworks that apply to learning are often found in the fields of educational or cognitive psychology, sociology or technology education. The importance of having a theoretical framework is explained in other references (*7–9*).

Researchers should also be familiar with similar research that can inform the proposed research design. References to similar research can be thought of as an application framework. New research should be proposed after a thorough review of other work on the question or related questions has been completed. Without a thorough review, the proposed research could be repeating what has been done or make the same experimental design mistakes that other authors have already identified.

An informal survey of research articles published in the same two month period, primarily from two journals—*Journal of Chemical Education* and *Journal of College Science Teaching*—show a range of studies that include both theoretical and application frameworks to those that include mostly application frameworks. The continuum of articles that have both theoretical and application frameworks (*10–16*) to those that have mostly application frameworks (*17, 18*) are included to provide the reader with some examples of the differences.

In the online HW study described above, the theoretical framework of the expected benefit of online HW was based on the theory of the importance of immediate feedback (*19, 20*) and the Transfer-Appropriate Processing (TAP) approach (*21, 22*). The TAP literature discusses whether it is important to have the homework (or practice) question format match the format of the test question. Another theoretical framework that could explain the effectiveness of online HW is the information processing view of learning that equates true learning with a change in schema organization (*23*). According to this theory, if there is no change in a student's schematic organization of knowledge then learning has not taken place. This theory has implications for the effectiveness of online HW. If online HW does not engage students in the processing of new knowledge with previously held knowledge, then it is doubtful that online HW will have a significant effect on learning that would result in increased test achievement.

The application framework for the online HW study included articles that reported a significant difference in achievement between paper vs online HW in a large enrollment physics class (*24*) and those that showed no significant difference between paper vs online HW (*25, 26*). Another study reported on student perception of the benefits of online HW (*27*). A second application framework considered included the effect of parameters that can be set in how an online HW assignment is conducted. For instance, if the number of tries that are set by the instructor in the online HW are too numerous, then guessing rather than thinking about the material may result. Kortemeyer (*28*) investigated how males and females reported using multiple tries in an online HW environment. Student responses about how they used multiple tries included "to submit random stuff or guess" as well as "having many tries allows me to try out my own approach without the stress or worry about grades". If multiple tries encourage guessing, then little change may take place in students' schemas and as a result, online HW may have little effect on achievement. If multiple tries encourage students to attempt different understandings of a concept, then online HW should increase student achievement.

Revising the Question

Consideration of the theoretical and application frameworks often leads to new ideas or new ways of looking at a situation. This can result in a revision of the original question or the development of subquestions that can be more easily measured than the overall question. In the online HW study, the theoretical and application frameworks led to a revision of the original question—does the use of online HW affect student achievement in chemistry? It was tthought that a revised question could address why some studies in the application framework report significant effects on student achievement of using online HW while others do not. The hypothesis was that not all online HW questions are the same. It was hypothesized that conceptual and traditional questions might have a differential effect on student achievement. Nothing in the literature showed that this aspect of the influence of online HW on achievement had been previously investigated. It was further hypothesized by the researchers that if the effect of online HW on achievement existed but was small or if the use of conceptual and traditional questions had opposite effects on achievement, then the overall result of online HW might be masked. The original question in the online HW study was modified to directly investigate the effect of conceptual vs traditional online HW questions on student achievement. Revised Question: Do conceptual and traditional online HW questions have a differential effect on student achievement? Revised take-home message: If there is a differential effect due to question type, teachers may purposefully choose which type of questions to include in their online HW assignments.

A treatment vs control design is an effective way to control many intervening variables at once. By identifying and controlling these variables, there is more confidence that the results of the research are valid. In traditional science experiments, treatment and control studies are fairly straightforward to set up. However, in studies with human subjects where the researcher often does not have the administrative ability to randomly assign students to one section or another or to split a given class into a treatment or control group, this experimental design is often difficult to initiate. In this situation, the best option for human subject researchers who want to do classroom research may be to use a one-class design where different levels or variations within the same class are compared. For instance, one can look at the effect of a particular teaching or learning innovation on high vs low achievers in the same classroom; male vs female students; or students who invest a certain amount of time into studying and those who do not; etc.

In the online HW study, a preliminary "learning laboratory" type study was planned to see if there was a difference in how students solved conceptual vs traditional questions in terms of where they looked on the computer screen for an online HW question and how often they correctly solved each type of question in an online environment. This study was based on the underlying theoretical framework of the mind-eye theory (*29*) that says people are thinking about the object they choose to look at. In this study, that meant it was important to investigate whether students looked at a different part of the problem on the screen (question, answers, and hint) or spent a different amount of time on the question, answers, or hint for conceptual vs traditional online questions. This data addressed the question of whether students solved conceptual and traditional questions differently in terms of what part of the question they concentrated on and for how long. The second question addressed in this preliminary study was whether students were successful in solving conceptual online HW questions at a different rate than traditional questions.

The mechanics of the the eye tracking technology used in this study included measuring the gaze patterns, which included where students looked on the screen *and* the amount of time spent on each of the components (question, answers, and hint). Eye tracking, like think-aloud interview protocols (see Herrington and Daubenmire (*30*)) is typically performed in a one-on-one interview format. The screen on the eye tracker in this experiment displayed the problem including question and answers with a box labeled "hint". If students chose the hint, a new screen appeared with the question, answers, and hint displayed. The eye tracker measured where and for how long students looked within each of these pre-designated Areas of Interest (AOI's)—question, answers, and hint. Students' answers to the questions were also evaluated for correctness.

Often the research tools we need in chemistry education research do not exist and must be either created or modified from existing tools. Regardless of whether tools are created or modified, issues of the validity and reliability of the tool(s) must be addressed. Validity is defined as the extent to which a tool measures what it is supposed to measure (*4*). There are three different types of validity, namely, content, criterion, and construct validity. Content validity deals with how well the instrument measures the appropriate content. Criterion validity addresses the question of how well the instrument measures the same variable as an established instrument. Construct validity compares an instrument to the underlying theory. Reliability, by contrast, is defined as a measurement of how free an instrument is from random error (*4*). It involves the stability of an instrument and its internal consistency. Researchers often use tools that have been used for other purposes and modify them to fit the current research. Examples of this include research using modifications (*18*) of the established online Student Assessment of Learning Gains (SALG) (*31*) and tools (*25*) originally developed for other purposes. When tools are modified or used with a different population, the validity and reliability should be re-established.

The questions used in this eye tracker study were modified from online HW questions already assigned, or if no appropriate question could be found, then a question was created. The questions and hints were designed to fit a 2x2 array which included Conceptual Question–Conceptual Hint (C-C); Conceptual Question–Traditional Hint (C-T); Traditional Question–Traditional Hint (T-T); and Traditional Question–Conceptual Hint (T-C). In some cases, new hints had to be constructed to fit the cell within the array. A traditional question was defined as testing recall or algorithmic knowledge. A traditional hint was a hint that gave specific directions for solving the problem at hand. A conceptual question was defined as a question that required higher order thinking such as analysis or problem solving. A conceptual hint was one that provided help with the underlying concept and not directly with the specific problem. The validity of the questions (appropriateness for the students enrolled in the course) was reviewed by the course instructor. The validity of the conceptual/traditional designation of questions and hints was determined from the ratings of two chemistry education research reviewers who were not otherwise involved in this research.

In the online HW study, ten students were recruited from a nursing chemistry class that had used online HW all semester. Each student was asked to solve 4 problems on the eye tracker that were randomly selected to include one question from each cell of the 2x2 array described previously. Students solved the problems and data were collected on both student achievement and gaze patterns for the questions assigned.

Triangulate Data

When interpreting data in an experiment, it is important to interpret it objectively. There is a temptation for the researcher to make the data say what is expected or hoped for. If the experimental design is not tight enough or the questions investigated not specific enough, a lot of variability can exist in data interpretation, which could lead to an incorrect conclusion. One way to handle this situation is to triangulate data that cannot be measured as accurately or as specifically as would be hoped. Triangulation is defined as an effort to measure a phenomenon by several independent measurement routes (*32*). Triangulating data can help control the variability in data interpretation through the development of a more extensive experimental design. In terms of planning for the possibility of nonsignificant results, the more data collected to address the same question and the more perspectives investigated, the stronger the case that nonsignificant results are valid results.

Triangulation can be a combination of qualitative and quantitative data to address the same question. Combining qualitative and quantitative data is the basis for a "Mixed Methods" design. Qualitative measures have the advantages of being able to help (1) identify pertinent variables that might not be otherwise known in a study, (2) bring the research question into sharper focus with the measurement or control of these pertinent variables and (3) support the accuracy of interpretation of quantitative results. Qualitative studies often involve smaller sample sizes, which limits the generalizability of the results. Qualitative studies *do* give a more indepth picture of what is actually happening in student learning by addressing questions of what kind of learning and why or how learning takes place. By contrast, quantitative studies typically include larger sample sizes and mark trends or combine results that increase the generalizability of the study. Quantitative studies often do not address the questions of why or how something takes place.

In the online HW study's preliminary eye tracking experiment, a qualitative approach involving ten students was used to inform the next part of the study that examined the effect of conceptual vs traditional online HW questions on student achievement on tests.

Building in Backup

With all the effort that goes into setting up a human subjects' research project, it is wise to redefine the original question into a series of subquestions that can each be investigated with individualized methodologies and tools. This approach increases the chances of finding significant differences in one of the several subquestions rather than hinging the entire project on the statistical results of one main question.

In the online HW study, the revised research question (Do conceptual vs traditional online HW questions have a differential effect on student achievement?) was further subdivided into several more specific questions that each had its own methodology and results to report. For the qualitative eye tracking experiment, the following two questions were proposed:

- Are students more successful solving one type of online HW question (C-C, C-T, T-C, or T-T) than another?
- Do students have different gaze patterns in areas of interest (question, answers, and hints) for each of these question types?

For the whole class study, the following questions were proposed:

- Does student achievement on conceptual or traditional online HW problems affect

    - overall student achievement on tests?
    - student achievement on conceptual vs traditional questions on tests?

- Does student achievement on test questions with an online HW antecedent question (conceptual or traditional) differ from achievement on test questions without an antecedent online HW question?

The addition of these finer grained questions requires different methodologies and collection of different data sets. By developing these questions as partial answers to the overall research question, we both increase our ability to more fully address the overall research question and increase our chances of obtaining significant results for at least one subquestion.

Pilot vs Preliminary Studies

Pilot studies are small scale studies that help researchers evaluate the feasibility of a larger study. Pilot studies are also very useful in trying out experimental designs and instruments. Many a problem, including nonsignificant results, can be averted when a particular experimental design is tested on a small scale. During a pilot study, one might find that the method of measuring differences is not detailed enough or the plan for interviewing students does not attract an adequate number of volunteers. Other important lessons that may be learned from pilot studies are that students do not volunteer to complete online surveys in a sufficient number or that the number of questions asked during an interview are too numerous and take too long to answer. Pilot studies can help the researcher avoid substantial squandering of time, effort, and money in the investigation of a research question that results in nonsignificant differences.

Preliminary studies are also done to determine the reliability of tools such as surveys or think-aloud questions. By conducting preliminary experiments, the researcher can determine the reliablility and validity of instruments before launching a full study.

The main difference between pilot and preliminary studies is that pilot studies normally include a majority of the components of the main study whereas preliminary studies are usually focused on a small part of the larger study. Both offer important information dealing with efficacy of the experimental design and analysis of the data for a planned larger study. It is highly recommended that either or both pilot and preliminary studies be conducted before launching a bigger study to insure meaningful research results.

In the online HW study, we did not do a pilot study. Instead we did a preliminary study using a limited number of students in an eye tracking learning laboratory situation. The results of the preliminary study suggested that the students did not treat conceptual and traditional questions differently in terms of the time spent or components (question, answers, hint) analyzed. The results of the preliminary study shaped our subsequent larger study in the sense that we no longer deemed it necessary to have specific types (conceptual or traditonal) hints available for every online HW question we included in the study.

Explain Conclusions within Theoretical Framework(s)

Although it is unusual to talk about conclusions before the data are analyzed, conclusions are included in this experimental design and theoretical framework section of the chapter because they involve some of the issues involving theoretical frameworks described here. When writing conclusions it is important to remember that conclusions must be based on the data presented and not the data the authors wish they had collected. In addition to relating to the data, the conclusions should refer back to the theoretical framework and be explained within that framework. The conclusions should be compared to the research results from the application framework and either expand conclusions in the research already published or confront them with explanations based on both the experimental design and the theoretical framework.

Table I presents a summary of both the questions asked and the subsequent results in the online HW study. The purpose here is to demonstrate how conclusions (based upon the data analysis) should be related to the theoretical and application frameworks found in the literature.

Conclusions about these results can be explained in terms of the theoretical framework of cognition theory that says that unless the activities (in this case, doing online HW) cause students to modify their existing schema, learning will not take place (*23*) and the Transfer Appropriate Processing (TAP) Model (*22*), which says that similarity of format of study and test question can affect achievement. In this study, similar online HW and test formats were used in this research that investigated whether having an antecedent online HW question affected achievement on *corresponding test questions*. The data for the research question of this study regarding whether having an online HW question in general affected achievement on *total* test scores also showed no significant differences. The conclusion then, based on the theoretical framework of schema modification and TAP, is that similar format of online HW and test questions is not sufficient to bring about schema modification resulting in increased learning as measured by achievement on tests in this study.

The application framework, as opposed to the theoretical framework for this research, included studies that showed no significant effect of online HW on achievement (*25–27*). Our online HW study produced similar results even when the variable of online question type (conceptual vs traditional) was included in the methodology. The conclusions of this research then indicate that the failure to prove an effect of using online HW questions on student test achievement is not explained by differentiating the type of online HW question assigned (traditional or conceptual) or matching the format of the online HW and test questions.

Table I. Overview of the questions and results of the Online HW study

Question	*Summary of Results*
Are students more successful solving one type of online HW question (C-C, C-T, T-C, or T-T) than another?	There was no statistically significant difference in achievement among any of the 4 types of questions.
Do students with different self-reported course grade levels have different gaze patterns or time spent in the Areas of Interest (AOI's) (question, hints, answers) for each of these 4 types of questions?	Yes, self-reported high achieving students (B+ or higher) spent a statistically significantly longer percentage of overall problem solving time on the question AOI for all 4 types of questions then self-reported lower achieving students (C or lower).
Does student achievement on conceptual or traditional online HW questions affect overall achievement on tests?	No, there was no statistically significant difference on overall test achievement based on achievement on conceptual or traditional online HW questions.
Does student achievement on conceptual or traditional online HW questions affect achievement on conceptual or traditional questions on tests?	No, there was no statistically significant difference on conceptual or traditional test questions based on student achievement on online HW conceptual or traditional questions.
Does student achievement on test questions with an online HW antecedent question (conceptual or traditional) differ from achievement on test questions without an antecedent online HW question?	No, there was no statistically significant difference in achievement on test questions that had an online HW antecedent question, regardless of type, vs those that did not.

These nonsignificant results led to both a re-examination of the theoretical and application frameworks previously identified and to an expanded search of other frameworks that might help explain the situation. For instance, the theoretical framework involving the change in cognitive structures might necessitate students reflecting on a problem after it was solved. Solving online HW questions does not necessarily require this self reflection. Online HW problems can be solved by students using notes, textbooks, web, or peers to help "get the right answer". In addition, if the number of tries for an online HW question is not limited, guessing rather than learning could be taking place. Graf and Ryan (*22*) describe the importance of conceptually engaged student study for increased retention on tests. If students are not engaged with the concept of the questions in online HW but rather are spending effort only on getting answers, there may be little or no schema modification or learning occurring and thus no significant difference in achievement on tests. Kang et al. (*20*) conclude that the more demanding the processing in the study situation, the greater the benefit on the test. If students circumnavigate the demands of online HW by making the answer a higher priority than the logic to understand the underlying concept, a diminished effect would be expected on achievement.

In an effort to re-examine the nonsignificant results of this study in light of other studies in the application literature, the reason for obtaining mixed results on the effect of online HW on achievement was hypothesized to be class size. If classes are small, the effect of online HW on test achievement might be minimized due to more individual mentoring or the teacher recognition and response to individual student confusion about concepts. In a large class with a high student to teacher ratio, this response might not be feasible. An examination of the literature with small (n = 45-64) (*25*) and large (n= 97-381) (*24, 26, 33*) classes did not bear out this hypothesis. Research with small and large classes both reported no significant results for the impact of online HW on achievement. Thus, this effort to explain the nonsignificant results was not useful.

The Kortemeyer study (*28*), which suggests that a large numbe of multiple tries for online HW may affect how students approach the completion of questions, did not explain the results either since students were given only two tries for each online HW question in this study. The limited number of tries in this study partially controlled for random guessing.

After investigating both the theoretical and application frameworks, there did not appear to be a ready explanation for why nonsignificant results were obtained in this study. So having convinced ourselves through a re-examination of both our theoretical and application framework and our methodology including tools used and data measured, our confidence in the validity of the nonsignificant results increased. Now it was time to conduct a post hoc analysis of the data from this study to extend the examination of the validity of our results and conclusions.

Summary of Planning for Nonsignificant Results

To summarize the planning aspect of research designs that limit the generation of non-valid nonsignificant results, we can identify several key points in our re-examination of our research plan, namely, reviewing that the following has occurred:

- Experimental design is appropriate and well planned
- Theoretical and application frameworks are well-defined and adequately researched
- Tools used are valid and reliable
- Analysis is appropriate and complete

However, even if all of these experimental requirements have been met and it is determined that the results are indeed nonsignificant, it is likely that reviewers and editors will look more closely at these parameters than normal before agreeing to accept the research for publication. A post-hoc analysis of data should also be completed to assure that the data analysis was appropriate, correct, and complete.

Post-Hoc Analysis

As has been discussed previously, even when the "rules" of good research design are implemented, nonsignificant results can occur. Before these results are declared valid and the study submitted for publication, it is important to engage in a post-hoc analysis to see if there are other results that can inform the findings or conclusions.

Analysis of Data

Formative evaluation of a research study should occur as the study proceeds. This includes modifying the research question(s) to make them more specific; building in back-ups in the experimental design; and using both the theoretical and application frameworks to interpret results. Post-hoc analysis, on the other hand, deals primarily with the original data analysis and possible additional analyses that can be performed on that data.

Double Check the Data and Statistical Output Files

Although it might seem trivial to double check your data file, it is surprising how often data entry errors are made. In addition to just entering data incorrectly, it is also true that data can be coded incorrectly when constructing and using transformed data in a spreadsheet. Thus, it is important to look at the data with "new eyes" asking yourself if the data appear to be error free and correctly transformed. Rechecking data entry and transformations should be part of the experimental design but checking it one more time during post-hoc analysis is recommended.

Once you have checked the data, look at the statistics used. It is worthwhile to review the assumptions that each statistical test requires. For example, all parametric tests require the following: the dependent variable is a continuous (vs categorical) variable; the sampling was random; the measurements are independent of one another; the population from which the sample was drawn has a normal distribution; and there is homogeneity of variances within the population (4). If these assumptions are violated, then either the choice of the statistical test, or interpretation of that test's results must be re-examined. The next step is to run the statistical analysis again to make sure that both the data were input into the statistical test correctly and that the statistical test itself was correctly run. It is more common than might be expected that errors in analyzing the data are made at this point in the analysis. It is always better for the author(s) to find these errors rather than the reviewers, editors, or readership. Such mistakes can be embarrassing no matter how novice or expert you are as a researcher (34).

Most researchers strive to run analyses on large samples (n=100 or more participants) but often this is not possible. If your sample is small, then you should take into account the statistical power of the analysis. Power is a measure of Type 2 error, which means that we fail to reject the null hypothesis when it is false (4). Three factors can contribute to insufficient power including: a small sample size, small effect size (a measure of the influence of the independent variable on the dependent variable), and the alpha level set by the researcher. Online calculators exist that can help you determine how large your sample size must be to detect the effect size with sufficient power that you want (35, 36).

The interpretation of your data in a manuscript should include a statement of whether the data met the assumptions of the statistical test used. In addition, the effect size and the power of the reported results should be included in the manuscript. Once you have checked the data and the analysis, look at your conclusions. Do they match the data you have or have you interpreted what the data say in order to draw the conclusion(s) you want? Ask colleagues to review your conclusions to see if they find the data convincing for the conclusions you suggest.

Revisit the Research Question

Review your study and ask yourself the following questions: Do the data collected and analyzed address the question(s) asked? If not, can you modify the question to match the data or do you have other data that were collected that address the question more directly? It is necessary that there is a good fit between the question asked and the data collected. If not, something has to be modified—either the data used, the analysis selected, or the question asked.

If the question asked and the data collected and analyzed are a good match, then ask yourself if there are any other questions that can be asked of the data collected. In the online HW study, the original intent was to investigate the effect of conceptual and traditional online HW questions on overall test achievement. When no significant effect was found, we asked other questions of the data we had collected (Table I), i.e., if there was no overall effect on test achievement, then was there a differential effect of conceptual and traditional online HW questions on achievement of conceptual and traditional test questions? When there was no significant effect of conceptual or traditional online HW questions on the conceptual or traditional questions on the test, was there a differential effect on achievement for those test questions that had an antecedent online HW question vs those that did not? All of these questions resulted in no significant differences. However, one nonsignificant result in the study led to more and more specific questions, each one of which also had nonsignificant results. In this case, the additional questions asked and the analyses performed were completed on data that had already been collected. These questions, although more specific than the original question, were related to it. With the results of the original question (overall test achievement) and the subsequent questions (conceptual or traditional test question and test questions with or without antecedent online HW questions), the nonsignificance of the analyses all point in the same direction, i.e., there was no effect of online HW on test achievement.

Another way to ask additional questions is look beyond statistical analysis. Sometimes it is possible in a mixed methods' experimental design (qualitatitve and quantitative data collected) to address a new question using aspects of the qualitative data already collected in a new way. In the first part of the online HW study, ten students were eye tracked as they solved online HW problems that had been selected and/or modified to fit a 2x2 array of Conceptual Question–Conceptual Hint (C-C); Conceptual Question–Traditional Hint (C-T); Traditional Question–Traditional Hint (T-T); and Traditional Question–Conceptual Hint (T-C). Areas of Interest (AOI's) were set up on each eye tracking screen for the question, answers, and hint areas. The number of times and the duration of each gaze in an AOI were recorded by the eye tracker. At the beginning of the eye tracking session, students were asked what their grade-to-date was in the course. Although we used this preliminary study to determine if there was a significant difference in the score on each of the questions or in the amount of time spent in each AOI in the 2 x 2 array, we then went back to the eye tracking data to address a new research question, i.e., is there a differential amount of time spent in each of the three AOI's for self-reported high achieving vs low achieving students? High achieving was defined as a student reporting a grade of B+ or higher while a low achieving student was one who had achieved a C or lower in the course to date. The eye tracker provided a visualization of where the students looked in the the three AOI's (question, answers, hints) as well as how often and how long they spent in each AOI. This eye tracking visualization consists of numbered circles showing where the student looked first and where they looked next. The size of the circle or the density of circles indicates how long and how often the student looked in a particular AOI. The visualization is accompanied by numerical data on duration of gaze within each AOI. This visualization of time of occurrence and duration of gazes is shown in Figure 1 and Figure 2. In each figure, the gaze pattern of a high achieving student is compared to that of a low achieving student.

From Figure 1 and Figure 2, it is obvious that both high and low achieving students in this sample did spend time analyzing the question before turning to the answers. This supports the idea that both high and low achieving students are attempting to solve the question by thinking about what it is asking. Notice in Figure 2 that both the high and low achieving students spent a considerable amount of time reading the hint (on the right hand side of each page within the figure) in connection with narrowing the choice of answers. This is the behavior that we would expect to see demonstrated in a problem solving situation. Even though our total eye tracking sample of 10 students was small, it does offer some data that students do use online homework the way it was intended. i.e., thinking about and trying to solve the problem presented. This supports the idea that students, at least in a learning laboratory experiment, know how to use online HW for analysis and problem solving as opposed to just searching for a right answer.

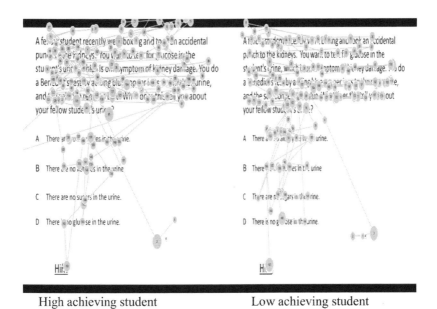

High achieving student   Low achieving student

Figure 1. Gaze patterns for high and low achieving students when solving a conceptual online HW question before a hint is requested.

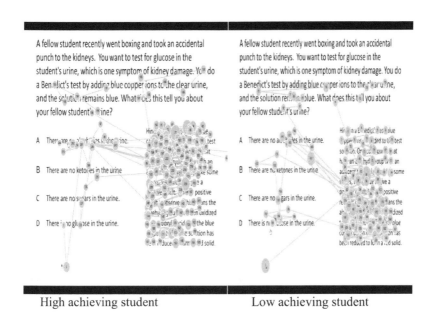

High achieving student   Low achieving student

Figure 2. Gaze patterns for high and low achieving students after hint when solving a conceptual online HW question with a conceptual hint.

The results of this new investigation did not yield a significant difference in the gaze pattern of high and low achieving students but the associated numeric metric did show a difference in the duration of gaze in the *question* AOI for the high achieving vs low achieving students The analysis offered insight into how both types of students (high and low achieving) analyzed online HW questions, answers, and hints. The eye tracking data revealed that both self-reported high and low achieving students spent a good deal of time analyzing the question given (as opposed to immediately turning to the answers) (Figure 1), however high achieving students spend significantly more time in the question AOI. The effect on gaze pattern and time spent in AOIs after the hint was displayed (Figure 2) demonstrated that students, both high and low achieving, spent considerable time analyzing the hint as well as the question. There was no significant difference in either gaze pattern or time spent in a particular AOI for high and low achieving students after the hint was displayed. The analysis of high and low achieving students was not part of the original research questions or analysiseven though such data were automatically collected in the preliminary study. This post hoc analysis provided information on the difference in problem analysis between high and low achieving students by demonstrating the increased time (and attention) that high achieving students spent on reading the question of an online HW problem.

Revisit the Theoretical Framework

When you have checked your data, analyses, and have looked at other questions that your data might allow you to investigate, it is time to turn back to the theoretical framework to help interpret your results.

In the case of the online HW study, our theoretical and application frameworks suggested that if too many tries were provided in the online HW, students would tend to guess or enter random answers. In this study, the number of tries was set at two and the eye tracking and grades for the online HW do not suggest that students were guessing or entering random answers for the online HW. The eye tracking data also suggest that students attempted to solve the online homework questions by analyzing the question, looking at the answers, and using the hints to narrow down their choice of answers. This behavior demonstrates a good problem solving approach to online HW problems in a learning laboratory situation by both high and low achievers. A second part of the theoretical framework, namely, the question of difference between the question format of thc online HW and the test was addressed in two different parts of the online HW study. The study of the effect of online HW on overall test achievement included multiple choice questions in the online HW and open ended questions on the test. The study of the effect of conceptual or traditional questions in the online HW on conceptual or traditional questions on the test, used multiple choice questions for both the online HW and the test. In these two studies, both with and without similar question formats, there were no significant results. Since we could not find an obvious reason for the nonsignificant results from the theoretical framework, we looked for other intervening variables that might explain the nonsignificant results found in this research.

Are the Nonsignificant Results Valid?

If you have checked the experimental design, analysis, and theoretical framework and can find no obvious threat to the validity of the nonsignificant results, then it is time to consider that they may indeed be valid results. As was mentioned earlier in this chapter, research is an iterative process. One experiment rarely addresses the entire question adequately and often leads to the development of a second or third study that builds on what was learned initially. It may be that nonsignificant results can point you or other researchers in a different direction for subsequent research.

In the online HW study, after looking at several questions from various angles and finding nonsignificant differences at each step, we concluded that the results for this study are valid even if they are nonsignificant. However, it is important to ask if our sample is representative of the much larger population of general chemistry students. The sample in this study consisted of nursing students whose chemistry course stressed conceptual learning over mathematical determinations. Would online homework have a similar nonsignificant effect on students in other general chemistry courses that incorporated a larger mathematical problem solving component? Our sample for the online HW study also represented a small class of approximately 50 students. In classes of this size, the instructor can get to know the students fairly well and to some extent, can more easily keep them engaged in class. When class size increases to 100, 200 or more, student anonymity becomes a bigger factor. Would the use of online HW have a significant effect on achievement in a larger class where direct student-teacher interaction decreases due to class size? Literature describing the effect of online homework on achievement included both large and small classes with no clear-cut difference in research results but perhaps this would be an important variable to take into consideration in future studies.

Examining the experimental design used in a study producing nonsignificant results, including the variables measured and controlled, is an important step in planning for the next experiment. In the case of the online HW study, there are a few things we would do differently in a subsequent study. These include the following:

1. Interview students shortly after they had completed the online HW to discuss what they thought they had learned from completing it.
2. Interview students after the test to discuss how and what they felt they applied from the online HW to the solving of the test questions.
3. Interview or survey students about their study methods preparing for the test to see if a review of the online HW problems and solutions was included.
4. Include classes of differing sizes in the study to test for the effect of class size.

5. Include students enrolled in a general chemistry course with more emphasis on mathematical problem solving.

Summary of the Two-Pronged Approach of Planning and Post-Hoc Analysis

The best way to deal with nonsignificant results is a two-pronged approach that includes both planning and post-hoc analysis. If adequate care is taken in planning the experiment and the analyses are done to support the validity of the nonsignificant results, then nonsignificant results can be publishable.

To help reduce the risk of nonvalid nonsignificant results, care must be taken in the planning and execution of the experimental design including attention to the following:

1. The research question asked is adequately addressed by the data collected.
2. The experimental design used is multi-variable and has adequately controlled for intervening variables.
3. The sample size of the research is large enough for the number of variables measured so that effect size and power of the statistics is adequate.
4. The assumptions of the statistical tests used have been met or the researcher has made valid adjustments to the test or the interpretation of the results.
5. The tools used (surveys, tests, achievement questions, rubrics, etc.) have been shown to be valid, reliable, and precise enough to accurately measure the variables needed to address the question.
6. The theoretical and application frameworks are extensive enough to explain why a significant result has or has not occurred.

If nonsignificant results are generated even with the implementation of good research design (planning), then a post-hoc analysis should be undertaken to include the following:

1. Double check the accuracy of the data used in the analyses and the appropriateness and correct implementation of the chosen analyses.
2. Ask new but related questions of the data already collected.
3. Use the theoretical and application frameworks to guide the conclusions drawn from the research results.
4. Construct new experimental designs for future studies based on lessons learned from the current study.

It is true that it may be more difficult to convince editors and reviewers that a study with nonsignificant effects is worthy of publication, but it is better to be transparent about the experimental design used and the results found, then to try to interpret the data in a way that is unsupported by the data presented. Editors and reviewers can quickly identify when an author is interpreting data to support his or her own purposes rather than letting the data speak for themselves. As an author, you do not have the authority to change the data, but you do have the ability to interpret it in light of the limitations of experimental design and theoretical frameworks. You will be judged on how thorough and logical your argument is. There is a fine line between presenting results objectively and trying to support an argument that you as the author believe to be true regardless of what the data say. As you write your conclusions, it is a worthwhile exercise to take stock of the pressure you are under to publish your research. Ask yourself if the pressure resulting from your situation (dissertation, tenure, or promotion) has an undue influence on your interpretation of the data. Keep in mind that if you start your research too close to your professional deadlines, your judgment and objectivity about what the data really say may be clouded. It is always best to have a colleague double check the logic of the argument you make in the conclusions of the manuscript submitted for review.

The result of this discussion on how to deal with nonsignificant results is clear. Nonsignificant results can be valid but they require that the author make an especially strong argument based on the principles of planning and post-hoc analysis to convince the editors, reviewers and readers that the results are indeed true.

Acknowledgments

Ashlie Wrenne, Matthew Tomney, and Regis Komperda, Chemical Education Graduate Program--The Catholic University of America, provided essential support in writing this chapter. Their help is gratefully acknowledged.

References

1. Jaykaran, D. S.; Yadav, P.; Kantharia, N. D. *J. Pharm. BioAllied Sci.* **2011**, *3*, 465–6.
2. Metcalfe, J.; Kornell, N.; Son, L. K. *Eur. J. Cogn. Psychol.* **2007**, *19*, 743–768.
3. Begg, C.; Cho, M.; Eastwood, S.; Horton, R.; Moher, D.; Olkin, I.; Pitkin, R.; Rennie, D.; Schultz, K. F.; Simel, D.; Stroup, D. *JAMA, J. Am. Med. Assoc.* **1996**, *276*, 637–639.
4. Pallant, J. *SPSS: Survival Manual*, 3rd ed.; Open University Press/McGraw-Hill Education: New York, 2007.
5. Bauer, C. F. Ethical Treatment of the Human Participants in Chemistry Education Research. In *Tools of Chemistry Education Research*; Bunce, D. M., Cole, R. S.; ACS Symposium Series 1166; American Chemical Society: Washington, DC, 2014; Chapter 15.
6. Bunce, D. M. Constructing Good and Researchable Questions. In *Nuts and Bolts of Chemical Education Research*; Bunce, D. M., Cole, R. S., Eds.; ACS Symposium Series 976; American Chemical Society: Washington, DC, 2008; Chapter 4.

7.   Abraham, M. R. Importance of a Theoretical Framework for Research. In *Nuts and Bolts of Chemical Education Research*; Bunce, D. M., Cole, R. S., Eds.; ACS Symposium Series 976; American Chemical Society: Washington, DC, 2008; Chapter 5.

8.   Williamson, V. M. The Particulate Nature of Matter: An Example of How Theory-Based Research Can Impact the Field. In *Nuts and Bolts of Chemical Education Research*; Bunce, D. M., Cole, R. S., Eds.; ACS Symposium Series 976; American Chemical Society: Washington, DC, 2008; Chapter 6.

9.   Bodner, G. M. *Theoretical Frameworks for Research in Chemistry/Science Education*; Pearson/Prentice Hall: Upper Saddle River, NJ, 2007.

10. Carmel, J. H.; Yezierski, E. *J. Coll. Sci. Teach.* **2013**, *42*, 71–81.

11. Pentecost, T. C.; Barbera, J. *J. Chem. Educ.* **2013**, *90*, 839, 845.

12. Sanger, M. J.; Vaugh, C. K.; Binkley, D. A. *J. Chem. Educ.* **2013**, *90*, 700–709.

13. Kendlhammer, L.; Holme, T.; Murphy, K. *J. Chem. Educ.* **2013**, *90*, 846–853.

14. Harshman, J.; Lowery Bretz, S.; Yezierki, E. *J. Chem. Educ.* **2013**, *90*, 710–716.

15. Cotner, S.; Loper, J.; Walker, J. D.; Christopher, D. *J. Coll. Sci. Teach.* **2013**, *42*, 82.

16. Hazari, Z.; Sadler, P. M.; Sonnert, G. *J. Coll. Sci. Teach.* **2013**, *42*, 82–91.

17. Bruck, A. D.; Towns, M. *J. Chem. Educ.* **2013**, *90*, 65–693.

18. Gron, L. U.; Bradley, S. B.; McKenzie, J. R.; Shinn, S. E.; Teague, M. W. *J. Chem. Educ.* **2013**, *90*, 694–699.

19. Lee, Y.; Palazzo, D. J.; Warnakulasooriya; Pritchard, D. E. *Phys. Rev. ST Phys. Educ. Res.* **2008**, *4*, 0101021–6.

20. Kang, S. H. K.; McDermott, K. B.; Roediger, H. L. *Eur. J. Cogn. Psychol.* **2007**, *19*, 528–558.

21. Roediger, H. L.; Gallo, D. A.; Geraci, L. *Memory* **2002**, *10*, 319–332.

22. Graf, P.; Ryan, L. *J. Exp. Psychol. Learn.* **1990**, *16*.

23. Shea, C. H.; Wulf, G. J. *Motor Behav.* **2005**, *37*, 85–101.

24. Bonham, S. W.; Deardorff, D. L.; Beichner, R. J. *J. Res. Sci. Teach.* **2003**, *40*, 1050–1071.

25. Fynewever, H. *Chem. Educ.* **2008**, *13*, 264–269.

26. Cole, R. S.; Todd, J. B. *J. Chem. Educ.* **2003**, *80*, 1338–1343.

27. Charlesworth, P.; Vician, C. *J. Chem. Educ.* **2003**, *80*, 1333–1337.

28. Kortemeyer, G. *Phys. Rev. ST Phys. Educ. Res.* **2009**, *5*, 010107.

29. Just, M. A.; Carpenter, P. A. *Behav. Res. Meth. Ins.* **1976**, *8*, 139–143.

30. Herrington, D. G.; Daubenmire, P. L. Using Interviews in CER Projects: Options, Considerations, and Limitations. In *Tools of Chemistry Education Research*; Bunce, D. M., Cole, R. S.; ACS Symposium Series 1166; American Chemical Society: Washington, DC, 2014; Chapter 3.

31. Student Assessment of their Learning Gains. http://www.salgsite.org/ (accessed 2014).

32. Frechtling, J.; Sharp, L. *User-Friendly Handbook for Mixed Method Evaluations*; Directorate for Education and Human Resources, NSF (RED 94-52965): Washington, DC, 1997.
33. Arasasingham, R.; Martorell, I.; McIntire, T. *J. Res. Sci. Teach.* **2011**, *40*, 70–79.
34. Monaghan, P. Chronical of Higher Education. http://chronical.com/article/UMass-Student-Talks/138763/ (accessed 2013).
35. Power and Sample Size Calculator. Statistical Solutions, LLC. http://www.statisticalsolutions.net/pss_calc.php (accessed 2014).
36. Calculators: Statistical Power. DSS Research. https://www.dssresearch.com/KnowledgeCenter/toolkitcalculators/statisticalpowercalculators.aspx (accessed 2014).

Chapter 14

Doing Chemistry Education Research in the Real World: Challenges of Multi-Classroom Collaborations

Jennifer E. Lewis*

Department of Chemistry, University of South Florida,
Tampa, Florida 33620, United States
*E-mail: jennifer@usf.edu

The focus of this chapter is the challenge of conducting a complex research project, for example a project that requires data collection from students in real classrooms at multiple institutions. Two project phases drawn from project management practices are discussed, the planning phase and the monitoring and controlling phase. Planning for specific issues that are likely to arise, such as attrition, changing circumstances for collaborators, and difficulties with human subjects review processes, is recommended. Data screening and regular contact with collaborators are suggested as pivotal monitoring strategies. The chapter concludes with a section on real world research failures that nevertheless produced research publications. The chapter is written primarily for researchers who are preparing to conduct their first large-scale project.

Introduction

One of the most daunting aspects of beginning research in chemistry education can be the relentless narratives of success offered by experienced researchers in the field. Not only is hindsight 20/20, it also tends to wear rose-colored glasses. If you have caught yourself thinking that a speaker extolling the many virtues of a successful large-scale research project sounds suspiciously like Voltaire's Candide—"All for the best in the best of all possible worlds!"—this chapter is for you. The more complex the research project, the more likely it is that at least one aspect of the planned work went seriously awry at some point. By the time you are hearing the final narrative of the project, however, that problem may have been addressed so cleverly that it now seems like a planned element—a feature—rather than a drawback. How do experienced researchers make that happen?

For the purposes of this chapter, let's set up a few parameters without getting too deeply into specifics. First, assume that you have a great set of insightful research questions and that your planned data collection methods (observation schemes (*1*), interview protocols (*2*), and/or quantitative measures (*3*)), sampling strategy, and research design should provide exactly the right information to address your questions if all goes well. While these things are not trivial, they are usually addressed thoroughly in standard research methods texts. What goes unmentioned is the practical side of doing research, in other words, the project management side. There are five basic phases to any project: initiating, planning, executing, monitoring and controlling, and closing (*4*). For this chapter, we'll think about a reasonably complex project by assuming that your research design requires collecting data from multiple classrooms, at least some of which are not at your own institution, and we'll discuss two of the middle phases: planning and monitoring/controlling. Finally, we'll discuss some failures that turned out to be salvageable.

Planning

Without planning, crisis mode takes over. Every time you turn around, there seems to be some new deadline for the project, and you have to struggle to get things in place to meet it. It is possible to achieve your goals by working efficiently to address each crisis in turn as it comes up, but staying in crisis mode leaves you unable to plan for real emergencies. The more complex your project, the more difficult it will be to simply address each crisis as it occurs, since inter-relationships between project elements will cause a cascade effect. In other words, you will experience a rapid increase in the frequency of crisis points as your besieged project gets further and further away from the possibility of success. Of course, before you execute a project is the best time to make a plan, but you may also realize after execution has begun that your initial planning wasn't sufficiently detailed to keep you out of crisis mode. Whenever you decide to plan, whatever your particular tools are—pencil and paper, sticky notes and a whiteboard, electronic spreadsheets or Gantt charts—the first step is to create a basic timeline for the project. What needs to happen when? Once you have the basic timeline, work backward from each deadline. If you need to collect data at a particular time, what other work needs to happen in advance? Who needs to be involved in each phase of the work? What resources will be required? How much time will each step take? Don't forget to account for potential differences in academic calendars among your collaborators. Also make sure to uncover and incorporate the "hidden work"—for example, shipping scantron forms to participating sites, or testing audio equipment under data collection conditions—as well as more obvious items, such as Institutional Review Board (IRB) approvals (*5*) and travel arrangements. Given the detailed information, how must each element be sequenced so that the overall project work can move forward smoothly?

This detailed planning process, which is more comprehensive than what would normally be presented in a grant proposal, is formally called developing a *work breakdown structure* (*6*). Because it is a plan for you to use rather than a plan for reviewers to evaluate, the end product can be as rough or as polished as suits your organizational style. A set of index cards can work, as long as you are able to see the project laid out and determine a logical workflow that you can incorporate into the calendar system you use to organize your work time. Don't let the phrase "calendar system" put you off; a physical appointment book and a collection of to do lists with deadlines can work just as well as a sophisticated app, but you do need a system for organizing your time. Since we're talking about the real world in this chapter, note that, no matter when you do this work breakdown exercise, the results will probably indicate that you should have started months ago and are already behind. Still, it is helpful to have the work laid out in front of you. Once you have your detailed plan, do a feasibility check. Which features of the work are essential, and which would be nice if all worked out well? For the essential features, can you think, right now, of how to cope if they don't happen as planned? This is when you determine what would constitute a real crisis, create a monitoring system, and list potential backup plans. Four elements of the planning process for complex research projects in chemistry education deserve special mention: attrition, pilot studies, multiple IRBs, and collaborator goals.

Plan for Attrition

If your data analysis will involve significance testing, power analysis provides an estimate of how many data points are necessary for a particular statistical test to have sufficient power to detect the result you expect to observe. Elsewhere in this volume, S. Lewis (*7*) highlights the importance of power analysis by observing that failure to conduct power analysis can mislead researchers with respect to the interpretation of non-significant findings, and Bunce (*8*) recommends the use of tables or software to assist with power analysis. G*Power (*9*) is helpful free software that covers many common statistical tests. In order to do an a priori power analysis, you need to know how large you expect the effect to be, which typically can be estimated based on previous studies, and which statistical test you plan to use to detect the effect. Once you have conducted a power analysis, ensure that your recruitment plan includes a margin for no-shows and refusals as well as true attrition, increasing the margin if your study design is longitudinal. For more complex analyses, such as a nested multi-level modeling design with multiple classes at different institutions or structural equation modeling, simulation studies can provide insight into necessary sample sizes (*10, 11*). When searching the literature, keep in mind that our understanding of how well statistical methods work under various real-world conditions is constantly being updated, and your particular data structure may not yet have been examined. Regardless, make your best guess from current knowledge, expect attrition at all levels, and start with more than the minimum! If your data collection does not involve significance testing, you still need to compare your recruitment plans with your sampling strategy. Have you created sufficient redundancy such that your research goals are protected if, for example, a few participants fail to attend their scheduled interviews? At what level will no-shows begin to cause a problem? Is there a particular group of interviewees that you may need to over-sample because you expect attendance to be an issue? Regardless of your data collection and analysis methods, having thought about likely sources of attrition ahead of time will enable you to monitor the situation and make adjustments when attrition rises above expected levels.

Incorporate Pilot Studies

It is unwise to treat the research reports from others working in similar settings as if they are pilot studies for your particular research project. These reports can be indicators that the data collection methods used by others may work for you, but there is no substitute for a researcher's own pilot test with the study population. For example, given the state of the art of measurement in chemistry education research at present (3), many instruments from our literature lack the support of extensive validation studies. Take the time to pilot an instrument with a sample from the study population and investigate whether you can make a case for valid inferences from the pilot data before you proceed with the full-scale study. In one situation, this simple step derailed the planned research project entirely, and the researchers instead decided to focus on suggesting possible modifications to the instrument itself so that other researchers would not be similarly derailed (12). While this particular example concerns an instrument, piloting is similarly important for observation schemes and interview protocols. Both small things like deciding where to sit for a classroom observation and big things like whether interviewees understand key terms are informed by pilot testing. You may not have the latitude to abandon your original research goals if things don't go well in the pilot phase, so choose your data collection methods thoughtfully and test them in advance! If the pilot work does indicate that changes are necessary, it is far better to know in advance than to find out only after you have collected a great deal of uninterpretable data. If your pilot work turns out to be extensive, consider sharing it with the community in a separate publication before you have completed your main study. We still know far too little about effective data collection methods for chemistry education research.

Expect Confusing IRB Rulings

When you plan to collect data at multiple institutions, each institution may require that your protocol be submitted to its own Institutional Review Board (IRB) rather than trusting your home institution's IRB to handle the review process (5, 13). Your carefully written and highly detailed research protocol, which is modeled after the approved protocol for a similar study you conducted previously at your home institution, is likely to emerge from the initial review processes at multiple external institutions with requested changes that are not compatible with one another. For example, consider a project for which you plan to video record instructors from different institutions teaching the same lesson, capturing their interactions with their students. One institution has requested that you rearrange seating in the classroom the days you plan to record, to place students off-camera if they have not consented, while another institution prefers that you record the class as a whole to protect students from the instructor's knowledge of who has or has not consented, as long as you store the data securely and do not analyze any video segments from non-consenting students. Both requests are reasonable, but unless one institution changes its mind, you'll have to be attentive to these differences not only during your data collection but also during your analysis.

Similar differences can occur with respect to the accepted norms for access to student data that are applied at each institution. For example, one institution may require that an instructor create a code list in order to send only de-identified student data to an outside researcher, while another institution may allow an outside researcher to see identifiable data in order to maintain privacy regarding consent status. These sorts of reasonable but different rulings have to do with the fact that each IRB creates its own interpretation and application of the relevant laws (14). What can be frustrating is that, although you may believe your original protocol does a better job of protecting human subjects in important ways as compared to the requested revisions from a specific institution, the IRB members have the power to make that determination, not the researcher. Compounding the difficulty of understanding differences in institutional norms is that the IRB members may not be willing to communicate directly with an outside researcher about whether they would be amenable to different interpretations, preferring to speak only to employees of their own institution as a matter of policy. For multi-institution projects, I have found it to be a useful practice to create a master document laying out the details of the protocol, from which I cut and paste verbatim into each institution's specific IRB protocol submission forms, adding specific notes in the master document regarding changes made for each specific institution as the review processes unfold. If my collaborator at an institution is the one communicating directly with the IRB, at least there is a full record of what is happening at other places that can be used to support a conversation about other options for achieving the intent of requested revisions. With sufficient lead-time, it can also be helpful to sequence IRB protocol reviews, beginning with the institution that has the most well-developed IRB infrastructure, for example, the institution with a medical school.

Ask Your Collaborators Why

What's in it for your collaborators? What is their incentive for working with you? When you pitched your multi-institutional project to instructors at a variety of institutions, no doubt you explained what you believed would be the benefits of participation. For example, you may have promised to provide data about student learning in a format that could be easily inserted into an annual review portfolio as evidence of engagement in scholarly teaching. Or perhaps you suggested that your classroom observations would not only address the research questions but could also be used to create specific suggestions to inform your collaborator's teaching. Maybe you proposed regular discussions among participating instructors, or support in learning a specific data analysis technique. It would be unusual if every collaborating instructor who agreed to participate did so for the same reasons. In some cases, when you have a follow-up discussion about participation before the project begins, you will uncover new motives for participation, different from those you originally suggested as options, particularly if there has been a time lag between the original agreement to participate and the start date for the project. Learning what is most attractive about participation for each collaborator as the project begins will allow you to prioritize that specific deliverable as you create your detailed project plan.

Monitoring and Controlling

Once you have begun executing the research project, it is important to have procedures in place that allow you to find and fix problems as they occur. Good monitoring strategies enable you to detect when one aspect of your complex project starts to go in a direction that will move the project as a whole off track. Control strategies are techniques for dealing with specific problems regarding aspects of a complex project in order to get the project as a whole back on track. This section of the chapter, therefore, describes monitoring strategies for detecting some common problems and suggestions for control strategies that may help if the problems occur.

Screen Data as It Is Collected

Prompt data screening is an extremely valuable monitoring strategy. Data screening simply involves reviewing the data to catch errors or other threats to validity. This means things like looking for outliers or unexpected responses, checking to see that there is variability in responses, and noting whether there are potentially problematic patterns of missing data. While data screening is often described in terms of quantitative data, the principles can be applied to qualitative data as well. For example, one easy to observe "potentially problematic pattern" of missing qualitative data would be a high degree of no-shows associated with interviews at one institutional site. A pattern that would require a greater degree of scrutiny to detect would be a consistent refusal by students to respond to a particular interview prompt or survey question. Focusing on prompt data screening is pivotal for success. A research project can recover from minor hiccups, but widespread validity issues are a serious ailment. If problems with validity are uncovered while the data is still being collected, they can be addressed before the project itself is threatened. Below are two specific benefits of prompt data screening that illustrate its value.

Prompt data screening allows detection of technical problems. You don't want to notice at the end of a long day of interviews that your audio recorder stopped working midway through the morning. If possible, leave a little time at the end of each interview to satisfy yourself that your equipment was working, sufficient time that, if it wasn't, you can augment your interview notes to include key ideas from the interview while you still recall them. If you were planning a close analysis of interview transcripts, you will not be able to salvage the interview, but the notes can still inform your thinking. Similar issues apply to any technologically-enhanced data collection method. "The camera wasn't on" is something you want to catch in the initial stages of data collection, not at the end. Data collection via rubric falls into this category as well. If classroom observations include a checklist observation protocol, plan to check inter-rater reliability throughout the study, not only at the beginning and end. Regardless of initial training, observers may change their interpretation of a protocol as they are exposed to real classrooms and can see what real variability looks like across the study participants. The same need to check inter-rater reliability at multiple points is true for graders working on open-ended test items as well, as anyone who has ever graded a large stack of exams knows. The level of "drift" that may be acceptable for one of many classroom assessments (particularly when students are free to come in and argue for more points) is larger than that acceptable for a key data point in a research project. If inter-rater reliability drops below an acceptable level, you need to schedule a discussion and retraining session before data collection continues.

Prompt data screening can uncover deviations from data collection protocols. Despite good intentions, in the press of day-to-day activities, collaborators may undermine a research project by failing to follow data collection protocols. If you are asking collaborators to send data to you for processing immediately after it has been collected, you can catch these problems before they snowball. For example, you may note that students at one institution did suspiciously well on the simple proportions section of the pre-test for a study. When you inquire, you learn that the instructor did not recall that students were not supposed to have access to a calculator during the test. Uncovering this problem early in the study allows you to decide on an appropriate specific control strategy—repeat the pretest for the deviant class if memory effects are likely to be minimal; modify the simple proportions section of the pretest and give it again across all participants; or, at the very least, ensure that this particular collaborator is reminded of the correct protocol for subsequent data collection and knows why this aspect of the protocol is important for achieving the research goals. Note that the control strategy of modifying the simple proportions section and repeating it across study participants could be elevated into a "feature"—a repeated measure can provide a more robust measurement strategy for math ability than originally planned. As suggested in the introduction to this chapter, one insight into research in the real world is to recognize that study design elements expressed as special features in final research reports may in fact have been developed as control strategies to address problems uncovered via monitoring.

The need to provide sufficient detail for data collection protocols is not limited to situations in which you will not be present to collect the data yourself. For example, consider classroom observations. Depending on the institutional norms, it may be inappropriate for an instructor to neglect to introduce an observer to a class. The details of the introduction, and, even if there is not a formal introduction, what sorts of answers the instructor will give to questions about the observer, need to be worked out in advance. Consider this unintentional sabotage: you are sitting in to observe a class session, and the instructor introduces you to the class as "reporting back to me about how well you work in your groups today," thereby ensuring that the students will become tense and artificial when you try to observe them. It is important not to deceive students about the basic reason for an observer's presence in the class, but they will certainly react to you differently as an observer if you are introduced as "the author of the activities we are doing today" as opposed to "a colleague who is helping me think about my teaching". Make sure to script an introduction that will enable you to accomplish the needed data collection.

Note that repeated deviations from data collection protocols may signal the need for a general control strategy to remind collaborators about the protocol details before each collection point. A cover sheet, a timely e-mail, a phone call to check in—one size does not have to fit all. Select strategies that will fit into your collaborators' preferred interactional modes. If respectful reminders don't work, another control strategy would be to consider whether your data collection protocol could be modified to fit better with your collaborators' normal instructional practices without compromising the research. Again, this modification could become a "feature" when you describe your finished project. Regardless, simply staying in touch with your collaborators is also an important monitoring strategy.

Keep Regular Contact with Collaborators

You established motives for participation in the planning phase and took care to deliver on your promises, but participation in the project may still have failed to meet collaborators' expectations. Have benefits been sufficiently valuable, or has project participation created more hardships than benefits? Are the anticipated benefits only slated for the long term, making it hard for collaborators to stay motivated to do the necessary project work right now? Have any new circumstances emerged that render the collaborators' original motivations less salient? Alternatively, collaborators may develop ideas for different types of involvement in the project that are still aligned with the research goals. Checking in with collaborators regularly will help you to stay on top of their changing circumstances and understand how their perceptions of the project evolve over time.

Keeping regular contact will forestall the "oops-I-forgot" problem. "What do you mean, you forgot to collect the data?" "Why couldn't you fit the target assignment into your class this term?" "What has changed?" These are questions you don't want to have to ask after a planned data collection period has ended. The best reminder strategies in the world do not work if a collaborator no longer values the benefits of participation in the project and has not had the opportunity to ask whether project participation can be reconciled with new concerns. Awareness of changing motivations through regular contact with collaborators can buy needed time to discuss possible adjustments. If the project truly cannot make sufficient accommodations to keep the collaborator engaged, your sincere attempts to consider the possibilities create a better exit scenario than waiting to hear "oops-I-forgot". Even if the exiting collaborator is not able to assist in recruiting a replacement, you have advance notice in which to develop a recruitment strategy, informed by what you have learned in the project thus far. If you have noticed a problem with your original sampling strategy but had limited resources to address it, perhaps this is an opportunity to make the project stronger (another "feature") with a new collaborator who can fill a specific gap. If you planned for attrition and your sampling is still sound, it may be that you can manage without the need for any replacement at all, diverting the resources to another aspect of the project that would benefit from them.

Making It Work

"A certain witty advocate remarked: 'One would risk being disgusted if one saw politics, justice, and one's dinner in the making (*15*).'"

Fellow researchers will understand that research is messy. Despite your best efforts, you will encounter limitations, outright failures, and unexpected findings, both positive and negative. What do you do when events conspire to create a situation in which your original research plan must be abandoned? The answer should not be to abandon all hope and switch careers. Typically there is a way to make something out of what you have managed to accomplish, despite the demise of your original plan. While you may need a moment to mourn, take heart that other researchers have faced similar situations and moved forward nonetheless. The chapter on non-significant results in this volume provides some suggestions for post-hoc analysis in that specific case (*8*).

Because the only research dilemmas I am privy to are my own, this next section focuses on a few examples from my research group to demonstrate that it is possible to recover from serious setbacks in the research process. In some cases, we developed a backup plan during the feasibility assessment phase of the planning process, but in other cases our research was halted until we could sort out what to do. Three ways of thinking about failure are presented below.

Turn Constraints into Affordances

A limited budget meant that a project could only supply sufficient resources to implement a curricular change a few times during the project period, in a complex setting that involved multiple faculty teaching multiple sections of a course, but each with only one section per term. Our original plan was to use historical comparison, asking the faculty who would have the opportunity to implement the change during the project period to also teach at least one section without the change in the project period, so that we would essentially have one quasi-experiment per participating instructor. This plan almost immediately proved intractable, as teaching assignments (not under our control) did not work out that way. Rather than giving up, we applied multi-level modeling techniques and expanded our data collection to be able to include all of the sections of the course taught during the project period. The result (*16*) was a more robust study than the original plan would have produced. The disadvantage was having to publish a more complex analysis and discussion than was typical for the chemistry education community at that time. Multilevel modeling has since grown in popularity and additional chemistry education researchers have begun to publish work that benefits from its use (*17*).

Consider What You Can Say with the Data You Have

Our original intention was to investigate whether formal reasoning ability as measured by the Test of Logical Thinking (TOLT) (*18*) would have a moderating effect on the influence of a curricular change. In other words, to address the question of whether the change might work better (or worse) for those with high reasoning ability. This intention was inspired by work on grouping strategies informed by formal reasoning ability (*19*), early results from which were being presented at professional meetings while we were planning our study. In the end, despite assiduous data collection with a large number of students for several years, we were on the low end of the necessary power to detect what appeared to be a small effect. Moderator effects were known to be difficult to demonstrate with the techniques we were using (*20*), so, although we might have explored analytic alternatives, we instead decided to explore whether we might use the data for another purpose. The result (*21*) was a straightforward comparison of formal reasoning (TOLT) and general achievement (SAT subscores) as predictor variables for detecting students at risk in first semester general chemistry, with the resulting recommendation that TOLT is a useful measure for this purpose. From there it was a small step to collect additional data (*22*) that further supported the TOLT in comparison to a similar instrument in use at that time, the GALT (*23*). This study was far from the original project, but still a contribution to the field of chemistry education. Likewise, as mentioned previously, after encountering repeated problems piloting a variety of instruments, we conducted a review and published our findings to promote the gathering and sharing of validity evidence (*3*).

View Mistakes as Opportunities

Sometimes others find your mistakes before you do; if you are lucky that will happen during the review process and you will have a chance to fix the problem before the final study report is presented or published. Occasionally, mistakes are not caught before they appear in print, particularly if you are trying to work on the "cutting edge" of the field. Don't be afraid to critique your own published work. If you realize that you now would have done something differently, it is likely that others have made and will continue to make the same mistake unless you also publish your new thinking. In a publication about a simple quasi-experiment, we mistakenly interpreted a non-significant p-value as evidence for equivalence between groups on a measure of interest. When we realized our mistake, we redid the comparison to include a true equivalence test and published the result as a follow up (*24*). We were fortunate that the appropriate analysis supported our original interpretation, but if it had not, we still would have been under an obligation to correct the record. When others catch your published mistakes before you do, be gracious! It means that they read your publication thoroughly and believe your work is important enough to be worth correcting. The more complex your project, the more likely it is that you will have made decisions that even you would make differently now that time has passed and knowledge has grown. As mentioned earlier, hindsight is 20/20. At least try to keep the use of rose-colored glasses to a minimum by providing sufficient detail that readers can judge for themselves whether they think your decisions were reasonable. If they would have made some decisions differently, so much the better. The best gift you can give another researcher is a gap to be addressed in a subsequent study.

Conclusion

In this chapter I have laid out several real world considerations for research in chemistry education. Any researcher who has completed even one study would be able to provide similar advice, and several of the topics I touched upon are also addressed in other chapters. What I hoped to do in this chapter was to make it clear that the normal processes of research are far from smooth, particularly for complex projects, regardless of the polished descriptions established researchers provide in most public presentations of their work. At the end of the day, there is no such thing as the perfect study. Researchers must make decisions in light of an unfolding sequence of unexpected events within any given project. Acknowledging and preparing for that reality is your best hope for being able to meet research goals. While becoming a full-fledged professional project manager is not necessary, attention to planning and to monitoring and controlling will bear fruit for research projects just as for any other project. If you are a beginning researcher reading this volume (25) in its entirety, you will have noticed that this chapter has a different tone and style than the others. That choice is deliberate: if I could also manage to get "the words DON'T PANIC inscribed in large friendly letters on its cover"(26), I would! Failing that, my intent is that this chapter will have started you thinking about problems that could occur during the large-scale research studies of your dreams, without turning those dreams into nightmares. Although some problems may initially seem fatal to a project, there is usually a way to share something from your work that will be of interest to the chemistry education research community. As a contribution to our growing understanding, a completed "good enough" project is much better than an abandoned perfect one.

References

1. Yezierski, E. J. Observation as a Tool for Investigating Chemistry Teaching and Learning. In *Tools of Chemistry Education Research*; Bunce, D. M., Cole, R. S., Eds.; ACS Symposium Series 1166; American Chemical Society: Washington, DC, 2014; Chapter 2.
2. Herrington, D. G.; Daubenmire, P. L. Using Interviews in CER Projects: Options, Considerations, and Limitations. In *Tools of Chemistry Education Research*; Bunce, D. M., Cole, R. S., Eds.; ACS Symposium Series 1166; American Chemical Society: Washington, DC, 2014; Chapter 3.
3. Arjoon, J.; Xu, X. *J. Chem. Educ.* **2013**, *90*, 536–545.
4. *A Guide to the Project Management Body of Knowledge (PMBOK® Guide)*; Project Management Institute: Newtown Square, PA 2013.
5. Bauer, C. F. Ethical Treatment of the Human Participants in Chemistry Education Research. In *Tools of Chemistry Education Research*; Bunce, D.

M., Cole, R. S., Eds.; ACS Symposium Series 1166; American Chemical Society: Washington, DC, 2014; Chapter 15.

6. Haugan, G. T. *Effective Work Breakdown Structures*; Management Concepts, Inc.: Vienna, VA, 2002.

7. Lewis, S. E. An Introduction to Nonparametric Statistics in Chemistry Education Research. In *Tools of Chemistry Education Research*; Bunce, D. M., Cole, R. S., Eds.; ACS Symposium Series 1166; American Chemical Society: Washington, DC, 2014; Chapter 7.

8. Bunce, D. M. A Two-Pronged Approach to Dealing with Nonsignificant Results. In *Tools of Chemistry Education Research*; Bunce, D. M., Cole, R. S., Eds.; ACS Symposium Series 1166; American Chemical Society: Washington, DC, 2014; Chapter 13.

9. Faul, F.; Erdfelder, E.; Buchner, A.; Lang, A.-G. *Behav. Res. Methods* **2009**, *41*, 1149–1160.

10. Jackson, D. L. *Struct. Equation Model.* **2003**, *10*, 128–141.

11. Maas, C. J. M.; Hox, J. *J. Methodol. Eur. J. Res. Methods Behav. Soc. Sci.* **2005**, *1*, 86–92.

12. Heredia, K.; Lewis, J. E. *J. Chem. Educ.* **2012**, *89*, 436–441.

13. Menikoff, J. *N. Engl. J. Med.* **2010**, *363*, 1591–1593.

14. Abbott, L.; Grady, C. *J. Empir. Res. Hum. Res. Ethics* **2011**, *6*, 3–19.

15. de Chamfort, N. *The Cynic's Beviary: The Maxims and Anecdotes from Nicolas De Chamfort*, Selected and Translated by William G. Hutchison; Elkin Mathews: London, 1902. Bartleby.com (accessed 2011).

16. Lewis, S. E.; Lewis, J. E. *J. Res. Sci. Teach.* **2008**, *45*, 794–811.

17. Pazicni, S.; Bauer, C. *F. Chem. Educ. Res. Pract.* **2014**.

18. Tobin, K. G.; Capie, W. *Educ. Psychol. Meas.* **1981**, *41*, 413–423.

19. Cooper, M. M.; Cox, C. T. J.; Nammouz, M.; Case, E.; Stevens, R. *J. Chem. Educ.* **2008**, *85*, 866–872.

20. McClelland, G. H.; Judd, C. M. *Psychol. Bull.* **1993**, *114*, 376–390.

21. Lewis, S. E.; Lewis, J. E. *Chem. Educ. Res. Pract.* **2007**, *8*, 32–51.

22. Jiang, B.; Xu, X.; Garcia, A.; Lewis, J. E. *J. Chem. Educ.* **2010**, *87*, 1430–1437.

23. Roadrangka, V.; Yeany, R. H.; Padilla, M. J. Paper presented at the *Annual Meeting of the National Association for Research in Science Teaching*, Dallas, Texas, 1983.

24. Lewis, S. E.; Lewis, J. E. *J. Chem. Educ.* **2005**, *82*, 1408–1412.

25. Bunce, D. M.; Cole R. S. Using This Book To Get Started on Your Own Research. In *Tools of Chemistry Education Research*; Bunce, D. M., Cole, R. S., Eds.; ACS Symposium Series 1166; American Chemical Society: Washington, DC, 2014; Chapter 17.

26. Adams, D. *The Ultimate Hitchiker's Guide to the Galaxy*; Gramercy Books: New York, 2005; p 6.

Chapter 15

Ethical Treatment of the Human Participants in Chemistry Education Research

Christopher F. Bauer*

Department of Chemistry, University of New Hampshire, Durham, New Hampshire 03824, United States
***E-mail: chris.bauer@unh.edu**

This is a pragmatic overview for researchers regarding the ethical treatment of human participants from the point of view of an experienced chemistry education researcher. The fundamental ethical principles, purpose, and process of institutional review are explained. Examples and advice provided are informed by discussions with other colleagues in the field, with novice investigators, and with my local Institutional Review Board office director.

Our students and colleagues are doing us a favor. By "us" I mean the community of chemistry education researchers, and by extension all educational researchers. The "favor" is that they are the source of the behaviors, thoughts, and emotions that we call data, from which we hope to draw inferences about how people learn chemistry and how we might improve how they learn. Having human data allows our investigations to be empirical – to be "evidence-based", to use a term that is popular now in educational circles. In the absence of evidence, one can speculate and build theories, but theoretical guidance alone will not move our understanding forward. No evidence means no progress in making sense of the complex world of human learning. Our students and colleagues, by allowing us to study them, are making a generous contribution to the effort, and make it possible to do what we as researchers are so interested in doing. It is a privilege to be able to engage them as participants in the research.

This chapter is a pragmatic introduction to the ethical study of human learning, with particular reference to chemistry education research. The intended audience is anyone who is thinking about classroom experimentation, whether you are engaging in formal theory-based design research, in activities that fall in the realm of the scholarship of teaching and learning, or in changes that you as an instructor might be considering for your own classroom. These comments come from an author who moved into chemistry education research (and a need to be concerned about human participants) after an early career in traditional chemistry. This chapter can only provide general guidance because each research study and study population will have unique characteristics, and because each institution establishes its own process and requirements.

Guidelines for conducting research with human beings have substantial scholarly underpinnings in ethics and the social sciences. All of the relevant professional societies have published statements: the American Educational Research Association (1), the American Psychological Association (2), and the American Sociological Association (3). The American Chemical Society has a code of ethics (4) which includes statements about the treatment of students in classroom settings but not in research settings. The foundation for establishing explicit protections for human participants is the 1979 Belmont Report (5), which was commissioned after serious cases of unethical medical research were exposed in the middle of the last century. Those cases of abusive physical and mental mistreatment raised questions about other types of potential mistreatment, and thus drew attention to social, psychological, and educational research. This is not a stagnant issue. Scholarly investigations and policy discussions continue. There are journals devoted to research ethics for science, most notably *Accountability in Research* (6) and *Science and Engineering Ethics* (7).

The U.S. Department of Health and Human Services has established the guidelines for ethical treatment of human research participants, also known as the "Common Rule" (8). These guidelines require that Institutional Review Boards (IRB) be established for institutions conducting federally funded research. For good ethical and legal reasons, many institutions have extended IRB oversight to all human participant research even if not federally funded. That brings us to the issue of concern for chemistry education researchers. If you want to do research on teaching and learning, the IRB has the responsibility to take a look at what you intend to do to assure that the people involved are treated ethically.

Where Do I Start?

Do I Need IRB Review?

Yes. You picked up this book because you are interested in conducting a study and in sharing publicly what you find out. Whether the participants are students or faculty, you will be gathering information about them, analyzing that information, and telling a story about it. The purpose of an IRB review is to make sure that you treat your participants with respect, beneficence, and justice. You show respect by providing information to them about the study and giving them an opportunity to decide on their own whether to participate. You show beneficence by considering, managing, and informing them about the risks they might be exposed to, balancing that against direct benefits for them and against the broader benefits of answering your research question. You show justice by considering and avoiding deceptive or inequitable treatment.

I occasionally hear, though with decreasing frequency: "Why does someone else need to approve my research? I'm a professional and can make reasonable decisions on my own." This objection, I suspect, arises partly from the fact that chemists' traditional research participants (chemical substances) aren't people. So, having someone else look over your research plan is unusual for chemists. The objection also comes from the perception that educational manipulations are not particularly invasive or risky compared with biomedical research, so why should we have to do extra administrative work when the risks are no big deal. Current scholarly investigations about institutional review in fact support this point (9). It is important to reiterate what the role of review is: The IRB's role is not to approve your research. It is to make sure that the human participants involved in your research are treated ethically. Why should an independent group make this judgment? If the researcher were to make the judgment, that would be a conflict of interest, which increases the chances of biased decisions.

Scholarly arguments aside, it is valuable to put yourself in the shoes of participants and envision what they might experience in the educational research activity. *Am I expected to make extra time available and do work outside of my normal classroom effort? Will this affect my grade? Are my comments hidden among the crowd, or am I going to have to reveal them out loud to another person face-to-face? Will I look or feel stupid when talking with this stranger about chemistry? What is going to happen to all this data, including mine? Will what I say come back to haunt me?* Regardless of your good intentions, this is how participants might feel. The researcher has a responsibility to anticipate such concerns, communicate explicitly with participants about how they might feel before, during, and after the activity, discuss what might mitigate discomfort, and give them a chance to consider whether they want to participate. This happens in the consent process and continues through the research activity. Remember that participants are donating their time and effort to your research investigation – it is sensible to make sure they feel good about that.

Formally, it's about risks and benefits (who takes the risk and who gets the benefits) and the intended use of the information (for public sharing or for institutional decision-making). As the level of risk rises relative to benefits, and as the degree of public display increases (which may increase the risks), the so-called "level" of IRB review increases. This is discussed more fully below. The other aspect is that participants must be informed about the risks and benefits and must have the chance and the capacity to decide whether to participate or not. How consent is obtained depends on the risk/benefit ratio. Certain special populations require additional care in the consent process: minors (e.g. pre-college students), the cognitively impaired, and prisoners, among others.

The IRB review process requires you to explain your research goals, how data is to be obtained from or about participants, and what will happen to the data. I have often found that writing a concise IRB application forces me to think carefully through my study design and the specifics of how events will unfold. These are things I would have to do anyway. This review process saves me from last-minute protocol construction, which if poorly done could compromise my study and waste time. I would rather invest time earlier and have a second set of eyes review my design than to trust that I can do it right all on my own and perhaps at the last minute.

The IRB is not there to prevent research. It is there to protect human participants from harm. If something should go wrong, and you have stuck with your protocol, the IRB and the institution are expected to stand behind you. Sometimes IRB feedback may comment on the scientific efficacy of your design, i.e. whether your design will help you answer your question. If you get that sort of feedback, that's a bonus because it is coming from people with valuable expertise. You do not always have to heed the advice regarding research design, but it may cause you to reflect on your plans.

Understand Your Local Review Process

Mysterious, arbitrary, slow, cumbersome. These are descriptions that I have heard people use to characterize the IRB review process and which continue to be the subject of some controversy (10). In my experience, I have not found these to be apt characterizations.

If you are new to educational research, or have recently moved to a new institution, the best way to get started is to meet with the chair of your IRB or the compliance officer in your sponsored research office. This should erase the image that there is a faceless institutional edifice standing in your way. Julie Simpson, the Director of Research Integrity Services in the University of New Hampshire (UNH) Research Office, put this forward as her first piece of advice. It is a particularly important suggestion when you have moved to an institution engaged in medical research, which inherently involves higher levels of risk and consequently more scrutiny to protect participants. Another good strategy is to make friends with campus colleagues in education, psychology, or sociology whose research likely involves human participants. They will have models of local IRB applications, know the twists and turns of local procedures, and can offer support and help. There is no reason for the process to be a mystery.

By law there must be at least five members on an IRB, with each member contributing a different viewpoint. Backgrounds must vary to promote complete and adequate review of the types of research brought before an IRB. That often spans qualitative and quantitative methodologies and multiple disciplines. At least one member must have science expertise (scientist or physician), one must have non-science expertise (lawyer, ethicist, clergy member), and one must be unaffiliated with the institution by appointment or family relationship. Membership must also be diverse in terms of gender and ethnicity. If prisoners are participants, at least one member must be a prisoner or prisoner advocate. Membership terms and the selection process are determined by the institution. One way to get better insight about how the IRB works is to offer to serve as a member, but you may not be eligible until you have gained substantial experience in preparing applications. Given that the review process is well defined and that board membership has breadth and expertise by design, the structure of an IRB is hardly arbitrary.

Most college and university higher education institutions have Institutional Review Boards registered with the U.S. Department of Health and Human Services (HHS). Registered boards are recorded in a searchable HHS database (*11*). You may be surprised at the number and breadth of institutions with registered IRBs. Hospitals will be listed, and some institutions have multiple IRBs to handle different types of research. IRB information and forms are typically posted on the institution's public website. Larger institutions will have a research compliance office with permanent staff. At smaller institutions, like four-year colleges, the contact provided may be the faculty chair of their IRB. If your institution does not have its own internal IRB, there may be an agreement in place for an external organization to provide IRB services. If you are part of a distributed educational system (e.g. community colleges or school districts), the system administrative office would be the place to contact for information. It is possible that some chemistry education researchers find themselves in a situation where no IRB oversight is available to them. In that case, you could spearhead the establishment of one, or you can seek a collaborator at another institution who has an IRB.

Each institution is charged with creating its own review process, so the phrase "location, location, location" is as pertinent here as it is in real estate. Different sets of humans operating in different institutional environments end up with different local procedures. For inexperienced researchers, it may be disconcerting that the IRB application process and the requirements for consent and risk management for participants can be so different across institutions. This means that advice from colleagues at other institutions may not translate to your institution. If all of your work is at a single institution, you will grow to understand and optimize the efficiency of that process. Be aware however that expectations have been changing over time, so copying the format of an old IRB application may not be satisfactory. In recent years, more explicit language has been expected in application materials and consent documents. The federal regulations have not changed. However, procedures are evolving as the research community gains more experience and refines ethical practices.

Most new researchers are concerned about how long the process will take. Your IRB will meet with a frequency consistent with the institution's level of research activity. Knowing the meeting schedule will tell you when to submit and how quickly you may get a response. At my institution, the board meets biweekly during the academic year and monthly in the summer. So, I can expect that a couple of weeks may pass before I have an initial response. At my campus, most applications are approved within about three weeks. Some research does exist on IRB response rates. For example, an on-line survey of researchers conducting classroom studies in economics (*9*) indicated that 56% of the researchers had received a decision from their IRB within 2 weeks, and less than 11% took more than a month. I would not characterize this as "slow". (The authors of this study took the glass-half-empty position because of the 44% experiencing longer response times.) The likely reason that it may seem slow to some is that the investigators waited until the last minute to submit their IRB application. The survey did not ask about procrastination. Just remember that poor time management on your part does not constitute an emergency on the part of the

IRB. Lastly, it is appropriate to reiterate a statement printed in big letters in most IRB materials: "Do not start collecting data from participants until you have complete IRB approval." Respect the process and your participants by allowing enough time to prepare the application and bring it through to approval.

Obtaining federal funding now uniformly requires evidence that students and faculty engaging in research have training in research ethics, a component of which includes training in human subjects research. Your institution or school district may have workshops, on-line learning modules, or both. The Collaborative Institutional Training Initiative (CITI), available through institutional subscription, is a widely used on-line source for initial and continuing professional development regarding human research (12). A certificate of completion, paper or electronic will be provided. Keep this in a safe place as you may need it in the future to confirm that you completed this training.

Writing the Application

This is a new genre of writing. It is not a research paper (a scientific argument), not a proposal (a persuasive argument), not a dissertation (a comprehensive detailed study), and not a research abstract (a summary of work done). It is description (who, what, when, where, why). It is possible that no one with chemistry expertise may read your application. Remember the purpose is to ascertain how the human participants in your research will be treated. The board will not find it valuable to read the full rationale that led you to be interested in why students draw evaporation as consisting of molecular dissociation for some molecules but not others, or the nuances of student language used when writing about equilibrium. They want to know what you are asking your participants to do, how they are to be invited or motivated to do that, and what will happen to the information gathered.

Consequently, your challenge is to write for an educated layperson in terms of the disciplinary content of your study. Using the example above about understanding the particulate nature of matter and evaporation, it may be enough to describe the specific aim of your study as: *Students will be given molecular drawings and asked to describe out loud the meaning that they attach to them.* It's not important that the IRB know by name every molecule to be used and why those in particular were chosen, or the expected content of their responses. What's important is that the task is clear: students are given something visual to inspect and then verbalize what they think about it. Because the task and actions are clear, it is easy to consider the potential risks involved.

Generally, an application is expected to include the following components:

- rationale and significance – the broader scope of the research
- specific aims of the research – specific tasks to be accomplished
- research protocol – describes the setting and population to be studied, the materials or approach by which data will be obtained from participants, how participants will be informed about the research and its risks and benefits, and how consent will be requested

- copies of participant recruitment materials, sample instruments or examples of activities, copies of consent forms and associated printed or announced information
- information about the researchers, including evidence of human research training
- nature of data collected, anticipated analysis to be done, storage and security plans
- explicit description of risks and benefits

Each institution devises its own format for IRB application. Some formats are narrative with a suggested structure. Others have specific itemized questions. Whatever the format at your institution, your application will be reviewed most efficiently if you follow the prescribed format. The board will look for information in familiar places in the documents. Putting it somewhere else, or not including it, will cause confusion and raise questions about your research. This advice is no different from advice for writing a successful research grant proposal. Confusing the reviewers will not work to your benefit.

Concise description should be your goal. This helps the reviewers and also helps you. The need to be very concise forces you to think very clearly about your research questions and hypotheses, the population you are studying, and your investigative design. Julie Simpson, my compliance officer at UNH, recommends using active voice. Instead of the typical passive voice of the research report, e.g. "Interviews will be invited during the lab period", say instead "The researcher will approach individual students during the lab period and invite them to be interviewed." Active voice makes it clear who does what and to whom. Another piece of advice when doing research in the classroom is to make it clear to the IRB and to students in the consent process what part of the activity is required as part of the class versus any part that is "only for research".

At UNH, the most frequent reason for a long review process is failure to use the consent form template provided, necessitating revision and re-review. Consent form information must include necessary boiler plate assurances. Although this can make the consent form long, I have not found this to discouarge students from reading it and making an informed decision. In our general chemistry classes, about 90-95% of students agree to allow their routine course and academic information and survey responses to be included in studies.

In the last five to ten years, one of the biggest changes for human participant protection is the use of electronic communication media and the associated risks of loss of anonymity and confidentiality. For instance, some survey software collects IP addresses automatically, in part as a means for controlling who receives the survey or to provide for programmed follow-up emails or subsequent surveys. An IP address can be an electronic conduit to that participant, so that data is no longer anonymous. Interviews conducted using video conference products are also not necessarily private, and so on. Typically, a statement reminding participants that electronic communication is not necessarily private is included in consent documents for studies involving any on-line exchange of personal information.

Normally, student researchers (graduate or undergraduate) must have mentors who co-sign their IRB applications, and these students must go through the training required by the institution. Training must be completed prior to that student conducting the work.

The Nature and Scale of Risks

One of the novelties of the thinking that goes into an IRB application is explicit consideration of how your research investigation might pose risks for your participants. It is your responsibility to identify those risks and benefits for participants as a consequence of engaging in your research. It is the IRB's responsibility to determine whether you have appropriately identified and managed those risks and balanced them against benefits for the participants and the research question. The goal is a win-win scenario for you and your participants. In the following section, I comment on the IRB judgment about level of review. Here I want to raise awareness of the types of risks that may exist. I am starting with the premise that your intention is to conduct an investigative study and publicly share the results outside of your home institution. Academic assessment data to provide intramural feedback to administrators, faculty, and students does not require IRB approval at most institutions as long as confidentiality is maintained (but again, check with your institution). This includes aggregate data released to accrediting agencies for program assessment.

Examples

Table I is a list of research scenarios involving students as participants, ordered from lower to higher risks (top to bottom). This is not intended to be an exhaustive list of research settings. The specifics of the suggested risk management strategies will change with context, and ultimately depend on what your local IRB requires. I had several issues in mind as I organized this table. (1) What are the chances that an individual's identity will be associated with the data they provide? Strategies in this case include removing identifying information, coding for anonymous tracking, and being wary of small data sets where indirect identification might occur. (2) What are the chances that the information if released might cause embarrassment or damage to reputation? Audio and video records (voice and image) make a guarantee of anonymity more difficult. Particularly with video, people may decline to participate because of possible embarrassment about self-image and concern that images could end up in public social network sites. There could also be the extreme case of a student in witness protection not wanting to compromise his/her anonymity. (3) Is the research opportunity presented in a coercive manner? Beware of possible instructor pressure and peer pressure to participate in a study. When bonus points are involved or an extra assignment provided strictly for research purposes, everyone in the class must have a fair shot at the bonus, and must be able to take a path to the bonus that does not involve the research task. The alternative task should be equivalent in time and effort, not overly onerous nor just a simplistic by-pass. In my experience, when students understand the purpose of the research, most will opt for the research pathway rather than the alternative. (4) Are emotional or health risks possible? For example, if students are to be handling chemical substances or laboratory materials, they should be reminded of the normal laboratory risks and safety practices, but also asked about allergen risks (e.g. latex) or chemical sensitivity. If random students are brought together to work in pairs, what if the student pair has a bad relationship history? Consequently, it is necessary to be clear about what you are expecting them to do and who they might be working with.

Table I. Levels of risk (increasing down table) and possible risk management strategies (your institutional IRB assessment may vary)

	Goal of research investigation	Risk management strategy
1	Compare learning outcomes from short surveys, exams, and computer homework from one organic chemistry class with those from another class.	Remove names and identification from database. Identity is not important for research question.
2	Compare learning outcomes for the same students over time from general through organic chemistry, coordinating several outcome measures and course data.	Remove names and identification from data after merging files, or give students a unique code and maintain a code key in a secure location to link student records until files are merged.
3	Same as previous but in a 15-person course, and where student written comments will be presented verbatim.	Chance of indirect identification. Describe population in general: institution type not name, course topic not title, pseudonyms for students, approximate date of course.
4	Two students in each lab section of a large class volunteer to write about their interactions with teaching assistants and other students weekly for bonus credit.	Students must be able to respond to the invitation privately and participate confidentially. A non-research bonus opportunity must be offered to non-consenters.
5	Hour-long audio-recorded interview or eye-tracking session while working on a chemistry problem.	Compensate volunteers (e.g. gift card, raffle chance) and keep list of recipients (tax records).
6	Video/audio record students in classroom groups, transcribe and analyze conversations.	Students must be able to consent privately and independently. Have the ability to edit out or hide people who decline to appear in public images.
7	Conduct lesson interventions in high school chemistry classes to test for enhanced student learning.	Student confidentiality and assent required. Parental consent needed or statement from school administration that it is not needed. Student records identified by code.

Again, it helps to put yourself in the shoes of your participants. What would you want to know about what you are being asked to do? How long will it take and how challenging will it be? What is the importance of the global research question? What are the chances that the information provided will embarrass you now or in the future or lead to negative consequences for your reputation? Will you feel obligated to participate because of your relationship with the instructor or presence of peers? If you start to feel embarrassed or concerned during an interview, are you welcome to voice that concern and stop further participation?

In honing your research question, think about what you really need to learn from participants, and consider whether there are less intrusive means than your original plans to obtain the information in which you are interested.

Incentives

On occasion, an incentive may be offered to encourage and reward participation. Typical incentives consist of course points, a monetary award, or a lottery chance for prizes. A national survey of NIH investigators and IRB chairs (*13*) determined that incentives are intended to create a win-win scenario: The perceived "win" for the research is boosting the number of participants; and the perceived "win" for participants is recognition that their time and intellectual commitment are valuable. Use of incentives, however, is not without controversy in terms of ethics and the potential for biasing research results. An inspection of the literature on this issue suggests that most scholarly treatments about compensation are focused on medicine, health, and social welfare as opposed to teaching and learning. Grant and Sugarman (*14*) provide a thorough philosophical argument suggesting that incentive use is generally not problematic unless participants have a dependency relationship with the researcher, risks are substantial, the research is degrading, or a large incentive is offered to overcome a strong aversive response. In terms of chemistry education research, the first concern might arise if research participants are enrolled in a course taught by the researcher, which is commented on later. Another potential issue is whether course credit is so substantial that it becomes too good to resist (becomes coercive). Psychology departments have well-established routines for awarding psychology course credit to create a steady pool of student participants to support multiple ongoing investigations. These colleagues would be valuable to consult with.

There seem to be fewer empirical investigations than one might think to support sound guidelines about incentives. The AERA Code of Ethics (*1*) has merely two sentences on this issue, essentially saying "be judicious". Lottery incentives have raised more concerns (*15*) because there can be only one or a few winners, often no information is provided on the probability of winning, and there is substantial evidence that humans overestimate that probability. The absence of specific research-based guidelines leaves the issue to the local investigator and IRB to figure out. In general, an incentive should not be so large as to be coercive (consider the socioeconomic status of participants) and it can be pro-rated for extent of participation but cannot be contingent on completing the research activity. Consider the following suggestions:

- Is an incentive really necessary? When the research involves a single survey event with low time commitment and low risk, you may garner enough participation without an incentive.
- If you seek an extended time commitment (e.g. multiple surveys over a semester) then course points for each event could be awarded. This automatically adjusts for different levels of participation.
- If you seek a longer interview or talk-aloud protocol in a single meeting then gift cards would be appropriate.
- If suggesting a lottery, provide specific information on the estimated chances to win and indicate who and when winner selection will be accomplished and documented. Not all IRBs will be comfortable with the lottery approach.

One last issue on compensation – this is taxable income. Awards disbursed from university funds currently require IRS reporting and, based on value, issuing a 1099 form and letter to participants reminding them that the income is taxable. Because of this policy, gift cards have replaced cash.

Faculty Participants

There is a tendency to think of students as the only participants who might be assuming risk. Instructors can be research participants, too. Whether you are studying students in one particular class or in a cross-institutional sample, the instructors are potentially at risk because the information that emerges has the possibility of affecting their professional status. Consider for instance if a particular curricular intervention you are studying in another instructor's class goes badly, and course evaluations for that person become more negative. At some institutions, this could affect tenure, advancement, or salary. So, although our colleagues may be willing to assist us in our studies by offering their class as part of our research, we should be sensitive to the need to consider risk/benefit and formal consent from them as well. Furthermore, we should be willing to write a letter of explanation for that person's professional file as a record of participation and as a potential backstop to any negative consequence.

Another potential area of risk is for an institution. Who signs consent for an institution? Actually, this is not an IRB issue. The IRB mandate applies only to individuals, not to institutions. However, in recognition of the potential for negative consequences, the simplest ethical strategy is keep institutional names anonymous. It is generally sufficient to describe the institution's characteristics, e.g. "a large public PhD-granting university in the Midwest". Institutions may be more concerned about this if you are an outside investigator, and thus they may insist on institutional anonymity or require a local faculty member to co-sponsor your work.

Experiments Including One's Own Students

Some institutions explicitly forbid you from including students in a study if those students are concurrently enrolled in a course you are teaching. Other institutions look at the risk/benefit balance, and the potential for coercion or for evaluation bias. The underlying issue is the power relationship between you, in the dual role of instructor and researcher, and your students, in the dual role of class members and research participants. Strategies for addressing this issue include working with colleagues and students in a course other than your own, asking someone besides yourself to introduce the research opportunity to students, or for the "sign-up" or initiation of the research opportunity to be remote from the presence of the instructor (e.g. on-line at home vs during class). Another protection is delaying analysis of student data until after the course grades have been recorded.

Minors

At the college level, the number of students likely to be younger than 18 years old is few. Your IRB may or may not consider these students to be adults. If such students are not defined as adults, you could attempt to get parental consent, or to get an IRB waiver of parental consent, but the easiest thing to do is to exclude the few students who are younger than 18 years old from the research. If course bonus options are part of the protocol, these students would be given the non-research option.

Studies of pre-college student learning necessarily include minors. Who gives permission for the study to be conducted? How are students involved in the study and how are they provided the opportunity to understand the study and to give assent to participate? Is parental permission for student involvement required? (This decision may be based on local policy.) How are protocol fidelity and participant protections maintained if implemented across many classrooms or schools? My own experience with this is narrow. There are many nuances to the ethical treatment of minors and to working in K-12 schools. Consequently, if I were to consider engaging in research in pre-college environments, I would talk with colleagues in chemistry education research who have this experience or with campus colleagues engaged in K-12 teaching and learning research.

International Studies

The same ethical principles apply to international studies, but in another country rules may be different or may not exist. You may need to be particularly cautious about issues of consent and coercion because of different cultural norms regarding authority, about the risks being assumed by any instructor involved in the study, and about government concerns about how their institutions might be perceived. Work with your international collaborators to understand their local process, secure approval of your protocol in the normal way at your home institution, and then proceed by being very cautious about protecting anonymity for students, faculty, and institutions at the international location.

Levels of Review

The levels of oversight (review) vary depending on the assessed levels of risk. Assuming that the IRB determined it has oversight, the increasing levels of review are Exempt, Expedited, and Full. In the past, it has been the case that many educational investigations have fallen into the Exempt and Expedited categories. This may be changing at some institutions.

Exempt Review

The U.S. Health and Human Services regulations (8) specifically refer to educational studies and the characteristics that make them exempt from higher levels of review. Here is the appropriate excerpt from section 46.101:

(1) Research conducted in established or commonly accepted educational settings, involving normal educational practices, such as (i) research on regular and special education instructional strategies, or (ii) research on the effectiveness of or the comparison among instructional techniques, curricula, or classroom management methods.
(2) Research involving the use of educational tests (cognitive, diagnostic, aptitude, achievement), survey procedures, interview procedures or observation of public behavior, unless: (i) information obtained is recorded in such a manner that human subjects can be identified, directly or through identifiers linked to the subjects; and (ii) any disclosure of the human subjects' responses outside the research could reasonably place the subjects at risk of criminal or civil liability or be damaging to the subjects' financial standing, employability, or reputation.

"Exempt" means that the levels of risk are minimal and annual IRB updates of status are not necessary. At some institutions, Exempt means exempt from IRB review while at others, including UNH, exempt is a level of IRB review. It is very important that you understand the procedure for such studies at your institution early on. The top three studies in Table I are likely to be classified as Exempt because the level of risk is minimal: anonymous surveys, records from typical classroom work, public behavior, and absence of differential treatment. You often just need to provide a final report when the study has been completed. The only exception is if you alter your protocol. In that case, you must request a modification and re-review from your IRB.

Expedited Review

An Expedited level review also requires that the risk presented by the study is minimal. It often is required when anonymity is not guaranteed, when audio and video recording are planned, and when differential treatment of members of the study population (e.g. a class) may occur. The fourth and fifth studies in Table I might be placed in this category. Expedited reviews are typically conducted by the IRB members who have the most pertinent expertise. Approval at this level extends for a year after which an annual report and request for approval renewal is required to continue. This oversight continues even if the participant enrollment and data collection processes have concluded and the research is entirely in the analysis and presentation stages. The renewal asks for an update on whether the study is still actively collecting data, and if not, where in the analysis and reporting process the project is. It also asks for verification of whether protocols are being followed as originally planned or whether they have been modified.

The reports are not long, but the renewal process may become somewhat overwhelming if one has multiple studies in progress. One method for helping to manage the reporting is to have continuing or umbrella protocols. For example, we have an ongoing effort to study student attitudes and performance in general chemistry, and with each new semester we ask student consent to use their course performance information and institutional data in support of a variety of specific hypotheses. Other colleagues have an umbrella consent process for interview studies, in which the protocol for inviting, rewarding, and conducting the interviews is the same, but the specific chemistry content changes because different graduate students are investigating different topics. Thus, modifications can be made by plugging new content into a single protocol without affecting the risk and consent structure.

Full Review

A Full Board review (involving all members of the board in the discussion at a convened meeting of the IRB) results when the risks are greater than minimal, when the study is invasive, or when protected populations are involved. Studies in K-12 science may fall into this category, as well as extensive video/audio recording, such as in the last two cases in Table I. Full Board reviews can only progress at the pace of the board meeting schedules, so these are likely to take the longest to move through the process.

Continuing Review and Modifications

What are your responsibilities once you have approval? Three things. First, do what you said you were going to do. Secondly, if things change (and they often do), you may have to request a modification. Recall again that the IRB's role is to protect human participants and the IRB will be concerned about changes to the levels of risk and consent procedures. Changing your consent form to 14 pt from 12 pt font size, or initiating an activity on Thursday instead of Monday, probably won't trigger concern. But if you find that one approach to recruiting student volunteers for interviews is not drawing participants, and you decide you need to offer a $50 gift card as an incentive, that modification requires review because of the potential for coercion and inequitable treatment. This would be considered "a problem", in the language of the federal legislation (8). When in doubt, contact your IRB chair or compliance officer, explain the situation and that person will let you know what to do. Needing to make modifications is common. Your main concern is to make clear how the modifications compare with the project as previously approved. I typically send a copy of the original documents with the editing marks shown, plus a clean copy with the changes completed.

Finally, you have a responsibility to inform the IRB if "unanticipated adverse events" occur. What this means is that if participants respond poorly to the research conditions, and these outcomes were not anticipated and addressed in the IRB review, then clearly participants are experiencing a new risk. Whether this new risk entails protocol changes must be considered. Thus, halting the research activity, alerting the IRB, and taking the time to review the event is necessary. The federal regulations (8) describe a number of case studies as examples, all of which involve physical and mental health issues, not teaching and learning. "Anticipated adverse events" should have already been considered in the protocol. You will be asked when closing out Exempt studies and in annual reviews of Expedited and Full studies whether there were any "problems" or "adverse events".

The Department of HHS does check up on institutions to verify compliance. A detailed process is described in the regulations regarding why and how this would occur (8). It could involve inspection of IRB minutes, review of study protocols, meeting with individual investigators, and the like. Serious breaches can result in suspension of the research privileges of individual investigators or of all federally funded research at that institution.

Projects Involving Multiple Institutions

Consider assessment of a curriculum intervention occurring at multiple institutions. Generally, the researcher will need approval at his/her own institution since he/she is conducting the research, and approval by the collaborating institutions will be needed because participants reside at the collaborating institution. Because of the variability among processes at different institutions, the specter of completing multiple IRB applications and annual updates could be a substantial burden for multi-institutional projects. If you have collaborators at the target institutions, an efficient way to proceed is:

(1) Obtain approval at your home institution.
(2) Use the internet to find the point person for the IRB at the other institutions. This will be an IRB compliance officer at larger institutions or the IRB faculty chair at smaller institutions.
(3) Send electronically a cover letter indicating in brief that you are collaborating with one of their faculty members in a multi-institution study, and attach the application submitted to your home institution, the approval letter, and documentation that you have completed human subjects training.
(4) Request that your home institution approval be considered a sufficient review. Some institutions will accept your institution's approval without further documentation, others may ask for a local faculty sponsor, and others may conduct their own review.

In my single experience with this issue, I found that in most cases my inquiries were addressed promptly and efficiently, even at a time of year when you might expect a slower response. This was for a project categorized as Exempt by my institution. A project based on a higher level of review might take longer.

Working in the K-12 environment is different. Unless you are working with a very large school district, there may not be an office devoted to assessment or institutional research with a contact person specifically familiar with human participant protocols. If so, you will need to work with the school district's Superintendent to get the appropriate permissions to proceed, including classroom and student access. As mentioned before, parental permission might be required. I think it is particularly important to consider risks for teachers since their local evaluation may be tied to student performance. If your institution has a school of education, your IRB will already be familiar with protocols for pre-college educational settings. My suggestion is that you find a colleague in education with this sort of experience as a starting contact.

When Does Review End?

At many institutions, you close out Exempt studies yourself by providing a final report. It may be enough to check off "it is done" and to indicate any publication or product that emerged from the work. If you don't remember to do it, your research office may clean house occasionally and remind you to look at the status of your exempt projects.

Expedited and Full studies have annual report requirements. There is no statute of limitations on a study that continues to be active at some level. If you fail to provide your annual report, the IRB is required to withdraw approval. That means that if you were still engaged in data analysis on that project, you would have to stop. It is possible to get a project reinstated. Should you run into this issue, contact your IRB chair or compliance office for guidance.

Ethical Educational Innovation

I have heard colleagues express the opinion that they are not in favor of "doing experiments on students". This concern is well placed in that it acknowledges that students have little choice regarding the courses and instructors they encounter in their required curriculum – so any innovation they experience is essentially coerced. They are not voluntary participants in a case like this. This concern is certainly an expression of the spirit of the Belmont report (5) which was triggered in part by medical experiments on captive audiences (i.e., patients, prisoners). On the other hand, the statement also implies that *any* attempt to manipulate the learning environment is unethical. But isn't this what every instructor does everyday using intuition, on-the-fly convenience sampling of student responses, and assessments that are not subjected to validity or reliability testing? Isn't every day in the classroom an experiment?

Investigations that invite students to participate outside the classroom – clinical or *in vitro* studies – where they have the ability to choose to participate, do not have this issue. Investigations that engage students while in the classroom – *in vivo* studies – in principle have a higher ethical bar to clear. What we instructors do every day is guided by our own professional experience, knowledge of practices, and sense of responsibility to our students. "I want to try this" is something one hears all the time in conversations among teachers. What one doesn't hear is whether there are ethical issues embedded therein. Consider three scenarios:

(1) You implement an innovation that affects all students, and use that to inform the next run of that course, but do not report publicly.
(2) You implement an innovation that affects all students, and report publicly outside the institution on the outcomes.
(3) You implement an innovation that targets some students, and you may or may not report publicly on the outcomes.

Case 1 is the typical classroom situation. It is entirely possible that no one but the instructor will be aware of this pseudo-experiment and its outcomes. Activities such as this range in the degree of rigor or theoretical justification applied. Consequently, the results may be meaningful for the instructor but will lack external validity. Outcomes will be tossed into the large bin of "anecdotal classroom stories". Instructors are generally given broad leeway under the principle of academic freedom to do what they think best, both in terms of content and instructional approaches. Case 2, which is identical to Case 1 in design, requires IRB review because of the public reporting. As a result of the IRB oversight, it is probable that the innovation will be given more thought in terms of design, implementation, and assessment. Furthermore, the public reporting provides opportunities for assessing the external validity and generalizability of the outcomes.

Case 3 may arise when it is desired to do a trial on a subset of a course population, or when there are resource constraints. This means there is a treatment group and a non-treatment group. This might be a quasi-experimental design, for example, when groups are "formed" simply by selecting different course sections, or it might be a true controlled experiment with groups decided by a random assignment. Regardless, this is analogous to a pharmaceutical drug effectiveness study. Assuming you can acquire participant information quickly and reliably during implementation, what will you do if the treatment group starts to show substantial positive outcomes relative to the control group (or the reverse)? In a drug study, at some point when the risk/benefit ratio tilts strongly one way or the other, the study is halted. If the benefits are strong, some action is taken to make sure the control group is not losing out on a benefit. I have not encountered an educational study where a similar result has occurred. I suspect it is primarily because the data collection and processing is not accomplished in real time but only after substantial analysis. Nevertheless, one way to anticipate this in a design is to balance the treatment so that it flips from one group to the other, each group now having an equal chance at "exposure". This may or may not be practical in your context, but it is something to consider.

Finally, let's reconsider Case 1. If the educational research literature in your field provides strong evidence that a particular type of instructional approach results in reliable and substantially-documented positive learning outcomes, what is your professional obligation (a) to be aware of that research and (b) to implement the approach in your teaching? In medicine, at some point ignoring the literature on improved treatment practices and failing to implement improved treatments is considered at least substandard practice and sometimes malpractice. At some point, will we be in a position of talking about "educational malpractice"? In the last few years, it is common to hear reference to "best practices". I get nervous when I don't know who decided what those practices are, what criteria were used to decide on what was "best", and who did the deciding. Perhaps "best" can be described as "grounded in modern cognitive, social, and educational theories" and informed by the existing literature as described, for example in the Disciplinary-Based Education Report (DBER) (*16*). We are still some distance from assuring educational outcomes in the same way that medicine might assure the efficacy of treatments, but the courts have already experienced this argument (*17, 18*). With the current pressure to demonstrate "value" for the cost of higher education, this issue is likely to grow in prominence.

Final Words

Consideration for the ethical treatment of students and faculty involved in our studies should be an expected and normal component of research. In principle, the investigator should be cognizant of these issues and incorporate ethical review into planning and execution. But having an independent group give specific attention to the ethical aspect assures that we do not forget it.

The views expressed here are those of the author and not his institution, and the advice may be generally informative but will not be definitive for a particular situation or institution. Talk with your campus representatives responsible for human participant research. If you want to expand your reading, I suggest a few recent books on the subject (*19–21*).

Acknowledgments

I acknowledge the kind assistance of Julie Simpson, Director of the Research Integrity Services Office at UNH, who was happy to share her wisdom from the standpoint of someone inside the IRB process, and who has patiently assisted me with IRB applications and processes over many years. The assistance of chemistry education graduate student Julia Chan is recognized for helping to edit this manuscript.

References

1. American Educational Research Association. Code of Ethics. *Educ. Res.* **2011**, *40* (3), 145–156.
2. *APA Ethics in Research with Human Participants*; Sales, B. D., Folkman, S., Eds.; American Psychological Association: Washington, DC, 2000.

3. *Code of Ethics*; American Sociological Association: Washington, DC, 1999. http://www.asanet.org/images/asa/docs/pdf/CodeofEthics.pdf   (accessed August 2013).

4. *Academic Professional Guidelines*, 4th ed.; American Chemical Society: Washington, DC, 2008. http://www.acs.org/content/acs/en/careers/profdev/ethics/academic-professional-guidelines.html.

5. *The Belmont Report*; The National Commission for the Protection of Human Participants of Biomedical and Behavioral Research, U.S. Department of Health and Human Services, April 1979.   http://www.hhs.gov/ohrp/humansubjects/guidance/belmont.html (accessed August 2013).

6. *Accountability in Research*; Taylor and Francis.

7. *Science and Engineering Ethics*; Springer.

8. *Code of Federal Regulations. Title 45: Public Welfare.* Department of Health and Human Services. Part 46: Protection of Human Participants. Revised January 15, 2009. (accessed November 2013). http://www.hhs.gov/ohrp/policy/ohrpregulations.pdf

9. Logus, J. S.; Grimes, P. W.; Becker, W. E.; Pearson, R. A. *J. Empir. Res. Hum. Res. Ethics* **2007**, *2* (3), 69–77.

10. American Association of University Professors. Regulation of Research on Human Participants: Academic Freedom and the Institutional Review Board. *Bull. AAUP* **2013**, *99*, 101–117.

11. U.S. Department of Health and Human Services, Office for Human Research Protections, Registered Institutional Research Boards.   http://ohrp.cit.nih.gov/search (accessed November 2013).

12. Collaborative Institutional Training Initiative. https://www.citiprogram.org/, page 5 (accessed November 2013).

13. Ripley, E.; Macrina, F.; Markowitz, M.; Gennings, C. *J. Empir. Res. Hum. Res. Ethics* **2010**, *5* (3), 43–56.

14. Grant, R. W.; Sugarman, J. *J. Med. Philos.* **2004**, *29* (6), 717–738.

15. Brown, J. S.; Schonfeld, T. L.; Gordon, B. G. *IRB: Ethics Hum. Res.* **2006**, *28* (1), 12–16.

16. *Discipline-Based Education Research: Understanding and Improving Learning in Undergraduate Science and Engineering*; Singer, S., Nielsen, N., Schweingruber, H., Eds.; National Research Council, National Academy Press: Washington, DC, 2012.

17. Fossey, R. *Educational Malpractice*, 2010. Higher Education Law. http://lawhighereducation.org/49-educational-malpractice.html (accessed August 2013)

18. DeMitchell, T. A.; DeMitchell, T. A. *BYU Educ. Law J.* **2003**, 485–518.

19. Coleman, C. H.; Menikoff, J. A.; Goldner, J. A.; Dubler, N. N. *The Ethics and Regulation of Research with Human Participants*; LexisNexis: Dayton, OH, 2003.

20. *Belmont Revisited: Ethical Principles for Research with Human Participants*; Childress, J. F., Meslin, E. M., Shapiro, H. T., Eds.; Georgetown University Press: Washington, DC, 2005.

21. *The Ethics of Educational Research*; McNamee, M.; Bridges, D., Eds.; Wiley-Blackwell: Hoboken, NJ, 2002.

Chapter 16

Preparing Chemistry Education Research Manuscripts for Publication

Keith S. Taber,*,[1] Marcy H. Towns,[2] and David F. Treagust[3]

[1]Faculty of Education, University of Cambridge,
Cambridge CB2 8PQ, United Kingdom
[2]Department of Chemistry, Purdue University,
West Lafayette, Indiana 47907, United States
[3]Science & Mathematics Education Centre, Curtin University,
Western Australia 6845, Australia
*E-mail: kst24@cam.ac.uk

This chapter offers an overview of the process of publishing reports of research in chemistry education. The chapter considers both issues of what makes a submission suitable for publication in a research journal and practical issues of what authors need to do, and what they should expect, at different stages of the process.

Introduction

This chapter has been prepared to offer guidance on the process of reporting chemistry education research (CER) in research journals. The focus of the chapter is publication in chemistry education journals - such as *Chemistry Education Research and Practice* (*CERP*) and the *Journal of Chemical Education* (*JCE*) – or science education journals – such as the *International Journal of Science Education* (*IJSE*), *Science Education* (*SE*), the *Journal of Research in Science Teaching* (*JRST*), or *Research in Science Education* (*RISE*). Much of the advice given here will also apply to publication in the many more general educational research journals, but our focus will be on the more specialist journals of particular interest to those working in chemistry education.

There are strong commonalities in the requirements and processes adopted by most research journals, and much of the advice offered here is generic in that sense. However, the chapter authors have particular associations with three of the journals that are most important to those who work in chemical education (KST: *CERP*; MHT: *JCE*; DFT: *IJSE*), and where there are significant differences in policy or practice between journals these will be described - presenting the journals in alphabetical title order (e.g. *CERP*, *IJSE*, *JCE*).

Thinking Ahead: Planning for Publishable Research

Writing up research is thought of as being something that happens at the end of a study, but the quality of a research report is only partly about the writing process. A high quality report depends upon a well-planned and executed study. The success of reports being submitted for publication therefore depends in part upon the quality of the thinking that goes into the design of the research. A poorly conceptualized study will not provide a basis for a report that will be judged able to make a substantive contribution to the research literature. Important and interesting findings cannot be published unless they are seen to be based upon a well designed study that provides a sound evidence base, and can be seen to have given consideration to the ethical treatment of research participants.

The Logical Chain Supporting Research Claims

Research reports make knowledge claims: claims to offer new scientific knowledge. 'Scientific' is used here in a broad sense that can encompass work in the social sciences (such as education) providing it draws upon systematic enquiry supported by an appropriate methodology (*1*). In terms of deciding whether a particular methodology is appropriate, it is important to consider the nature of what is being studied, and what kind of knowledge it is possible to acquire about it (*2*). (These issues are sometimes labelled as ontological and epistemological matters, but often in CER reports these terms are not explicitly used.)

So, for example, research might investigate the level of resourcing of school chemistry laboratories in terms of provision of glassware; the level of qualifications of college chemistry teachers; students' understanding of isomerism; teachers' beliefs about effective chemistry pedagogy; the use of formative feedback in high school chemistry classes; project-based learning in undergraduate chemistry; the nature of student dialogue in laboratory group-work, etc. These different foci concern very different *kinds of things*. It is relatively easy to identify and count test tubes and flasks, but teacher behavior (offering formative feedback) may be more difficult to define and observe. Different researchers are much more likely to agree on how many reflux condensers there are in a classroom than whether particular comments should 'count' as formative feedback. Features of other people's mental experiences - such as their knowledge, understanding and beliefs - are not directly accessible to us, may be quite labile, and can only be inferred indirectly by what they tell us (*3*).

Consideration of such ontological and epistemological matters, informed by a careful review of relevant research literature to find out how other researchers have understood the research foci, can help in the formulation of clear and viable research questions, and to understand the kinds of answers which will be feasible for those questions. Research questions are the starting point for producing a research design based on a suitable methodology. This research design will employ particular data collection tools, and just as importantly, data analysis tools, to be used within a particular sampling frame. Some designs (experimental, quasi-experimental studies, and surveys) will be planned in detail in advance; but the adoption of other methodologies (such as grounded theory for example) allows a more iterative approach where early data collection and analysis informs further stages of the research.

Selection of an inappropriate methodology can undermine a research study to the extent that it is not possible to collect the right kind (or amount) of data to answer the research question(s). One cannot answer a question about a general situation (e.g. the extent to which project work is used in undergraduate courses) by undertaking a case study in one institution; and similarly one is unlikely to understand the various factors and their interactions in complex situations (e.g. how understanding of key chemical vocabulary influences learning about core chemical concepts) by surveying students with a written questionnaire. We could give many more examples, but readers are recommended to review the other chapters of this book to see how different approaches are suitable for collecting certain kinds of data and answering particular kinds of questions.

Ultimately knowledge claims rest on the presentation of an argument supported by evidence that can be recognized to be the result of an appropriate analysis of data of a suitable kind and sufficient quantity or representativeness to address clear research questions motivated by a well-informed consideration of the nature of what is being studied.

The Importance of Research Being Seen To Be Ethical

All scientific research should be underpinned by ethical considerations. For example, committing substantial resources to a research project to answer a trivial question would be considered unethical. Ethical standards apply to such aspects as offering a balanced argument in a case for research funding, and to writing up research in ways that honestly report findings: not, for example, selectively reporting some results because these best support the researchers' preferred conclusions. These sorts of considerations are as important in CER as they are in research in chemistry or other natural science disciplines.

However, CER usually involves additional areas of ethical concern that do not apply in most research in the natural sciences. Most studies in CER involve human participants as 'subjects' of the research. The term subject has here been put into scare-quotes, as it implies something done to (rather than with) people, and that is generally considered an inappropriate way to conceptualize those who help us in our research. Research with people, unlike research on samples of non-living substances, requires us to show high levels of respect and concern for those who are involved in our studies.

Ethical Guidelines for Research with Human Participants

In many national contexts there will be widely available guidelines on the ethics of working with people ('human subjects') in research. In the United States, guidelines are published by the American Educational Research Association (*4*), in the United Kingdom by the British Educational Research Association (*5*), and in Australia by the National Health and Medical Research Council/Australian Research Council (*6*). Within national contexts there will be local procedures for getting approval of any research that involves human subjects, for example through Institutional Review Boards (*7*) or departmental ethics committees (see the Chapter by Bauer, this Volume).

Research journals will expect authors submitting manuscripts to have adhered to the guidelines that apply in their national and institutional context. Often authors may be asked to confirm they have followed such guidelines as part of the submission process. Good practice in research writing would include a statement confirming that ethical guidelines are followed in the text of the research report itself (see below). Precise expectations about what constitutes ethical research may vary from one context to another. However, journal editors and reviewers may expect that certain ethical issues should always be addressed by authors of manuscripts, regardless of the norms in the particular context where the research to be reported was undertaken. We briefly discuss some key issues here.

Informed Consent

It is widely considered that people who are involved in research projects should have given informed consent. This means they must have freely agreed to take part in the research, without duress or without the expectation of substantial reward or – should they decline participation – potential negative consequences, having been properly briefed about what they are agreeing to. It is accepted that mild deception may *sometimes* be acceptable in educational research on rare occasions when full disclosure of research purposes would undermine the research: but then full debriefing, and an opportunity to retrospectively withdraw, should be offered to participants.

When researchers approach an institution to seek permission to ask teachers or students to participate in research there will be 'gatekeepers' (such as institutional principals, heads of department or class teachers) who will decline permission if they have reservations about the research or the researchers. When teachers undertake research with their own students or colleagues, this safeguard could be by-passed, and so it is important to involve a more senior colleague as an informed person who could take on the gatekeeper role when there are no IRB procedures that need to be followed (*2*).

No Expectation of Harm or Disadvantage

An important feature of ethical research is that the inquiry could not be foreseen to do harm to participants. Harm need not be physical – it would include undue stress, ridicule, or educational disadvantage. This might, for example, become significant in experimental or quasi-experimental studies comparing two different instructional approaches. If there are already good reasons to believe that one approach is educationally superior to another, then it is questionable whether an artificial situation should be set up where one group of learners is deliberately subjected to instruction *expected* to be inferior. The importance of replicating studies to confirm findings, or testing out existing findings in new contexts, has to be balanced against reasonable expectations based upon the current research literature. As suggested below, research should address research questions or foci that are strongly motivated by a review of existing literature.

Anonymity of Participants and Institutions

A commonly accepted principle in educational research is that research data should be kept confidential and securely held, and that participants and participating institutions should be offered anonymity where they cannot be identified in reports of the research. This may raise particular issues where researchers are inquiring into their own professional and institutional contexts where full reporting often requires acknowledging the particular relationship of the researcher(s) to the context or where the methodology adopted in the research requires reporting a 'thick description' (8) that provides extensive details of the research participants and context.

Generally where individual teachers or learners are described in research reports in any detail, pseudonyms are given to personalize the account, and these are understood to substitute for the actual name. There may be occasions where it is appropriate, with permission, to identify a participant – for example if he or she occupied a unique role or had some particular distinction that was the basis for his or her selection as a participant. In general, reasonable effort should be made to provide anonymity.

Institutions also may be given assumed names in reports, and described in general terms (regional location, size, nature of student body), but this does not offer anonymity if authors are clearly reporting work in their own institution(s). The purpose of the anonymity principle is to offer protection for participants and institutions, and – as with all ethical guidelines – such a principle needs to be interpreted in the particular context and case being considered (see the Chapter by Bauer, this Volume).

Determining Authorship

The issue of authorship is of considerable importance to potential authors, and is also covered by ethical principles (9). The key issue here is that all those (but only those) who made substantive intellectual contributions to a study should be included in the author list (*10*). This leaves scope for discussion of precisely what comprises a substantive intellectual contribution to a particular study, and authors need not all have contributed at all stages, and may not all have been involved in drafting the report (although all authors need to agree to the final text before submission) (*11*). Whilst it may not be possible to finalize such judgments until a report is written, it makes sense for those working in a research team to have a clear idea from the outset of whether their involvement is likely to amount to authorship rather than be seen as purely a technical contribution. Expectations about ordering names in author lists may also vary (e.g. in terms of scale of contribution, alphabetical, seniority, cycling around the team over several reports, etc.), and it is sensible for teams of colleagues to agree upon an approach in advance.

Planning for Quality Research

Research journals use a process of peer review (described below) where editors ask others working in a field to read and evaluate submissions, and make recommendations on whether they should be published. It is well over a decade since Eybe and Schmidt published a set of quality criteria for CER (*12*), and most journals will ask reviewers to evaluate submissions on the basis of criteria of this type (*13*, *14*). The nature of strong manuscripts suitable for publication, and the process of peer review are discussed in more detail later in this chapter. Researchers are advised to be aware of the issues discussed below when planning and executing their research to ensure that they are in a position to write up their work in a form likely to be evaluated highly in peer review.

Preparing the Manuscript for Submission

Research journals tend to reject the majority of submissions they receive, so authors cannot assume their research report will be published by their preferred outlet. However, as journals tend to have their own particular style requirements, it is sensible to have an outlet in mind when writing up a study and checking upon the submission procedures and guidelines for authors. This information is nearly always available from a journal's website.

Selecting a Target Journal for Publication

When identifying a preferred outlet for your work, there are a number of issues to be considered. You should give attention to the match between your work and the journal; the intended audience for your work; and the prestige of the journal.

To take the last point first, academics are in part (and sometimes quite a considerable part) judged by their publications list, and all other things being equal it is better to be published in what is widely considered a leading journal in a field, rather than in a lesser-known outlet. This is not just a matter of kudos - your work is more likely to be noticed and influential in a 'top' journal as others working in the field will regularly access these journals to keep up with their reading, whilst generally only accessing less prestigious journals when following particular references.

Major Journals in the Field

A recent publication found that amongst researchers in the field the most prestigious research journals in CER and science education (presented in alphabetical order within categories) are (*9*):

(a) specifically for chemistry education: *Chemistry Education Research and Practice* (*CERP*) *Journal of Chemical Education* (*JCE*)
(b) for science education more generally: *International Journal of Science Education* (*IJSE*); *Journal of Research in Science Teaching* (*JRST*); *Research in Science Education* (*RISE*); *Science Education* (*SE*).
(c) Other journals that accept CER *International Journal of Science and Mathematics Education* (*IJSME*); *Studies in Science Education* (*SiSE*); *The Australian Journal of Education in Chemistry* (*AusJEC*); *The Chemical Educator* (*TCE*). *Biochemistry and Molecular Biology Education* (*BAMBEd*)

Both *JCE* and *CERP* are published by learned societies - the American Chemical Society and the (UK) Royal Society of Chemistry respectively. However, despite both being linked to national societies, these are international journals that publish work from all over the world, and have international membership for their editorial boards and colleges of reviewers.

Similarly, *IJSE*, *JRST*, *SE*, *RISE* and *IJSME* are international research journals. These are all published by major commercial publishers: Routledge/ Taylor and Francis (*IJSE*), Springer (*RISE, IJSME*) and Wiley (*JRST, SE*) and three have current or historic associations with learned/professional bodies (*IJSE*: European Science Education Research Association; *JRST*: the (US) National Association for Research in Science Teaching; *RISE*: Australasian Science Educational Research Association).

Three other journals that also publish CER are *AusJEC, TCE, and BAMBEd*. *AusJEC* is published by the Royal Australian Institute Inc. and includes aspects of chemistry content, technology in teaching chemistry, innovations in teaching and learning chemistry, research in chemistry education, laboratory experiments, chemistry in everyday life, news, and other relevant submissions. *AusJEC* has an irregular publication schedule. *TCE* is a peer-reviewed journal providing articles on current topics, experiments, and teaching methodology. Being an on-line journal, video clips of demonstrations and laboratories, animation, and full-color graphics are available to enhance the clarity and usefulness of articles. *TCE* also provides quality articles in the expanding field of CER. Studies published in this area provide concrete evidence and conclusions about techniques that improve teaching effectiveness. *BAMBEd* is published by the International Union of Biochemistry and Molecular Biology and seeks to improve teaching and student learning in biochemistry, molecular biology, and related sciences. CER that focuses on biochemistry is published in this journal as well as laboratory experiments, innovative pedagogical approaches, and reviews of emerging areas. *BAMBEd* is published six times per year.

There are many other journals that will consider submissions on themes related to chemistry education. Some of these are regional or national, and so tend to largely publish work from particular parts of the world. These journals will often accept submissions from anywhere in the world, and given the wide availability of internet search facilities, articles published in such journals can be accessed internationally. However, these journals often publish material that would not be accepted in the major international journals because it is considered insufficiently original or important (or not of high enough quality in terms of evidence and argument).

There are also increasing numbers of internet-based, international journals published by commercial companies, and often these journals publish material on a broad range of topics. The rigor of reviewing/editorial processes varies considerably among these journals. Some of these journals follow perfectly acceptable peer-review processes, but where articles on CER are published alongside material from a diverse range of other educational themes they are less likely to be noticed by others working in CER.

Publication Models

Most well-established research journals have both a print and on-line version (*CERP* is on-line only). In these cases the content of the articles in print and electronic versions are identical, although sometimes color is used in figures in on-line articles where grey-scale is used for the printed version. Publishing is a costly business, especially when it involves the printing and distribution of hard copy journal issues, and so has to be financed, which is usually done by charging readers or authors. Traditionally most research journals do not charge authors for publication, but instead charge subscribers (who receive or access the full journal) or occasional readers (for accessing specific articles 'pay-for-view'). Institutional library subscriptions are usually substantial amounts, but entitle all staff and students in an institution, such as a university, electronic access to the journal.

There have long been some journals which charge authors fees for the costs of administration and publishing, a model which is also used in vanity publishing (there are many commercial presses where anyone can have his or her work published as a book if he or she pays the publisher to cover the costs). The distinction here is whether there is meaningful peer-review, rather than just an agreement to publish as a commercial arrangement. With the advent of electronic journals, and the advantage of open access (where anyone with access to the internet can read published material), academic journals (e.g. *RISE*) are increasingly either employing open-access, or making it an option (so authors can choose whether to pay a fee and have their article freely accessible to non-subscribers when it is published on-line).

CERP is an exception as a journal that is free to access on the Internet, but does not make any charges to authors. CERP does not have a print edition (thus reducing costs significantly) and the costs of editorial and production work on the journal are covered by sponsorship of the Education Division of the RSC as part of the educational mission of the Society.

Ensuring Fit with a Journal

One key reason for submissions being rejected from research journals is that material submitted is not considered to fit the remit of the journal. Most journals have available guidance on the kind of material published by the journal.

For example *CERP* invites reviews, theoretical perspectives, and reports of empirical studies in chemistry education, and offers guidance on the nature of these three types of contributions (*13*). *CERP* regularly rejects submissions without peer review because they do not match this remit. The journal also receives suggestions for new ways of teaching topics that have not been evaluated in classrooms or are not explained in terms of educational theory and CER literature; details of new or modified laboratory activities with extensive technical information, but without evaluation in classrooms; and reports of authors' own educational practice lacking rigorous evaluation, for example relying on the author's own impressions of how things went, and simplistic student responses to course evaluation questionnaires. In each of these cases the submissions do not match publicly available information on the characteristics of contributions the journal publishes.

JCE invites ten different types of articles (*14*). Those that may be most closely aligned with the purposes of the CER community include activities, communications, demonstrations, laboratory experiments, and articles. The first four types of articles may have components of CER in them but are not formal research studies and may be missing specific information required for CER articles. Recently an editorial and guidelines specific to CER articles were published in order to clarify expectations for CER manuscripts (*15*). For example, it is expected that the manuscript will cite literature relevant to the study, have clearly described theoretical and methodological frameworks, findings that are based upon the analysis of the data, limitations (every study has them), and implications for CER and the learning of chemistry. Manuscripts that suffer from shortcomings in not meeting the guidelines are rejected and not sent out for review.

In addition to empirical research papers, *IJSE* accepts position papers, innovations, theoretical papers, and letters to the editor, but as noted elsewhere the majority of articles are not specifically related to CER.

One particular question authors need to consider is whether they wish to publish in a CER journal, a science education journal, or a more general educational journal. In general, chemistry education journals expect a strong focus on CER, where science education journals will usually expect submissions to have clear relevance across science education, even if reporting CER work. Similarly, high-ranking general education journals will publish work in chemistry education where it seems to have clear relevance to wider education issues beyond CER alone.

It has been suggested that there are three 'levels' at which research undertaken in chemistry teaching and learning contexts can be considered CER – inherent, embedded, and collateral (*16*). Inherent CER inquires into foci that are core concerns of chemistry education – for example learning difficulties in specific areas of chemistry. These themes are well suited to CER journals but may be seen as of too specific in interest to be considered by general education journals. Embedded CER studies explore wider issues in education (such as the nature of student-teacher classroom discourse) but are carefully contextualised within a chemistry teaching and learning context. These types of studies can also be suitable for CER journals as they relate the wider issue to the specifics of teaching and learning chemistry. The final level of study is labelled collateral CER. This level concerns studies that explore general educational questions and happen to be carried out in chemistry teaching and learning contexts because these have been identified as accessible sites for data collection. Work perceived in this way, even though relating to teaching and learning in chemistry, is less likely to be considered suitable for consideration by CER journals.

Readers may suspect that, to some extent, classification of work into such categories as inherent, embedded, or collateral CER will depend upon the presentation in the write-up. At this time, authors can decide on the extent to which they link their work to prior literature and what they emphasize in reporting their study. There is certainly *some* scope for such flexibility in writing up work, but the initial conceptualization of a project and the specific research questions (RQ) posed inevitably channel the direction a study takes (see above), and constrain how the work can be framed when reported. So the nature of the RQ may lead to work that is either unlikely to be considered suitable for a major CER journal, or alternatively might limit publication of the work to a specialist CER journal.

Specialist and Interdisciplinary Science Education Journals

As well as the major CER and science education journals we discuss above, there are also a number of other potentially relevant journals that researchers in CER might consider as outlets for their work and will be more suitable for some types of studies. Here we refer to some particular examples likely to be of interest to someone working in CER.

Journal of College Science Teaching (*JCST,* published by the National Science Teachers Association in the US) publishes research articles in its *Research and Teaching* column (*17*). These articles span the sciences and include those that are interdisciplinary in nature. Thus, if the study crosses disciplines or has an emphasis in biology, physics, or earth and atmospheric sciences and also relates to chemistry, then this journal may be an outlet to consider.

Studies in Science Education (*SiSE*, published by Routledge/Taylor & Francis, and formerly published by the Centre for Studies in Science and Mathematics Education at Leeds University, UK) publishes review articles that discuss particular areas of research and scholarship in science education, often from a particular perspective. Whilst some other journals also publish reviews, *SiSE* accepts in-depth reviews that would be considered too long to be considered by most journals. *SiSE* commissions many of its reviews and also considers material submitted in the usual way, so authors are advised to approach the editor informally before setting out on writing their review manuscript.

Science & Education (*S&E*, published by Springer, and associated with the International Philosophy, History and Science Teaching Group) publishes material related to science education on themes linked to the philosophy and history of science - for example, teaching about the nature of science (or nature of chemistry).

Foundations of Chemistry (*FOCH*, published by Springer) is set up as an interdisciplinary journal concerned with "conceptual and fundamental" issues in chemistry publishing "philosophical, historical, educational and interdisciplinary studies of chemistry" (*18*). It does not publish mainstream CER research, but its scope includes philosophical perspectives on chemistry education.

Theme/Special Issues

Some journals have issues that are completely or partially dedicated to a single topic. Usually these special or theme issues are edited by, or with support from, guest editors who have particular expertise in the topic. A call for articles will be issued well in advance of the publication of the issue setting out the remit for the issue; often the guest editors will be prepared to offer authors advice on whether a planned article is likely to fall within the scope of the special issue. A special issue brings together a number of articles on current research in a topic area and so arguably provides greater visibility among workers in that subfield for contributions included in the themed issue. (Arguably this is less of an advantage now that most journals publish material on-line where it can be readily found by Internet searches.) Often guest editors include an editorial introduction to the theme issue, contextualizing the place of each accepted article within the topic area. Some examples of journal theme issues in areas of chemistry education are presented in Table 1.

Table 1. Examples of journal theme issues in CER

Theme	Issue
Constructivism in Chemical Education	*FoCH* (2006) - Volume 8, Issue 2
Diagnostic Assessment in Chemistry	*CERP* (2011) - Volume 12, Issue 2
Sustainable Development and Green Chemistry in Chemistry Education	*CERP* (2012) - Volume 13, Issue 2
The Application of Technology to Enhance Chemistry Education	*CERP* (2103) - Volume 14, Issue 3
Physical Chemistry Education	*CERP* (2014) - Volume 15, Issue 3

Continued on next page.

Table 1. (Continued). Examples of journal theme issues in CER

Theme	Issue
Advanced Placement Chemistry	*JCE* – Announced for 2014
Teaching and Learning about the Interface between Chemistry and Biology	*CERP* – Announced for 2015

Reaching a Professional Audience

Journals generally have a strict rule that submitted manuscripts must be original work not published (or being considered for publication) elsewhere. However, this does not mean that authors cannot write more than one article about the same research study, as often studies have a number of different aspects that are best reported in discrete articles. Moreover, authors are allowed to write about aspects of their work reported in journal articles in other formats.

In particular, there are periodicals that are primarily intended for a professional audience (such as school and college teachers) rather than other researchers. For example, *Education in Chemistry* is a magazine for chemistry teachers at all levels published by the Royal Society of Chemistry. These publications do not publish formal research papers, but will consider articles that inform their readers about the findings of research projects. Generally such articles are shorter and less formal than research reports, omit detailed discussions of methodology, and offer a selected bibliography for further reading, rather than citing academic references in the text. Editors expect such articles to focus on implications for teachers, and to get to the point quickly. Colorful images and short quotations are often welcome, rather than a formal presentation of data.

The two main CER journals (*CERP* and *JCE*) are both aimed at teachers as well as researchers but there is value in looking to disseminate research results more widely, especially where those results have clear implications for practice. A sensible policy for authors is to first focus on writing the formal research report for publication in a research journal so that the study is reported to the research community. Once a manuscript is accepted for publication it may then be sensible to write a more popular account focusing on the implications for teachers to be placed in an educational magazine or newsletter, or practitioner journal with wide teacher readership (for example, *SSR*, the *School Science Review* published by the UK's Association for Science Education). This article will not offer sufficient details of the research to satisfy a research audience but can include a citation to the formal research report being published elsewhere for any reader interested in finding out more.

Indeed for major projects it is suggested there may be four levels of writing (*19*). The first is the full technical report containing all the information on the study (and for research students this would be their thesis). The second level is the research paper (or papers) published in research journals. The third level is accounts written for practitioners (or policy makers) and published in other types of publications. Finally, it may be appropriate to produce short press releases for distribution to the news media and for institutional newsfeeds.

Evaluating a Journal as an Outlet for Your Published Work

Given the proliferation of journals in recent years, it is sometimes difficult to know whether a journal that is not one of the major journals is a suitable place to publish work. Table 2 suggests some indicators to consider when evaluating unfamiliar journals as potential outlets for published work.

Table 2. Some indicators of journal quality

Indicator	*What to look for:*
Publisher /sponsor	Major academic publisher or learned society
Editor(s)	Well-known/respected in the field
Editorial/advisory/review board(s)	International coverage, including major contributors to the field
Establishment	Is the journal well established? (It *should* have an ISSN (International Standard Serial Number), although this does not ensure quality.)
Indexing	Is the journal indexed in major indexing services (such as the Social Science Citation Index, ERIC, and SCOPUS)?
Impact factors	Higher values indicate articles in the journal are more often cited in other published works.
Content	Authors of repute are publishing material there; published material appears to readers to be of a high standard (no obvious lapses and deficiencies that should have been spotted in peer review)

Other factors you may wish to consider are whether a journal is open access so anyone can readily obtain articles and whether it requires authors to assign copyright to the publishers (this issue is considered below). You may also be interested in knowing how quickly submissions tend to be processed and printed. This information is not normally publically listed as it can vary so much from article to article (depending on how quickly reviewers report, the extent of revisions needed, how quickly authors can undertake these, etc). However, published articles often include information on the date that a manuscript was received by the journals, when it was accepted, and when it was first published (usually on-line): for example "Received 09 Jan 2013, Accepted 05 Feb 2013 First published online 19 Feb 2013" (*20*).

Writing Style

The major journals described above only accept manuscripts written in English. It is understood that authors who are not native English speakers may lack precision in their use of English grammar. Normally editors and their reviewers will tolerate minor errors of English as long as these do not obscure the clear meaning of the text. Manuscripts that cannot be clearly understood, however, are not acceptable, and careful proofreading and advice on correcting errors should be arranged by authors *before* submission. Where word limits are provided in the guide for authors, these must be followed; for example, IJSE specifies no more than 8500 words or 35 pages (which includes abstract, references, tables and diagrams. Even if no word limits are given (for example *CERP*), editors and reviewers will expect manuscripts to be as concise as is consistent with making a thorough argument to support the knowledge claims made.

Writing in First or Third Person?

Often in scientific writing third person writing is used, i.e. "observations were made" rather than "we observed...". This is consistent with the current natural science tradition that research should be objective: that if research techniques are undertaken carefully then another qualified researcher could be substituted and the outcomes should be the same. This procedure is recognized as an ideal, because often some laboratory scientists are considered to have a special "touch" in carrying out difficult techniques (or using temperamental equipment) due to personal tacit knowledge that is not easily recognized and communicated in technical manuals (*21, 22*). In educational research, however, it is recognized that the same kind of objectivity is unrealistic in some types of studies. For example, in-depth interviews often depend upon the researcher developing a rapport with the research participant, and the use of semi-structured interviews involves the researcher in making real-time "on-line" decisions about how to respond to and follow-up on participant comments. The data produced are often considered a co-construction of the interviewer and interviewee (*23*). In these situations it may well be more appropriate for authors to write using 'I' and 'we' as long as their accounts are clear and factual.

Many CER studies report innovative practice undertaken and evaluated by chemistry teachers working in their own classrooms and institutions. Such studies allow researchers to bring particular insight into the research context of the study but may also bring bias (as it is difficult for researchers not to want and expect their innovations to have positive effects). Some would see overly formal - third person - accounts of research in such contexts as offering an inappropriately objectified report – and consider that first person accounts are more honest to the nature of the research.

Where authors are unsure how to frame their writing, it is important to check for guidance in the journal instructions to authors. Where no strong guidance is given, it may be useful to see how other authors have written about similar studies in the target journal.

What a Manuscript Should Include

A manuscript reporting empirical research in CER is expected to have certain components. (Not all of these features are necessary in reporting other kinds of research and scholarship such as a review or theoretical perspective). Submissions will be expected to include the following:

Indicative Title

A submission should be given a title that is clearly indicative of the nature of the research reported. Although some journals will accept titles which include cultural references ("The secret life of the chemical bond: students' anthropomorphic and animistic references to bonding" (*24*)), puns ("Conceptualizing quanta - illuminating the ground state of student understanding of atomic orbitals" (*25*)), alliteration ("Mediating mental models of metals: acknowledging the priority of the learner's prior learning" (*26*)), or quotations ("'I believe I will go out of this class actually knowing something': Cooperative learning activities in physical chemistry" (*27*)) etc, authors will be asked to modify titles that are considered obscure or misleading in regard of the content of the work. "Safe" titles are likely to be those that succinctly summarize what the research is about (e.g. "The timing of an experiment in the laboratory program is crucial for the student laboratory experience: acylation of ferrocene as a case study" (*28*); "Students' understanding of the nature of matter and chemical reactions – a longitudinal study of conceptual restructuring" (*29*)). There may be limits to the length allowed for a title (e.g. 50 words for *CERP*, 30 words for *IJSE*, authors are asked to be concise in forming titles for *JCE*)

Abstract

All scientific publications should have an abstract that summarizes all the key features of the work (such matters as the theoretical base or paradigm; type of methodology used; grade level or ages of learners; locations of research site; overview of findings etc). An abstract should give potential readers a good idea of what they will find out by reading the full article. Some journals give a strict word limit for the abstract but this varies across journals (e.g. 350 words for *CERP*, 250 words for *IJSE*, and 200 words for *JCE*). Additionally *JCE* requires a graphical abstract for each manuscript.

Keywords

Some journals ask the author to indicate a small number of descriptive key words that can be useful for indexing purposes. These may be required to match those specified in a list provided by the journal (*JCE* for example). Other journals no longer ask for key words given the ability of modern search engines to readily search full content of articles.

Review of Previous Literature

All studies published in research journals are expected to review previous published research relevant to the work being reported. The aim is not to cite every article that could be considered vaguely relevant, and generally submissions that make a series of statements each followed by lists of citations are not well received by reviewers. Rather, the purpose of the review is to show that the current study is informed by and builds upon previous research.

It is important to cite seminal articles relating to the theme of a submission and to acknowledge major differences in perspective or approach where these are represented in the literature to show that authors are aware of alternatives to the approach they have selected. In part, the literature review leads to the research questions by showing that the current study has potential to add to existing knowledge in the field. Also the review sets out a conceptual framework (*30*) for the present study that shows how the research focus is understood by the authors (see the comments on ontology above).

This section of the manuscript will therefore likely be selective in focusing on literature that develops the understanding of the research focus adopted in the study whilst acknowledging where there are different legitimate theoretical perspectives (*30*) that could be applied to the same focus. For example, a study considering factors influencing effective learning could look at teaching style, students' beliefs about learning, students' prior learning, teaching models employed, level of student-student interaction, the nature of teacher questions, and a good many other things. A particular study may only select one, or some, of these issues to explore (supported by citations to relevant research) but would acknowledge other areas not being considered in the study (supported by a small number of citations to seminal work or reviews of these topics).

Research Question

Usually an empirical study seeks to answer a specific research question or questions (RQ) motivated by the literature review presented. A RQ may be a precise hypothesis to be tested or might be much more open-ended depending upon the existing state of knowledge and understanding related to the research focus and context. The RQ needs to be stated before the methodology is described, and should be revisited when the findings are presented later in the manuscript and where an explicit evaluation of the extent to which the study offers answers to the RQ should be offered. Guidance on developing research questions in CER is available in an earlier volume in this series (*31*).

Many studies in CER are carried out by researchers who are following research programs that are iteratively addressing a research focus or question (*32, 33*). However, in papers reporting research undertaken in the authors' own professional context it is quite possible that a study might be motivated by an identified problem or issue in the authors' own practice, rather than a theoretical issue arising from the literature. However, when such "context-directed" research is reported in research journals it still needs to be contextualized within the current state of research literature to show that the study can make a contribution to knowledge in the field. If this is not possible then the research may well have value within the authors' own institutional context, but is not suitable for publication. In other words, evaluation of practice that is innovative within the authors' own context is not worthy of wider publication unless it is shown to have generalizable implications.

Methodology

A research report must include details of the methodology used and the research design used. A wide range of methodologies is acceptable in CER (*1*), and different types of methodology are appropriate for different studies (*2*). The methods section of a research report should include the following information:

Participants and Research Context

Authors need to describe the research context while being aware of the need to not identify institutions or individuals where anonymity has been assured. Factors that might be relevant are phase of education, national or regional context, characteristics of participants, nature of the course being taught or followed, etc. For some kinds of studies, such as a case study which looks at a single instance in some detail, readers expect "thick description" (*8*) that supports analytic or reader generalization (*23*) – that is sufficient context for readers working in other contexts to make a judgment about whether what they read in a research report "is likely to apply here".

Data Collection and Analysis

Sampling procedures must be described so that the reader understands the context as well as how and why a particular sample was chosen. At times it is a sample of convenience; however, this may not yield the strongest and most robust research design. For some studies (consider one that investigated underrepresented groups in chemistry), the sample is purposefully chosen to allow the research question to be addressed.

The procedures used to collect and analyze data must be clearly explained, allowing the work to be replicated. Data gathered using instruments to measure student learning, attitudes, or skills need to be evaluated for reliability and validity. Quantitative analysis procedures should be explained in enough detail to be replicable as well as to be understandable by the readers. Choices in analytical methods, such as parametric or non-parametric approaches to data analysis, should be justified (see the Chapters by S. Lewis and by Pentecost, this volume). In cases with voluminous data analyses, such as factor analysis, some figures of merit may be included in the main text of the manuscript while other analyses can be moved to appendices (e.g. *CERP*) or supplementary on-line materials (e.g. *JCE*).

In qualitative studies, data are collected in a variety of ways including interviews or videotapes. How the data are collected should be carefully described and protocols should be included (or placed in appendices or supplementary materials if overly long). Any software packages used to manage transcripts should be named and the analytical process used to code, analyze, and interpret the data must be described. The theoretical framework used to guide the work must inform this process. Manuscripts may include the coding scheme or a short coded vignette to help the reader understand how the data was coded and analyzed. Figures may help the reader understand how codes group into categories that support the assertions that emerged.

Finally, if a mixed methods approach is used, then care should be taken to describe how the qualitative and quantitative studies inform one another and how the data analyses are integrated to develop the findings.

Journals expect explicit detail of how ethical concerns were addressed to be included in the methodology section. For example, the Royal Society of Chemistry (publisher of *CERP*) includes the following in its ethical guidelines:

"In cases where a study involves the use of live animals or human subjects, the author should include in the Methods/Experimental section of the manuscript a statement that all experiments were performed in compliance with the relevant laws and institutional guidelines, and also state the institutional committee(s) that have approved the experiments. They should also include a statement that informed consent was obtained for any experimentation with human subjects" (*34*).

Findings/Results

The results of the study should be clearly reported. In studies using quantitative data analysis, the statistics obtained should be clearly presented in tables and the key findings highlighted and interpreted in the text. In studies using analysis of qualitative data with interpretive modes of analysis, sufficient quotations from the original data should be used to exemplify the themes and categories identified in the analysis.

Conclusion/Discussion/Implications

A research report should conclude by summarizing the main findings of the study in the context of the research reviewed at the start of the paper. Particular emphasis should be given to findings that are surprising in terms of prior literature, or illuminate current points of disagreement in the literature. This section should emphasize the contribution of the study to the field, and should indicate implications of the study – whether for policy, practice, or questions raised for further research.

This section should also acknowledge limitations of the study, bearing in mind that some limitations are inevitable even in the best studies. Case studies, for example, may offer excellent insights but have limited generalizability. Surveys may representatively sample broad populations but are usually limited to questions with pre-determined categories (either for responses themselves or how responses are coded during analysis), and so have limited potential for revealing new features of a research topic. Experimental studies are often especially challenging, normally relying on "natural" experiments (working with existing courses and teaching groups) and potentially open to a range of threats to validity such as the expectations of researchers and participants, students' responses to novelty (*35*), and teachers' lack of confidence and expertise when first teaching through novel approaches or with unfamiliar materials (*36*).

Focus and Overall Coherence

Although each section of the paper is important in its own right, authors also need to ensure that there is a clear argument that develops and flows through the manuscript. It is worth putting a draft aside, and returning to it afresh with a critical eye a few days later. It can also be very valuable to seek feedback from colleagues whose opinion you value.

Two key issues here are whether the paper has a strong focus, and whether there is a coherent argument throughout. The links between each section should clearly lead the reader through a cumulative argument for the knowledge claims being made. Equally it is important to be sure that there is no extraneous (and so potentially distracting) material that is not directly relevant to the argument being made. For example, where a paper reports one aspect of a larger project it is important to include all the details needed to make sense of the study being reported. It is also just as important to excise material which is pertinent to other aspects of the wider research but which has no direct bearing on what is being reported in the particular study.

Presentation of the Submitted Manuscript

Different journals have their own rules or norms in how articles are presented in the journal – again authors should refer to any guidelines offered by a journal and browse recent issues for current practice.

317

Headings and Sub-Headings

Some education journals have expectations about the particular sections included in a manuscript and the headings used for these sections. Generally CER and science education journals are quite liberal in these matters. However, if manuscripts are structured using the kinds of sections and headings presented in the previous section on what a submission should include then this helps editors, reviewers, and other readers find their way through a report and identify key information. It may often be sensible to adopt these or similar headings unless there are good reasons to take a different approach.

Use of Tables and Figures

In general, journals encourage authors to include tables and figures where these add clarity for readers. Tables and figures should be numbered consecutively (separately for tables and for figures) and each table or figure should have its own title. All tables and figures included in the manuscript must be explicitly referenced in the main text.

Acknowledgements

Authors should acknowledge funders who supported the work reported, and provide grant titles and numbers if available. It is also appropriate to acknowledge contributions made to the work that fall short of deserving authorship, such as: providing materials or advice, feedback on draft writing, providing access to research sites, technical work on data collection, analysis, or manuscript preparation, etc.

Appendices and Supplementary Materials

Journals generally allow authors to submit supplementary materials to be included as appendices at the end of articles or to be accessed separately via a website. These devices should not be used as a way of meeting word lengths by moving essential materials to appendices – all key information should be included in the main text of the manuscript but within the required word or page limit.

Examples of course materials developed or evaluated in a study, examples of how data were analyzed (e.g. factor analysis), examples of interview transcripts or observation notes to illustrate how the data were coded and analyzed, and other materials of these kinds may sometimes be considered suitable for including as appendices or supplementary material.

Journals require citations to literature to be formatted in the journal's style. Many education journals (including *CERP*, *IJSE*) use the name-date of the American Psychological Association or Harvard approach (*37*), whilst some (such as *JCE*) adopt the approach of successively numbering references in the text - in the style used in this book. Authors using bibliographic software packages will be able to choose the appropriate style for a particular journal and have the formatting done automatically but the output should be checked for accuracy. All sources cited in the text (and only these) should be included in the reference list.

Blinded Versions of Manuscripts

Some journals require the submitted form of a manuscript to be blinded for review. This mean that the authors' names are not included in the manuscript file, and that citations to their own previous work are replaced (temporarily for the purposes of peer review) by anonymized versions of the references that do not allow the source to be identified as in the following examples:

- Author (2009) Book. Dordrecht: Springer
- Author (2012). Chapter in edited book. Hatfield, Hertfordshire: *Association for Science Education.*
- Author (2013) Article in research journal.
- Author and colleague (In press). Chapter in edited book. Dordrecht: Springer

If a blinded manuscript is required, authors should check for any inadvertent indications that may be present in the main text ("...one of my colleagues in the Faculty of Education at Cambridge suggested...") or acknowledgements.

Some journals (e.g. *CERP, JCE*) do not blind authors to reviewers and so this type of anonymizing is not needed in preparing a manuscript. Some journals may ask authors to submit both an unblinded and a blinded version of the manuscript.

The Submission Process

Most journals now operate on-line submission and review processes. This means that anyone who wishes to submit manuscripts, or be considered as a peer-reviewer for a journal, needs to register with the journal by opening an account and providing basic information such as contact details, institutional affiliations, and – if wishing to be considered as a reviewer – research interests and areas of expertise.

The Author Center at the Journal Website

Once registered, logging into the account leads to a menu of options that will include an author center and (if asking to be considered as a reviewer) a reviewer center. The author center is a page with information and links relating to manuscripts submitted to the journal by that author. This will show details of any previously submitted manuscripts and decisions on them. It also has a link for starting a new submission. Once a new submission is started it will show at the author center, allowing the author to return at any point to continue work on the submission until ready to submit. Many articles in CER have multiple authors; however, one author has to take responsibility for administering the submission process through his or her log-in. This person will upload the necessary files, answer the submission questions needed for the editor and journal production office, and will approve the submission when it is ready to go to the editors. Until the manuscript is formally submitted, the journal's editorial staff has no knowledge of the planned submission, as it only becomes visible to them once the author authorizes the submission.

What You Need to Upload

When starting a submission, the journal website will take the author through a series of questions over several webpages and request the uploading of files. You should check the submission guidelines for a specific journal to check whether you need to upload information using a particular template or file format. Increasingly, journals accept manuscripts with the tables and figures positioned in the main text. Journals will have expectations about the quality of the original figures suitable for reproduction.

Many computer drawing programs give the user options for saving and exporting images (formats such as jpg, tiff) and often at different levels of resolution (quoted in terms of a metric such as "dots per inch", dpi). Higher resolution allows better quality reproduction, but increases the size of the image file. Figure 1 shows the same image from a drawing program saved at two different resolutions (50 dpi and 200 dpi), demonstrating the difference in clarity of the image.

Figure 1. Saving images at two different resolutions.

Information You Will Be Asked to Provide

Once you have logged into your account, entered your author center, and started a submission, you will then be asked to provide a range of information. Typically you may be asked:

- type of submission/journal section
- title of submission
- abstract (typed or pasted in or uploaded as file)
- details of co-authors (if these are not already registered with the journal, you will need to provide contact details)
- corresponding author (where several authors co-write an article, one is usually nominated to be designated as a contact about the study for readers of the published article. This is often the submitting author, but need not be.)
- some journals (e.g. *CERP, JCE*) will ask if you wish to nominate any reviewers or request that specific people should not be asked to act as reviewers
- comments/letter to the editor – if you wish to explain any context to the submission
- whether you wish your accepted manuscript published immediately as an accepted manuscript before a copyedited version is produced by the production office (if that option is available for the journal, e.g. *CERP, JCE*)
- whether you wish your article to be available open access if accepted (where the journal offers this as an option but it is not standard for all accepted articles)
- whether a previous version of the manuscript has been submitted (where a journal may sometimes reject a submission but allows authors to resubmit a reworked version of the same study: note a resubmission is different to submitting a revision – see below)
- permissions to use previously published materials or photographic releases for images of people
- the word count of the manuscript, and the number of tables and figures included

You may also be asked by some journals to confirm:

- that the manuscript is being submitted solely to that journal and is not published, in press, or submitted elsewhere
- that the research reported follows the ethical guidelines and norms in the context where it has been carried out
- that all authors have approved the submitted version of the manuscript and agreed to be named as authors

Once you have provided all the required files and information, you will typically be asked to download a PDF file of the uploaded materials for review to check, and asked to confirm the PDF is in order before confirming submission. Once you click on the "confirm submission" button you will lose the ability to make further changes to the submission and you will be sent an email confirmation that the submission has been logged for the editorial staff of the journal.

The Peer Review Process

The peer review process occurs in a number of stages.

Screening of Manuscripts

When your submission is received by the journal it will be checked and then subjected to a screening process. The check is to ensure that all the necessary files have been received in the correct formats. The screening is an initial review by a member of the editorial team to ensure that the submission has a *prima facie* case for being considered by the journal. At this point, editors will reject a manuscript that is not aligned with the remit of the journal, clearly does not meet submission guidelines (such as being much longer than the acceptable length), or does not seem to provide a suitable basis for evaluating research quality.

So, for example, if a submission did not offer a review of relevant prior literature, or did not include a section detailing research design and methodology, then the editor would judge that peer reviewers would have no basis for considering the manuscript as suitable for publication, and so would reject it without peer review. This is very unlikely to happen if the advice offered earlier in this chapter has been followed.

Peer Review

Peer review is considered an essential feature of the scientific process, and involves others who are considered to have expertise in the topic of a submission, making an evaluation of the merits of the manuscript. Peer reviewers will normally be others working in the field, and usually those holding doctoral degrees and having themselves published work in peer-reviewed research journals. Generally editors invite at least two reviewers to consider a submission and to make a report.

Selection of Reviewers

Editors have discretion, and may sometimes invite a reviewer from outside the immediate field where they consider this appropriate. So many articles in chemistry education draw upon perspectives from other areas of scholarship, and so an editor may decide to invite a reviewer who has a particular expertise (say in metacognition) who is not working in chemistry or science education. In general, however, editors of journals such as *CERP* and *JCE* will invite researchers working in CER to review. Editors of science education journals, such as *IJSE*, will normally invite reviewers working in science education, and will make a decision about whether this needs to be someone with a strong background in CER depending upon the subject matter of the submission. Editors of general education journals are less likely to invite reviewers from CER as the decision to submit to a general journal implies the manuscript is considered to be of wide interest among educational researchers.

Sometimes authors are invited to suggest preferred reviewers and editors may adopt these suggestions. However, this is always at the editor's discretion and preferred reviewers are unlikely to be selected if they are not considered to offer the necessary level of experience and expertise. Generally colleagues who have strong institutional or other associations with authors are not considered as reviewers as they will have a potential conflict of interest. Reviewers are assured of anonymity in their work, so editors will not reveal to authors who reviewed their work, even in a single-blind journal where reviewers are told who the authors are.

Editorial Input into Review

A reviewer makes a recommendation to the editor and provides a report on the strengths and weaknesses of the submission. Usually reviewers are invited to offer comments for communication to authors, and also are given the opportunity to offer confidential comments (for example if they suspect malpractice such as plagiarism, multiple reporting, or concocting data). The editor considers the reports, and the submission, and makes a decision taking into account the reports and recommendations. If reviewers do not reach a consensus, the editor may adjudicate based on his or her own judgment or may invite another senior colleague to act as an adjudicator.

Editorial Decisions

The precise range of decisions reached varies a little between journals but generally may include the following options:

- A submission may be approved or accepted for publication as submitted (although this is a rare outcome)
- A submission may be approved or accepted for publication subject to completion of minor revisions specified by the editor
- Authors may be told that their submission is not currently suitable for publication, but are invited to undertake major, specified, revisions, so that it may become suitable. In *JCE* this is noted as a major revision as specified by the editor.
- A submission may be rejected but with the invitation to consider resubmitting as a new manuscript at some point if issues raised in reviewer reports can be addressed

- The submission may be rejected with no invitation for subsequent resubmission. Sometimes this may be accompanied by a suggestion that the manuscript is more suitable for a different journal or type of publication.

Usually the editor's response includes copies of the reviewer comments and indicates the changes that are required or suggested before the manuscript could be accepted for publication. Often the editor will expect all changes recommended by reviewers to be carried out, but the editor has discretion to suggest some of the reviewer suggestions are less essential. The timescale from submission to an initial decision can be quite variable. When reviewers quickly accept and complete reviews, and when their recommendations are consistent, a decision may be forthcoming in less than a month. However, for some journals, a period of approximately three months is typical.

Responding to Review Decisions

If a journal asks for minor revisions, it is nearly always going to be possible to complete these satisfactorily and quickly. If a journal asks for major revisions (or rejects, but invites resubmission) then the authors need to decide whether they wish to continue to seek publication in that journal, or look to have their work published elsewhere. There is no obligation on authors to undertake revisions and they may instead submit their work to another journal rather than seek to respond to the revisions requested.

In considering such a decision the authors should consider whether they will be able to satisfactorily undertake the requested revisions; how much work will be involved; and whether this will change the nature of the report in ways with which they are uncomfortable. Often criticisms made by reviewers for one journal are likely to be raised by reviewers for another journal with similar standards. A sensible default position therefore should be that reviewer comments help authors to improve their work and revisions should be carried out if feasible. However, authors are at liberty to decide to disregard reviewer comments and seek publication in a different journal if they strongly disagree with changes they have been asked to make.

Editors carefully read the authors responses to the reviewers' remarks. Editors do not necessarily expect that authors will make every change that reviewers suggest. However, they do expect authors to describe why they may have chosen not to make a suggested change such as a change in analytical approach, etc. There is more scope for disagreement in some areas of work: for example the editor of the journal *Science & Education* often sends manuscripts to experts with a range of disciplinary backgrounds (e.g. history of science, philosophy of science, etc., as well as science education) and the editor acknowledges that there is often scope for dialogue with authors over making recommended revisions.

Revising Your Manuscript

Generally, when revising a manuscript in response to an editorial request for revisions, authors are asked to submit a 'tracked' manuscript, i.e. one that shows where the new submission has changed from the original submission, and a response to reviewers. That response should cover all the changes requested stating what the authors have done to respond. Often a table is most suitable for this, although journals do not usually specify a format. As suggested above, editors will consider arguments for why particular recommended changes were not appropriate, but may not accept these arguments.

There is usually a limited time for making revisions after which a journal will assume that the authors are not intending to continue to seek publication in that journal. Editors normally have discretion to extend this period, provided they are asked for an extension within the standard timescale. Usually authors are sent an automated reminder when a revision is nearly due.

When the revision is submitted through the journal website, the editor will decide whether (a) requested changes have clearly not been made; (b) requested changes have satisfactorily been made; or (c) substantive revisions have been made that should be sent to the reviewers for evaluation. A decision to accept at this stage without further involvement of reviewers is most likely when only minor revisions have been requested. If authors do their best to make it very clear that they have completed all required revisions, then it is more likely that the editor will feel able to move to an immediate accept decision, without the delay of further rounds of peer review.

Sometimes there are several stages of revision, each of which can take some time to complete, before the editor is satisfied that the manuscript is ready for publication. It is also not uncommon for authors to fail to make revisions to the satisfaction of the reviewers and editor, and so an invitation to revise a manuscript does not imply that publication will necessarily follow.

The adage that "the editor's decision is final" should not be taken as an absolute rule, as journals may have mechanisms for challenging editorial decisions. However, if an editor rejects a manuscript based on clear recommendations from reviewers, then it is extremely unlikely that decision will be overturned on appeal. There is necessarily a subjective element in the nature of peer review and each member of the research community brings different experiences and perspectives to evaluating submitted work – so ultimately if authors feel the merits of their work are not being recognized by one journal, they are free to submit to another journal instead.

However, journals are normally very strict about the rule that manuscripts can only be considered by one journal at a time. It is not acceptable to submit a revision to one journal and simultaneously ask another journal to consider the manuscript.

After Approval or Acceptance

Once a manuscript has been approved or accepted for publication, the editor will pass it to the journal's production office to prepare "proofs" – the final text in the correct format for publication. Most journal production offices carry out light copyediting to correct grammatical errors or modify writing style if it does not fit journal norms. This is particularly important when the first language of authors is not English. Journals may assess the level of copyediting needed and fast track those requiring only modest attention (e.g. *CERP* does this). *IJSME* pays specific attention to manuscripts written by authors whose native language is not English, and the editors have made arrangements for support in re-writing where appropriate.

Proofs will be sent (by email) to the corresponding author for checking, giving an opportunity to respond to any queries from the copyeditor and make any final checks (including on any changes made by the production office). Authors are not allowed to made substantial changes to their manuscript at this stage – for example adding new results or interpretations. At this point it is particularly important to check on such matters as the format of tables and the legibility of figures, the accuracy of equations, the formatting of quotations, and to be certain that the list of authors (and their institutions) is complete and correct. The normal means of reporting necessary corrections is to prepare a detailed list (giving page and line numbers) of the precise changes needed, or marking up electronically a copy of the PDF file with required changes. Usually only a few days are given for checking proofs; sometimes a lack of response is taken as approving the proof provided.

The production office will make the necessary changes, and prepare a final PDF file of the article for publication online. This will be assigned a DOI (Digital Object Identifier) code, and once this is published online, there is a definitive published version of the article that cannot be changed further. This process may take from a few days to several months depending upon the journal.

At this point the article will not be assigned to a journal volume and issue, although it is published and may be cited as such. Later it will be compiled with other published articles into a journal issue, at which point it will be given a new citation, and the PDF file of the electronic version will be changed to reflect this (but will also show the date of first publication). However, the actual content of the article, and its DOI, will be the same as the version originally published.

The timescale from original submission to publication clearly depends upon various factors, some of which are within the control of authors (following the formatting instructions for the journal, ensuring all tables and figures are included and correctly labelled in a submission, responding to revision requests quickly and careful documenting changes made, etc) and some of which are not (the speed with which reviewers evaluate assigned manuscripts, and whether reviewer reports are consistent or indicate the need for additional reviewer evaluation). Authors have experienced tortuous processes in some exceptional cases where the full process from submission to acceptance taking well over a year. However, the advent of electronic journals has tended to encourage publishers to seek to speed up the process, and the best-case scenarios can now be fairly short timescales. One of the authors has experienced both extremes. One paper submitted to an education journal in July 2007 was finally published after several rounds of revision in January 2010 (130 weeks later). The same author submitted a manuscript for publication in a CER journal on 9th January 2013. In response to reviewer comments, the author was asked to make revisions. These were undertaken and the revised version was accepted for publication on 5th February the same year. It was then published (*38*) two weeks later – less than six weeks after the original submission.

CERP and *JCE* are now offering authors the opportunity to have a preliminary version of their article published almost immediately after it is accepted. In *CERP* the author's final submitted version will (if the author chooses) be published on the website as a "just accepted" manuscript with the assigned DOI within a day or two of being accepted by the editor. The usual process of preparing and correcting a proof is then carried out and the copyedited version of the article (with the same DOI) replaces the author-prepared provisional version once it is ready. This version will in turn be replaced by the definitive version that includes volume, issue and page numbers once the article is included in a published issue. In *JCE,* the manuscript is edited, published as "ASAP" (As Soon As Publishable), and posted on the website with an assigned DOI. As with *CERP*, the usual production processes take place, then the final version of the article replaces the ASAP article carrying the same DOI.

Citing and Reproducing Your Work

Once your work is published, you will be able to cite it as appearing in the journal. Sometimes authors wish to cite work that is in the process of being prepared for publication or has been accepted for publication. This may take a number of forms as listed in Table 3.

Table 3. Citing work that has not yet been published

Form of citation	Implication
Author (forthcoming); Author (in preparation)	These forms are sometimes used for something that an author is in the process of writing or planning, but for which there is no complete manuscript available.
Author (submitted); Author (under review)	These forms may be used for writing which has been submitted for publication but which has not been accepted and is still in the review process – sometimes authors will make such work available to interested colleagues in manuscript form.
Author (accepted for publication); Author (in press)	These forms are used where an article has been accepted for publication but has not yet been published. Authors may make their version of the final submitted manuscript available to interested colleagues.

Assigning Copyright or Licensing Work

Publishers generally publish materials in exchange for exclusive rights to the material. They generally ask authors to either assign copyright to them (so they become the owners of the author's work and can control its use) or to sign a license allowing them exclusive rights to publish the material. However, if an open-access option is agreed upon, different arrangements may be entered into.

Open-Access Archiving and Article Repositories

Some funders require articles reporting funded work to be published under open access conditions or require the work to be made available to the academic community via institutional depositories. So called "Gold" open-access allows the author to make available freely the final published version of the article as it appears on the journal website. Journals do not normally allow this except where a fee is paid by the author (or their institution) for open access publication, as the publisher will lose revenue from pay-for-view access.

Alternative "Green" models of open-access allow the published article to be made available after a period of embargo (e.g. from 6 months to 2 years) so that the publisher may collect fees for earlier access. However, publishers often allow authors to make publically available their own PDF versions of the final (accepted) version of the submitted manuscript (i.e. before formatting for the journal by the production office) provided it includes an acknowledgement and citation to the published version. Sometimes publishers also provide a PDF of the final version that an author can provide to interested colleagues. Practice in this area varies between publishers and is in flux, so authors should check the requirements and practices for particular publishers at the point of article submission.

What You Can Do with Your Work

Usually publishers use licenses where the authors retain certain rights over the use of their work or ask for copyright on the basis of offering authors certain rights. In general, authors are allowed to use their published work, or extracts from it, in their teaching or talks. They are usually also allowed (subject to due acknowledgements) to republish a journal article as part of a collection of their own writings in a published book. Publishers will normally allow authors to reuse diagrams and tables from published works with acknowledgement but without charge. However, practice can vary between publishers and should be checked in particular cases. Sometimes publishers will allow the authors to produce a translation of their article into another language for publication in a different place, but again this should be checked with the specific journal concerned.

Citation of Your Work

Although the copyright for your writing will be retained (if you license the work) or held by the publisher, others can refer to your work. If they do this in academic writing, then they will be expected to provide a full citation to your published work. Usually journals allow the preferred form of citation for published articles to be downloaded, often in formats that can be directly imported into common bibliographic software.

Other authors are allowed to copy short sections of your writing in their academic (but not commercial) work as long as it is shown as direct quotes, under what is known as "fair use". However, normally other authors will not be able to copy figures and tables from your published work (even with acknowledgement) without express permission of the copyright holder; sometimes publishers charge for this if they hold the copyright.

In time you will be able to find citations to your work through indexes and using such applications as Google Scholar or Scopus. Publishers often include links on their webpages to where published articles are cited in other articles in that journal (or in any other journal on their platform).

Conclusion

It is clear that there is a great deal to think about when seeking to publish reports of CER in the research literature. The overall process of publishing a CER study moves through a number of phases, extending well beyond writing and submitting the manuscript. Successful publication is likely to depend upon the quality of the thinking informing the design of the study to ensure that serious flaws are not revealed only at the writing stage – or when a journal editor returns reviewer comments and a decision. Moreover, as different journals have different expectations and preferred formats, it makes sense to select a preferred target journal when commencing the write-up, so that matters such as maximum article length are taken into account. As suggested above, it is quite rare for any manuscript submitted to a research journal to be published precisely as first submitted, so authors always have to be open to reviewers' recommendations for improving a manuscript, and be ready to undertake one or more revisions of their original submission.

The processes and procedures described in this chapter have become established to ensure that material published in prestigious journals in the research literature is of high quality and offers originality. As users of research – that is, as readers of articles in research journals – we all benefit from the editorial and peer-review processes that both select submissions of merit and support authors in improving their reports. Editors and reviewers are human beings, and most authors will occasionally find editorial decisions that they consider ill judged – just as most readers will occasionally spot what seems a clearly flawed study that has managed to get into print.

Despite the occasional "outlier", the peer review system generally helps authors publish research of potential interest to the rest of the CER community, whilst preventing the journals being diluted with work that is either of poor quality or limited interest to colleagues. Setting out on publishing CER may seem daunting to the novice, but publication of a study in a widely recognized peer-reviewed research journal is an important indicator of work that is judged to make a genuine contribution to the CER community.

References

1. *Scientific Research in Education*; National Research Council Committee on Scientific Principles for Educational Research: Washington, DC, 2002.

2. Taber, K. S. *Classroom-Based Research and Evidence-Based Practice: An Introduction*, 2nd ed.; Sage Publications, Inc.: London, 2013; pp 355 + xii.

3. Taber, K. S. *Modelling Learners and Learning in Science Education: Developing Representations of Concepts, Conceptual Structure and Conceptual Change to Inform Teaching and Research*; Springer: Dordrecht, 2013.

4. American Educational Research Association Code of Ethics. *Educ. Res.* **2011**, *40* (3), 145–156.

5. *Ethical Guidelines for Educational Research*; British Educational Research Association: London, 2011.

6. *Australian Code for the Responsible Conduct of Research*; National Health and Medical Research Council, Australian Research Council: Canberra, Australia, 2007.

7. Bauer, C. F. Ethical Treatment of the Human Participants in Chemistry Education Research. In *Tools of Chemistry Education Research*; Bunce, D. M., Cole, R. S., Eds.; ACS Symposium Series 1166; American Chemical Society: Washington, DC, 2014; Chapter 15.

8. Geertz, C. Thick Description: Toward an Interpretive Theory of Culture. In *The Interpretation of Cultures: Selected Essays*; Basic Books: New York, 1973; pp 3–30.

9. Towns, M. H.; Kraft, A. The 2010 rankings of chemical education and science education journals by faculty engaged in chemical education research. *J. Chem. Educ.* **2012**, *89* (1), 16–20.

10. *Publication Manual of the American Psychological Association*, 6th ed.; American Psychological Association: Washington, DC, 2009.

11. Taber, K. S. Who counts as an author when reporting educational research? *Chem. Educ. Res. Pract.* **2013**, *14* (1), 5–8.

12. Eybe, H.; Schmidt, H.-J. Quality criteria and exemplary papers in chemistry education research. *Int. J. Sci. Educ.* **2001**, *23* (2), 209–225.

13. Taber, K. S. Recognising quality in reports of chemistry education research and practice. *Chem. Educ. Res. Pract.* **2012**, *13* (1), 4–7.

14. Towns, M. H. New guidelines for chemistry education research manuscripts and future directions of the field. *J. Chem. Educ.* **2013**, *90* (9), 1107–1108.

15. Towns, M. H. Content Requirements for Chemical Education Research Manuscripts, 2013. American Chemical Society. http://pubs.acs.org/paragonplus/submission/jceda8/jceda8_cerguide.pdf.

16. Taber, K. S. Three levels of chemistry educational research. *Chem. Educ. Res. Pract.* **2013**.

17. Journal of College Science Teaching: Write for JCST. National Science Teachers Association. http://www.nsta.org/college/guidelines.aspx (accessed July 6, 2013).

18. Foundations of Chemistry. Springer.com. http://www.springer.com/philosophy/epistemology+and+philosophy+of+science/journal/10698 (accessed November 17, 2013).

19. *Good Practice in Educational Research Writing*; British Educational Research Association: Southwell, Nottinghamshire, 2000.

20. Royal Society of Chemistry. http://pubs.rsc.org/en/content/articlelanding/2013/rp/c3rp00012e#!divAbstract (accessed November 17, 2013).

21. Polanyi, M. *Personal Knowledge: Towards a Post-Critical Philosophy*, Corrected version ed.; University of Chicago Press: Chicago, 1962.

22. Collins, H. *Tacit and Explicit Knowledge*; The University of Chicago Press: Chicago, 2010.

23. Kvale, S. *Interviews: An Introduction to Qualitative Research Interviewing*; Sage Publications, Inc.: Thousand Oaks, California, 1996.

24. Taber, K. S.; Watts, M. The secret life of the chemical bond: students' anthropomorphic and animistic references to bonding. *Int. J. Sci. Educ.* **1996**, *18* (5), 557–568.

25. Taber, K. S. Conceptualizing quanta – Illuminating the ground state of student understanding of atomic orbitals. *Chem. Educ. Res. Pract.* **2002**, *3* (2), 145–158.

26. Taber, K. S. Mediating mental models of metals: Acknowledging the priority of the learner's prior learning. *Sci. Educ.* **2003**, *87*, 732–758.

27. Towns, M. H.; Grant, E. R. "I believe I will go out of this class actually knowing something": Cooperative learning activities in physical chemistry. *J. Res. Sci. Teach.* **1997**, *34*, 819–835.

28. Southam, D. C.; Shand, B.; Buntine, M. A.; Kable, S. H.; Read, J. R.; Morris, J. C. The timing of an experiment in the laboratory program is crucial for the student laboratory experience: acylation of ferrocene as a case study. *Chem. Educ. Res. Pract.* **2013**.

29. Oyehaug, A. B.; Holt, A. Students' understanding of the nature of matter and chemical reactions – A longitudinal study of conceptual restructuring. *Chem. Educ. Res. Pract.* **2013**.

30. Taber, K. S. Methodological Issues in Science Education Research: A Perspective from the Philosophy of Science. In *International Handbook of Research in History and Philosophy for Science and Mathematics Education*; Matthews, M. R., Ed.; Springer: New York, 2014.

31. Bunce, D. M. Constructing Good and Researchable Questions. In *Nuts and Bolts of Chemical Education Research*; Bunce, D. M., Cole, R. S., Eds.; ACS Symposium Series 976; American Chemical Society: Washington, DC, 2008; pp 35–46.

32. Taber, K. S. *Progressing Science Education: Constructing the Scientific Research Programme into the Contingent Nature of Learning Science*; Springer: Dordrecht, 2009.

33. Erickson, G. Research Programmes and the Student Science Learning Literature. In *Improving Science Education: The Contribution of Research*; Millar, R., Leach, J., Osborne, J., Eds.; Open University Press: Buckingham, 2000; pp 271–292.

34. Ethical Guidelines. Royal Society of Chemistry. http://www.rsc.org/Publishing/Journals/guidelines/EthicalGuidelines/index.asp (accessed November 17, 2013).

35. Clark, R. E. Reconsidering research on learning from media. *Rev. Educ. Res.* **1983**, *53* (4), 445–459.

36. Rosenthal, R.; Jacobson, L. Teacher's Expectations. In *The Ecology of Human Intelligence*; Hudson, L., Ed.; Penguin: Harmondsworth, U.K., 1970; pp 177–181.

37. *Publication Manual of the American Psychological Association*, 5th ed.; American Psychological Association: Washington, DC, 2001.

38. Taber, K. S. Revisiting the chemistry triplet: Drawing upon the nature of chemical knowledge and the psychology of learning to inform chemistry education. *Chem. Educ. Res. Pract.* **2013**, *14* (2), 156–168.

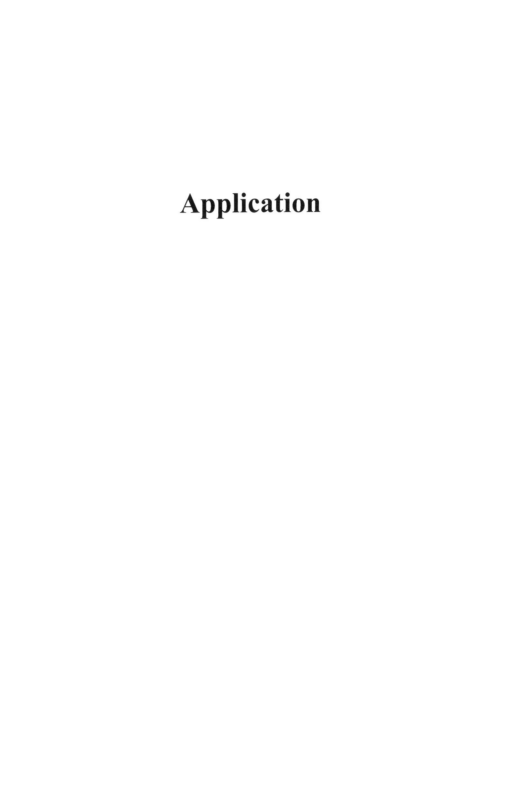

Application

.

Chapter 17

Using This Book To Get Started on Your Own Research

Diane M. Bunce[1] and Renée S. Cole[2]

[1]Chemistry Department, The Catholic University of America,
Washington, DC 20064, United States
[2]Chemistry Department, University of Iowa,
Iowa City, Iowa 52242, United States
*E-mail: Bunce@cua.edu

The purpose of developing this book is to provide both novice and experienced researchers with a resource for planning, executing, and publishing their research. When starting out doing research or implementing a new research methodology, the myriad of details can seem daunting. We hope that this book will help you get a handle on the things to consider when planning a research study and offer new ways for analyzing and reporting the data and conclusions. As an example of how we think this book might be helpful, let's consider two hypothetical research questions and how different information in this book might be used in the planning, execution and analysis of these studies. The two projects we will consider are the following: 1) Does lab have an effect on student understanding of chemistry concepts presented in lecture? and 2) Does achievement on multiple choice questions accurately measure a student's understanding of chemistry?

Does Lab Have an Effect on Student Understanding of Chemistry Concepts Presented in Lecture?

If this is the question that first drives the research, then look at the expected take home messages that you would like to be able to deliver (Bunce, Chapter 13). The take home message could be 1) experience in lab helps students more deeply understand the concepts presented in lecture or 2) lab experiences are not integrated with conceptual understanding emphasized in lecture and therefore, have little effect on student understanding.

Once you look at the question you want to address and the take home message you would like to deliver, it is time to start searching for the theoretical and application frameworks (Bunce, Chapter 13) that will help inform your investigation. In this case, the theoretical framework might involve the effect of experience on understanding; seeing the macroscopic effect of a concept and relating it to the particulate or symbolic levels; or whether discussing ideas with other students, as might happen between partners or within lab groups, helps students engage with a concept more fully resulting in a deeper understanding. Possible application frameworks (research done on similar questions) might include studies on the Science Writing Heuristic or POGIL labs where collaboration is paramount to the pedagogical approach or other studies that show whether lab experiences have any effect on student achievement.

The theoretical and application frameworks will most likely help you revise your original question and make it more specific and thus more easily measured. Understanding could be defined by either the quality of the actions taken by students in lab to execute the lab measured by observation of student behavior in lab (Yezierski, Chapter 2) or by an analysis of the discussion among students in lab (Cole, Becker, and Stanford, Chapter 4). These approaches could help determine one part of your research question, i.e., the quality of the lab experience. Since you are interested in relating this to student understanding in lecture, you must still determine how to measure student understanding in lecture both with and without the lab experience.

To measure student understanding of a chemistry concept within the confines of a real world chemistry course, you could measure student conceptual understanding before and after the lab experience. However, this general plan does not yet address how you would measure that understanding.

One way to measure understanding in lecture would be to devise open -ended conceptual questions to be used in interviews to probe understanding before and after the lab experience. The chapter on interviews (Herrington and Daubenmire, Chapter 3) would be useful in devising such interviews. An alternate way to measure understanding would be with chemical inventories (Bretz, Chapter 9) by either using one that is already developed and could be modified for use in this study or one that would be developed solely for the purpose of the current investigation. Alternatively, if your understanding of knowledge formation in students includes the change in schema development, then the Pathfinder tool (Neiles, Chapter 10) might be useful to measure the change in students' schema development of a chemistry concept before and after a lab experience.

To further investigate this question regarding the impact of the lab experience on student conceptual understanding, you could consider including different lab approaches such as directed labs, inquiry labs and the intermediate guided inquiry labs. This might require involving institutions other than your own. Expanding the research project beyond your own institution where you are physically close to the situation and can make in-course corrections or modifications more easily, has both benefits and drawbacks. The benefits include a more generalizable answer to your research question. The drawbacks deal with the organizational problems of managing a multi-institutional research project (J. Lewis, Chapter 14).

Since you are collecting data on human subjects, you will need to apply for IRB permission to conduct the study (Bauer, Chapter 15). Depending on your institution, you might be required to use data from classes where you have no direct supervision. Whether this is the case or not, you will need to devise and secure written permission from the students in this study if they can be identified through their participation in this study. In this case, you will probably need a signed permission waiver from any student in one or more of the following research activities: one-on-one interviews, observations that are video-recorded, or conversations among lab groups that are audio-recorded. To coordinate the data taken in lab with achievement or demonstrated understanding in lecture, you will need signed waivers from students completing chemistry knowledge inventories, pathfinder sessions, open-ended interviews, or achievement tests that will be included in the analysis. One challenge will be to collapse the myriad of signed permissions into a limited number of waivers that would cover both the lecture and lab components of the study. If student grades on achievement tests or SAT scores are needed in the study, then separate Family Education Rights and Privileges Act (FERPA) permission may be required of each participating student as well as an IRB waiver.

Based on the type of data collected, you might analyze it using multivariate methods where your measure of student understanding is the dependent variable and the score on your tools such as chemical inventories, interview rubrics or pathfinder analyses are your independent measures. The number of groups you enter into a multivariate design would depend on the number of different lab approaches you include such as directed, guided inquiry and inquiry. The configuration of the data for input and the overview of the analysis in a program such as SPSS or R is discussed in the Multivariate chapter (Pentecost, Chapter 6) and the chapter on the use of the statistical package R (Tang and Ji, Chapter 8). If your population is not evenly distributed or if you want to include data that is categorical you might want to analyze it nonparametrically (S. Lewis, Chapter 7). If the data you collected is qualitative including the analysis of transcripts, software programs that facilitate this type of analysis and knowing how to use them, will be very helpful (Talanquer, Chapter 5).

Before actually starting your study, it is wise to review some ways you can help insure that your results, even if nonsignificant, are still valid (Bunce, Chapter 13). Changes made to the planning at this stage will serve you well in the overall execution of the plan. It might also be helpful to review some of the organizational techniques for handling real world experiments (J. Lewis, Chapter 14) that would prove helpful in managing the generation and collection of data from multiple sources at various times during the project.

It is never too soon to think about publication. Knowledge gained from doing research is not useful if it is not shared. Sharing knowledge on important questions is best done through a peer-reviewed journal or other online source. Peer review is the key to establishing quality articles and reports. No matter how well you have done a study, if you can't present, explain, and defend your investigation to others who are knowledgeable in the field, then perhaps your study is not as convincing as you thought. Insights from the editors or associate editors of some of our most prestigious journals in the field of chemistry education (Taber, Towns, and Treagust, Chapter 16) should help you as you collect, analyze and write up the report of your study.

Does Achievement on Multiple Choice Questions Accurately Measure a Student's Understanding of Chemistry?

Our second example of a hypothetical research study shares many of the same issues as the first study. Once again you might start the investigation with a question that has been on your mind as a chemistry teacher and chemical education researcher for some time, namely, "Does achievement on multiple choice questions accurately measure a student's understanding of chemistry?"

It could be that at your institution, due to its size or custom, multiple choice questions serve as the measurement of choice for achievement in a chemistry course. You may have wondered if this tool is measuring students' test taking or mathematical skills rather than their understanding of the underlying chemistry. If you pursue this question, then it is important that if you have strong opinions either way on this issue, that you make these opinions explicit at the beginning of the study and devise your methodology to investigate the question objectively. Consider including triangulation in a mixed methods approach here (Bunce, Chapter 13) to safeguard against interpreting the data in a prejudiced manner. Above all else, you will need to convince editors, reviewers and readers that your conclusions are based on collecting the most convincing data and analyzing it in as transparent and thorough a means as possible. This self-examination might lead you to restate your original question in a way that is more easily measured such as, "Do multiple choice questions measure the same understanding of chemistry concepts by students as open-ended questions and if so, do they do it as well?"

This revised version of your original question restates the phrase "accurately measures" more explicitly as "measure the same under understanding" and "do it as well". This actualization of the phrase "accurately measures" will lead to the use of more specific tools and measurements in your study.

Just as in the previously discussed study, reflection on what you want your take-home message to be, should direct your reading in theoretical and application frameworks (Bunce, Chapter 13). In this case, your theoretical framework may involve information processing and how students solve problems using encoding of information in the questions and the access of stored knowledge in different testing formats (multiple choice or open-ended). The effect of time on student achievement might be another variable of interest since multiple choice tests usually cover a larger number of topics in a set amount of time requiring both more encoding and access of different stored knowledge than a more limited number of ope- ended problems in the same time period. On the other hand, open-ended questions may require a deeper understanding of a concept than a multiple choice question, therefore, differences in depth of understanding required by the two question formats might be important. Other things to consider include cognitive demand produced by either addressing many different topics on the same multiple choice test or many parameters on individual open-ended questions. Student problem solving ability metacognition and self efficiacy could serve as important intervening variables in such a study and should be included in the theoretical framework.

This type of research question is obviously large and could constitute a life's work on the topic. What makes the question so difficult is that there are a great number of variables involved and adequate tools for measuring each identified variable may not be available. Even if there are tools available, the number of tools needed and the time necessary for each student to complete the measures of individual variables might prove overwhelming. Thus, it may be that after looking at the theoretical framework, the research question should be restated to include a more limited number of variables. For instance, just comparing the amount of encoding that is necessary for multiple choice vs. open-ended questions may be a large enough question for a single study. Or determining the cognitive demand of multiple choice vs. open-ended questions on a single topic might be more than enough for a single study.

One of the problems with finding a relevant application framework for such a research question is that there are many studies that have been published which compare one or two variables involved in this question but do not acknowledge and/or control the large number of intervening variables. If you have developed a more theoretical view of student understanding needed to solve multiple choice vs. open-ended questions, then you probably are aware of how complicated the issue is. Overly simplistic experiments that compare student scores on multiple choice and open-ended tests without control of the intervening variables such as encoding, cognitive demand and time constraints as well as other pertinent variables, are not helpful in understanding the situation. As a result, the application framework might be populated with a good number of unhelpful and flawed studies. In a case like this, it may be necessary to better define a smaller aspect of the larger question to investigate and rely on the theoretical foundation rather than the application framework to more heavily direct the study.

Depending on which aspect of the original question you choose to investigate, you will need a way to determine the depth of the concept measured by both multiple choice and open-ended questions. If you decide to pursue the variable of encoding and how it might differ between these two types of questions (multiple choice and open-ended), you could consider the use of eye tracking (Havanki and VandenPlas, Chapter 11) as a means of determining how much encoding is required by each question type. You will also need a way to determine what knowledge stored in long term memory is accessed for each type of question. This issue of accessing long term memory knowledge and determining the quality of this knowledge could be investigated through interviews (Herrington and Daubenmire, Chapter 3). It could also be measured through determinations of student schema (Neiles, Chapter 10); the location as measured by eye tracking of where students look on a list of possible examples of requisite knowledge (Havanki and VandenPlas, Chapter 11) or how well students demonstrate and modify their knowledge through the BeSocratic program (Cooper, Underwood, Bryfeznski, and Klymkowsky, Chapter 12). Each of these tools delivers different types of information and each tool must be tailored and used in a way that is aligned with the variable and the qualities of that variable that you want to measure.

It is obvious that this study, even if it is limited to investigating the impact of only a couple of variables such as encoding or accessing knowledge from long term memory, is likely to generate a great deal of data. Planning for how this data will be collected and readied for analysis is key (J. Lewis, Chapter 14). In addition, a re-examination of the planning to help decrease the chance of generating nonvalid, nonsignificant results (Bunce, Chapter 13) becomes increasing important. The planning for choosing and handling incoming data should include not only what data is collected but also how it will be analyzed either quantitatively or qualitatively. For quantitative studies, identifying the main statistical method needed (parametric, nonparametric or both) and the right kind of data to collect, should be identified and planned for (Pentecost, Chapter 6), (Tang and Ji, Chapter 8), and (S. Lewis, Chapter 7). For qualitative data, having an overview of what you want to be able to glean from an analysis will make the use of qualitative instruments much more efficient (Talanquer, Chapter 5). It is much easier to modify a methodology at this point in the planning process than to try to do it while in the midst of an experiment or even worse, when the study has been completed and you no longer have access to the same population.

Planning for the IRB review for this study should be front and center in the planning process. If you want to interview students using think-aloud sessions as students solve multiple choice vs. open ended questions; eye track their behavior as they view these questions on screen; or evaluate the identity and quality of their knowledge accessed from long term memory through chemical inventories, BeSocratic interactions or Pathfinder sessions, you will need signed IRB permission waivers (Bauer, Chapter 15). If you want to analyze the achievement of students in classroom testing situations, you will need access to their grades. Such access will require IRB as well in some cases, FERPA waivers. IRB approval is needed If you plan to present the results of this study in an open meeting such as a departmental seminar or local, regional or national professional meeting or plan to publish the study in a journal.

If publishing the study is your end goal, then just as in the previous example, it is wise to review the insights of the editors and associate editors of journals in the field of chemistry education (Taber, Towns, and Treagust, Chapter 16) to become familiar with what editors expect in a finished manuscript.

The purpose of this book is to help both novice and experienced researchers meet the expectations of the field of chemistry education on different components of how to conduct quality chemistry education research. One reason for including a wide array of authors is to provide different perspectives on how to accomplish this. The emphasis of this volume is *how* to do it, thus the title of the book *Tools of Chemistry Education Research*. We don't assume that this volume contains all the tools, but we hope we have provided a starting point for the reader to become familiar with some tried and true and other up and coming tools and analytical techniques used in our field of chemistry education research. Now is the time to start planning that research project that you have been thinking about.

Editors' Biographies

Diane M. Bunce

Diane M. Bunce (Ph.D.) is a Professor of Chemistry at The Catholic University of America. She received her B.S. degree in chemistry from Le Moyne College, M.A.T. degree in science education from Cornell University and Ph.D. in chemical education from the University of Maryland—College Park. Her research deals with how students learn chemistry and the mismatch between the way we teach chemistry and what we know about how students learn. Diane received the ACS 2011 Pimentel Award for Chemical Education and the 2001 Helen Free Award for Public Outreach. She has served as editor or co-editor of two other ACS Symposium Series books.

Renée S. Cole

Renée S. Cole (Ph.D.) is an Associate Professor of Chemistry at the University of Iowa. Dr. Cole earned a B.A. in chemistry from Hendrix College, and M.S. and Ph.D. degrees in physical chemistry from the University of Oklahoma. Her research focuses on issues related to how students learn chemistry and how that guides the design of instructional materials and teaching strategies as well on efforts related to faculty development and the connection between chemistry education research and the practice of teaching. Dr. Cole is also an associate editor for the *Journal of Chemical Education*.

Indexes

Author Index

Subject Index

353

Printed and bound by CPI Group (UK) Ltd, Croydon, CR0 4YY